Crops and Vegetation Monitoring with Remote/Proximal Sensing

Crops and Vegetation Monitoring with Remote/Proximal Sensing II

Guest Editors

Kenji Omasa
Shan Lu
Jie Wang

MDPI

Basel • Beijing • Wuhan • Barcelona • Belgrade • Novi Sad • Cluj • Manchester

Guest Editors

Kenji Omasa
Faculty of Agriculture
Takasaki University of Health and Welfare
Nakaorui-machi
Japan

Shan Lu
School of Geographical Sciences
Northeast Normal University
Changchun
China

Jie Wang
College of Grassland Science and Technology
China Agricultural University
Beijing
China

Editorial Office
MDPI AG
Grosspeteranlage 5
4052 Basel, Switzerland

This is a reprint of the Special Issue, published open access by the journal *Remote Sensing* (ISSN 2072-4292), freely accessible at: https://www.mdpi.com/journal/remotesensing/special_issues/R1CWB0K300.

For citation purposes, cite each article independently as indicated on the article page online and as indicated below:

Lastname, A.A.; Lastname, B.B. Article Title. *Journal Name* **Year**, *Volume Number*, Page Range.

ISBN 978-3-7258-4717-4 (Hbk)
ISBN 978-3-7258-4718-1 (PDF)
https://doi.org/10.3390/books978-3-7258-4718-1

Contents

About the Editors vii

S. Sunoj, C. Igathinathane, Nicanor Saliendra, John Hendrickson, David Archer and Mark Liebig
PhenoCam Guidelines for Phenological Measurement and Analysis in an Agricultural Cropping Environment: A Case Study of Soybean
Reprinted from: *Remote Sens.* **2025**, *17*, 724, https://doi.org/10.3390/rs17040724 1

Noah Bevers, Erik W. Ohlson, Kushal KC, Mark W. Jones and Sami Khanal
Sugarcane Mosaic Virus Detection in Maize Using UAS Multispectral Imagery
Reprinted from: *Remote Sens.* **2024**, *16*, 3296, https://doi.org/10.3390/rs16173296 29

Wenchao Liu, Jie Wang, Yang Hu, Taiyong Ma, Munkhdulam Otgonbayar, Chunbo Li, et al.
Mapping Shrub Biomass at 10 m Resolution by Integrating Field Measurements, Unmanned Aerial Vehicles, and Multi-Source Satellite Observations
Reprinted from: *Remote Sens.* **2024**, *16*, 3095, https://doi.org/10.3390/rs16163095 54

César Sáenz, Víctor Cicuéndez, Gabriel García, Diego Madruga, Laura Recuero, Alfonso Bermejo-Saiz, et al.
New Insights on the Information Content of the Normalized Difference Vegetation Index Sentinel-2 Time Series for Assessing Vegetation Dynamics
Reprinted from: *Remote Sens.* **2024**, *16*, 2980, https://doi.org/10.3390/rs16162980 75

Konstantin Nahrstedt, Tobias Reuter, Dieter Trautz, Björn Waske and Thomas Jarmer
Classifying Stand Compositions in Clover Grass Based on High-Resolution Multispectral UAV Images
Reprinted from: *Remote Sens.* **2024**, *16*, 2684, https://doi.org/10.3390/rs16142684 101

Jialong Gong, Xing Zhong, Ruifei Zhu, Zhaoxin Xu, Dong Wang and Jian Yin
An Angle Effect Correction Method for High-Resolution Satellite Side-View Imaging Data to Improve Crop Monitoring Accuracy
Reprinted from: *Remote Sens.* **2024**, *16*, 2172, https://doi.org/10.3390/rs16122172 119

Lei Wang, Xiuzhen Han, Shibo Fang and Fengjin Xiao
Comprehensive Assessment of NDVI Products Derived from Fengyun Satellites across China
Reprinted from: *Remote Sens.* **2024**, *16*, 1363, https://doi.org/10.3390/rs16081363 138

Jihua Meng, Xinyan You, Xiaobo Zhang, Tingting Shi, Lei Zhang, Xingfeng Chen, et al.
Remote Sensing Application in Chinese Medicinal Plant Identification and Acreage Estimation—A Review
Reprinted from: *Remote Sens.* **2023**, *15*, 5580, https://doi.org/10.3390/rs15235580 154

Mishkah Abrahams, Mbulisi Sibanda, Timothy Dube, Vimbayi G. P. Chimonyo and Tafadzwanashe Mabhaudhi
A Systematic Review of UAV Applications for Mapping Neglected and Underutilised Crop Species' Spatial Distribution and Health
Reprinted from: *Remote Sens.* **2023**, *15*, 4672, https://doi.org/10.3390/rs15194672 176

Kesevan Veloo, Carlos Zúñiga Espinoza, Alberto Espinoza Salgado, Pete W. Jacoby and Sindhuja Sankaran
Multispectral, Thermal, and Hyperspectral Sensing Data Depict Stomatal Conductance in Grapevine
Reprinted from: *Remote Sens.* **2025**, *17*, 137, https://doi.org/10.3390/rs17010137 207

About the Editors

Kenji Omasa

Kenji Omasa received his Ph.D. degree in Engineering from the University of Tokyo, Japan. Kenji had been a professor of the Department of Agricultural and Life Sciences at the University of Tokyo from 1999 to 2016 after working as the head, etc., of Biotron and Environment Plant Science at the National Institute for Environmental Studies, Japan, from 1976 to 1999. Kenji had also been the president of the Agricultural Academy of Japan from 2018 to 2022 and an executive board member of the Science Council of Japan from 2014 to 2017. Kenji is currently a professor emeritus at the University of Tokyo, an emeritus researcher at the National Institute for Environmental Studies, the dean of the Faculty of Agriculture at Takasaki University of Health and Welfare, the vice president of Association of Japanese Agricultural Scientific Societies, and a board member of Japan Academy of Life Sciences. Kenji's interests are in imaging of the cell level to plants, remote sensing, smart agriculture, modeling of ecosystems, and analysis of global change effects on plants and ecosystems. To date, Kenji has contributed to 650 scientific publications, 250 of which are in journals with referees and 400 are in books, reviews, and other publications. Kenji received 40 academic awards, including a Medal with Purple Ribbon in 2013 and The Order of the Sacred Treasure, Gold Rays with Neck Ribbon in 2024, awarded by the Government of Japan.

Shan Lu

Shan Lu graduated with a M.Sc. degree in Physical Geoscience from the Northeast Normal University, China, in 2002 and a Ph.D degree in Biological Environmental Engineering from the University of Tokyo in 2007. She is now a Professor at the School of Geographical Sciences at Northeast Normal University, China. Shan Lu has participated in 20 scientific projects, being the coordinator in 10 projects. Her current research interests include hyperspectral remote sensing, multiangular remote sensing, polarized remote sensing, and their applications in biophysical parameter assessment of vegetation and soils. To date, she has contributed more than 50 papers, and 36 are in international journals with referees. She is also a reviewer for different papers in international conferences and journals.

Jie Wang

Jie Wang graduated with a Ph.D. degree in Ecological from the University of Oklahoma, USA, in 2017, and received the M.Sc. and B.S. degrees in Geography Information System (GIS) from the Chinese Academy of Sciences and Nanjing University, China, respectively. Now, she is a professor at the College of Grassland Science and Technology at the Chinese Agricultural University, China. Her current research interests include land use/cover change and carbon cycle, multi-source remote sensing, and their applications in biophysical parameter assessment of vegetation and soils. To date, she has participated in more than 10 scientific projects as PI or Co-PI. She has published more than 40 papers and contributed to 35 international journals as a referee.

remote sensing

MDPI

Article

PhenoCam Guidelines for Phenological Measurement and Analysis in an Agricultural Cropping Environment: A Case Study of Soybean

S. Sunoj [1,†], C. Igathinathane [1,*], Nicanor Saliendra [2], John Hendrickson [2], David Archer [2] and Mark Liebig [2]

1 Department of Agricultural and Biosystems Engineering, North Dakota State University, Fargo, ND 58102, USA; sunoj@illinois.edu
2 Northern Great Plains Research Laboratory, USDA-ARS, Mandan, ND 58554, USA; nickzs1@gmail.com (N.S.); john.hendrickson@usda.gov (J.H.); david.archer@usda.gov (D.A.)
* Correspondence: igathinathane.cannayen@ndsu.edu; Tel.: +1-701-667-3011; Fax: +1-701-667-3054
† Current address: Agricultural and Biological Engineering, University of Illinois Urbana-Champaign, Urbana, IL 61801, USA.

Abstract: A PhenoCam is a near-surface remote sensing system traditionally used for monitoring phenological changes in diverse landscapes. Although initially developed for forest landscapes, these near-surface remote sensing systems are increasingly being adopted in agricultural settings, with deployment expanding from 106 sites in 2020 to 839 sites by February 2025. However, agricultural applications present unique challenges because of rapid crop development and the need for precise phenological monitoring. Despite the increasing number of PhenoCam sites, clear guidelines are missing on (i) the phenological analysis of images, (ii) the selection of a suitable color vegetation index (CVI), and (iii) the extraction of growth stages. This knowledge gap limits the full potential of PhenoCams in agricultural applications. Therefore, a study was conducted in two soybean (*Glycine max* L.) fields to formulate image analysis guidelines for PhenoCam images. Weekly visual assessments of soybean phenological stages were compared with PhenoCam images. A total of 15 CVIs were tested for their ability to reproduce the seasonal variation from RGB, HSB, and Lab color spaces. The effects of image acquisition time groups (10:00 h–14:00 h) and object position (ROI locations: far, middle, and near) on selected CVIs were statistically analyzed. Excess green minus excess red (EXGR), color index of vegetation (CIVE), green leaf index (GLI), and normalized green red difference index (NGRDI) were selected based on the least deviation from their loess-smoothed phenological curve at each image acquisition time. For the selected four CVIs, the time groups did not have a significant effect on CVI values, while the object position had significant effects at the reproductive phase. Among the selected CVIs, GLI and EXGR exhibited the least deviation within the image acquisition time and object position groups. Overall, we recommend employing a consistent image acquisition time to ensure sufficient light, capture the largest possible image ROI in the middle region of the field, and apply any of the selected CVIs in order of GLI, EXGR, NGRDI, and CIVE. These results provide a standardized methodology and serve as guidelines for PhenoCam image analysis in agricultural cropping environments. These guidelines can be incorporated into the standard protocol of the PhenoCam network.

Keywords: image processing; PhenoCam; phenology; remote sensing; soybean; vegetation index

1. Introduction

A "PhenoCam" is a near-surface remote sensing system for collecting time-lapse images automatically for phenological studies, but the term also refers to a network of sites (e.g., https://phenocam.nau.edu/webcam/, accessed on 8 February 2025) deploying cameras to track phenological change [1–3]. PhenoCam networks have evolved significantly over the past two decades. Recent advancements in PhenoCam networks have demonstrated their versatility in monitoring vegetation dynamics across diverse ecosystems, from agricultural landscapes to natural habitats [4–6]. As highlighted by Richardson [7], PhenoCam technology has generated over 200 publications, establishing itself as a fundamental tool in understanding ecosystem responses to climate change.

The increasing adoption of PhenoCams for agricultural monitoring has underscored their potential for providing cost-effective, high-temporal resolution data for crop phenology assessment [5,8]. These methodological advancements have enhanced the reliability of PhenoCam-based vegetation monitoring and expanded its applications through improved image processing techniques [9,10] and standardized protocols for data collection [11,12]. Our study builds upon this foundation while extending its application to agricultural systems. The phenological growth stages in plants, such as the timing of emergence, peak greenness, flowering, seed production, and senescence, are crucial for making management decisions [13].

The phenological growth stages are typically displayed by changes in the greenness intensity of the canopy. Several studies have correlated the greenness intensity from PhenoCam images to phenological growth stage dates in forest ecosystems and rangelands [14–16]. As of May 2020, there were about 106 agricultural test sites with PhenoCams installed [17]; however, as of February 2025, the number of sites had increased to 839 of vegetation type I, 102 of type II, 39 of type III, 173 of agricultural test sites, (https://phenocam.nau.edu/webcam/gallery/, accessed on 8 February 2025), which suggests the growth and research interest in this field. However, it can be observed that agricultural applications have mapped the greenness intensity only to identify the start and end of the season [18,19]. Therefore, this study used PhenoCam images from two soybean (*Glycine max* L.) cultivation sites to analyze how the greenness intensity was related to various growth stages of field crops within an agricultural cropping environment along with other objectives.

The PhenoCam-based approach could provide an objective method to assess the crop phenological stages compared to the laborious and subjective manual crop scouting approach. Furthermore, it enables continuous assessment with increased image and temporal resolutions, thereby facilitating early detection of crop stresses by capturing subtle differences in canopy color, and storage of images for long-term assessment and analysis [1,2,15]. Insights into the phenological processes obtained through image analyses from PhenoCam systems, combined with the established agronomic practices, are expected to help improve management decisions [3,20].

However, PhenoCam applications in agricultural systems also face data acquisition and processing challenges, such as a lack of guidelines on suitable color vegetation indices (CVIs) to use for comparison, and appropriate image acquisition time and object position from the camera. Recent agricultural applications have shown promise in using PhenoCams for crop phenotyping [5,8], yield estimation [4,5], and stress detection [21,22], yet standardized protocols for agricultural implementations remain limited. Existing information in the PhenoCam network includes camera setup and installation protocols, archived data products, image analysis tools, and modeling packages, but lacks information on the above-mentioned aspects for agricultural systems. Therefore, it is desirable to use PhenoCam images from agricultural cropping environments and assess their suitability for the extraction of phenological growth stages compared to visual assessments to gain insight and develop guidelines for phenological image analysis.

Using the common red, blue, and green (RGB) channel values from images, CVIs are derived by mathematical expressions of pixel values from two or more color channels to describe canopy color [23,24]. PhenoCam networks have only widely used green chromatic coordinates (GCC) and excess green (EXG) CVIs for monitoring canopy greenness. Even though several other CVIs have been developed and used in different agricultural remote sensing applications [25,26], their applicability to phenological cameras has not been systematically evaluated. The selection of suitable CVIs for PhenoCam applications requires considering multiple factors, such as their existing usage in vegetation monitoring, demonstrated performance in agricultural applications, and mathematical combinations across diverse illumination conditions. RGB-based indices are particularly valuable for PhenoCam networks due to their compatibility with standard digital cameras and cost-effectiveness, while more complex indices incorporating multiple color space transformations may offer enhanced sensitivity to vegetation characteristics [3,7,27,28]. This study evaluates a comprehensive set of CVIs that represent both traditional and enhanced approaches to vegetation monitoring.

Image color intensity variation caused by illumination conditions in agricultural fields is the most common issue in phenological monitoring [29], and another source of variation is camera type [30]. Andresen et al. [31] reported that the GCC and EXG index values fluctuated due to changes in scene illuminations caused by weather and solar radiation. As the PhenoCam images are captured in open atmospheric conditions, abrupt fluctuations in the phenological curve are observed from the images. However, the actual phenological variation (curve) is expected to be smooth with only a slight variation as the crop transitions into different growth stages.

In most of the PhenoCam studies, using either an average or 90th percentile color value from a narrow time window around midday (10:00 h–14:00 h) is recommended to minimize fluctuations from different illumination conditions [32,33]. Koebsch et al. [34] used a broader time window (entire day—05:30 h to 18:30 h) and minimized the GCC fluctuations by a clustering approach, which determined a common GCC value over a 3-day moving window and assigned to the current time. However, these studies did not quantify the noise produced, nor did they evaluate the suitability of other CVIs. Therefore, the other CVIs might produce a smaller deviation than the GCC and EXG with reference to a smoothed trend representing natural variation.

Richardson et al. [35] suggested plant color evaluated from the pixel intensity values inside a region of interest (ROI) might vary as a function of distance (object position) from the PhenoCam, but it has not been quantified, analyzed statistically, or reported. In general, an ROI is identified within the field of view of the PhenoCam image, and the color information from the ROI is extracted for analysis [15]. Traversing through the PhenoCam image pixels from the bottom to top, the image objects translate to moving away from the camera in the field. Positioning the image ROIs from the bottom to the top represents the object's position from the PhenoCam. In recent years, PhenoCams have been applied in field phenotyping [5], where the color values from 88 experimental plots were employed in the study. However, the study did not account for the selection of ROIs for extracting color values, which may introduce a higher likelihood of error and potentially affect the accuracy of the inferences drawn. Therefore, analyzing the effect of object positions (through ROIs) on CVIs can be used to determine a suitable ROI position for PhenoCam studies. For this study, we hypothesize that (i) a smoother phenological variation (curve) than GCC will result from other CVIs if they are less prone to noise, and (ii) the image acquisition time and the object position (ROI selection or pixel location) will affect the CVIs describing the soybean phenology.

Given this background, this research aimed to determine a set of CVIs that best represents the natural smooth phenological curve and to evaluate the effects of image acquisition time and object position in the camera's field of view (through ROI) on the selected CVIs. The results from this research can be incorporated into existing protocols for PhenoCam image analysis for a consistent analysis of crops. The recommendations (CVI, time, and ROI) can be applied in future crop-based studies throughout the PhenoCam network. Therefore, the specific objectives for this research were:

1. Evaluate the variations in existing RGB-based CVIs and other novel combinations of alternative color space-based CVIs across image acquisition times (10:00 h–14:00 h at 30 min intervals for ≈140 days; 2018 growing season).
2. Determine a set of best CVIs with a lower variation than the GCC curve and validate the CVI curve against the visually assessed soybean phenological stages.
3. Assess the effect of image acquisition time on the selected CVIs within the PhenoCam's field of view.
4. Assess the effect of object position on the selected CVIs within the PhenoCam's field of view.

2. Materials and Methods

2.1. Site and Management Description

Two experiment fields were selected for the study, namely, Mandan-H5 (46.775°N, 100.951°W) and Mandan-I2 (46.761°N, 100.923°W) at an elevation of 593 m above sea level, located at the Area 4 Soil Conservation Districts' Cooperative Research Farm, Mandan, North Dakota, USA (Figure 1). The area of these fields were 19.8 ha and 22.1 ha, respectively. The site has an annual average temperature of 62 °C and receives annual average precipitation of 355–457 mm. During the growing season (April–October), the average temperature is 21–32 °C, with growing season precipitation averaging 190 mm. The soil is predominantly Temvik–Wilton silt loam in both fields. The experimental field had been under a corn–soybean rotation system for the past four years.

Soybean (Mycogen seeds 5B024, The Dow Chemical Company, Midland, MI, USA) was planted using a John Deere 1890 planter at 543,631 seeds/ha (220,000 seeds/ac) in both fields on 29 May 2018 and 31 May 2018 , respectively. Row and plant spacing were maintained as 0.19 m and 0.06 m. The soybean crops were harvested on 18 October 2018 and 17 October 2018, respectively.

The dataset employed in this study covered one growing season with a single crop species. It included the PhenoCam imagery with corresponding ground-truth measurements for field validation. Our analysis focused on the geometric relationships between ROI placement and sun position during image acquisition, which are the factors that remain constant irrespective of growing season, climate conditions, or crop type. Therefore, this dataset effectively supported our objective of developing generalizable guidelines for PhenoCam image analysis in agricultural cropping environment.

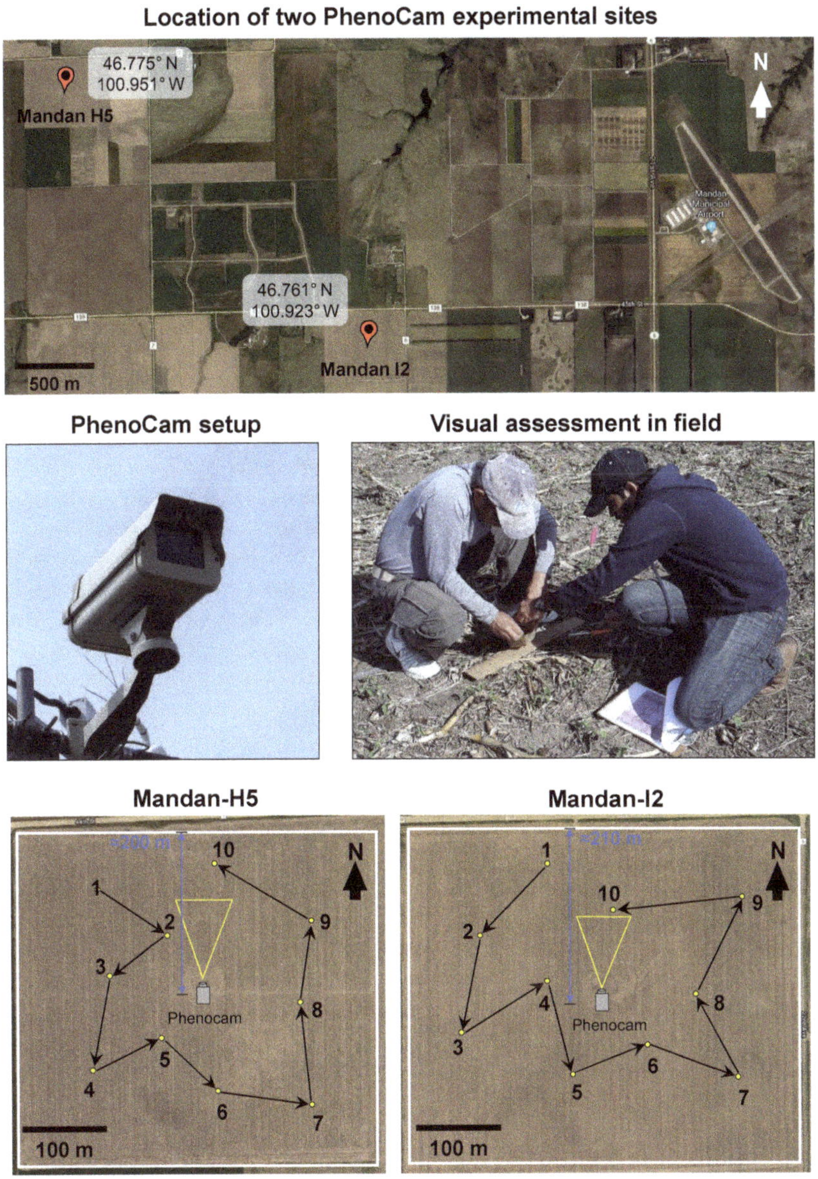

Figure 1. Location of PhenoCam study experiment sites in Area 4 Soil Conservation Districts' Cooperative Research Farm, Mandan, ND, USA; PhenoCam setup used in this study, phenology visual measurements in the field, and an overview of the two PhenoCam fields: Mandan-H5 (46.775°N, 100.951°W; ≈19.8 ha) and Mandan-I2 (46.761°N, 100.923°W; ≈22.1 ha). Ten yellow points in each field indicate the locations for the visual assessment of phenological stages (ground truth), and the yellow triangle represents the approximate field of view area (≈0.27 ha) for each PhenoCam. Note: The top panel aerial image was obtained from Google Maps to show the study sites' location and details.

2.2. PhenoCam Setup and Image Analysis for Crop Phenology

2.2.1. PhenoCam Images and Data File Collection

Both PhenoCam study sites were installed with a NetCam color digital camera (NetCam SC IR, StarDot Technologies, Buena Park, CA, USA). The camera was equipped with a 5-megapixel CMOS sensor. The camera provided a maximum spatial resolution of 2592 × 1944 pixels with frame transfer capability. The sensor featured RGB color channels with 8-bit radiometric resolution per channel (total 24-bit color depth). The camera's light sensitivity was 0.3 lx in color mode, and it included automatic exposure control, color balance, and digital noise reduction capabilities for optimal image quality. The digital cameras were housed inside a weather-proof enclosure (Figure 1) for continuous monitoring and were configured based on the instructions provided by the PhenoCam network [36]. The images (30 min interval) were downloaded from the PhenoCam website (https://phenocam.nau.edu/webcam/gallery/, accessed on 8 February 2025) for each field separately ("mandanh5" for the site Mandan-H5 and "mandani2" for site Mandan-I2) by specifying the date range and time span.

The total number of growing days between planting and harvest dates were 143 d for Mandan-H5 and 139 d for Mandan-I2. The images were collected for a time span between 10:00 h and 14:00 h on each day as this time span is commonly used in PhenoCam studies [32,33]. The total number of images collected for the entire growing season in Mandan-H5 was 1287 images (143 × 9/d) and 1251 images (139 × 9/d) in Mandan-I2. The details about downloading the images from the PhenoCam network are provided in (Supplementary Materials Section S1; Figures S1–S6).

To assess uncertainties in our data processing and analysis pipeline, we evaluated multiple sources of variation. Image acquisition uncertainty was quantified by analyzing the coefficient of variation for each DAP in color values across different times (10:00 h–14:00 h). ROI selection uncertainty was evaluated by calculating the standard deviation of CVI values across different positions. We evaluated the variability in CVI across different image acquisition times and growth phases.

Along with PhenoCam images, data files (provisional data) that contained several parameters, such as daily mean RGB values, mean GCC, 5–95th percentile RGB values, 3-day mean RGB, etc., also accessible through the PhenoCam website [1,37], were downloaded for analysis. In the present study, the daily mean RGB values from these data files were used to compare different CVIs, as these values were readily available without image processing and represented a broader time span (07:00 h–20:00 h). The procedure for downloading the data files and calculating RGB-based CVIs from the files using an R program [38,39] is presented in (Supplementary Materials Section S2; Figure S7). The images and the data files used in this study belong to the PhenoCam dataset v2.0 [40].

2.2.2. Color Value Extraction from PhenoCam Images

Although a free web-based tool called `xROI` was available for processing PhenoCam images [41], its capabilities were limited to extracting color values only from RGB color space at a single ROI. But for our study, it was desirable to have a tool that could extract color values from other color space (HSB, Lab) along with RGB at multiple ROIs simultaneously. Therefore, a user-coded plugin was developed using Fiji software (Ver. 1.52p, Rasband [42], Schindelin et al. [43]). The details on the plugin development and its operation can be found in (Supplementary Materials Section S3; Figure S8). The images downloaded from the PhenoCam network resided in a single folder by default. This series of images could be opened by dragging and dropping the entire folder into ImageJ.

The images were opened as a stack, and the user drew a polygonal ROI on any of the images in the stack, which was applied to all the images, to extract color information.

The plugin had two inputs (color space and ROI selection) and three color space (RGB, HSB, and Lab) options (Supplementary Materials Figure S8). The user could choose at least one color space and either "Single ROI" or "Multiple ROI" with the number of rows and columns specified. The plugin was developed in such a way that the color values were extracted for all the images in a single run, and the mean RGB, HSB, and Lab values of each sub-ROI (1–9) were generated as a comma-separated value (CSV) file for further analyses.

2.2.3. Color Value Extraction from Multiple ROIs

In this study, multiple ROIs with 3 rows and 3 columns (9 sub-ROIs) were used for color value extraction. Care was taken in specifying the whole ROI so it matched with the ROI specified by the PhenoCam network for Mandan-H5 and Mandan-I2 (Figure 2). To study the effect of time (10:00 h to 14:00 h at 30 min intervals) on the CVIs, three replications were considered by averaging the sub-ROIs along each column (replication#: sub-ROIs# $\Rightarrow$ (1: 1, 4, 7); (2: 2, 5, 8); and (3: 3, 6, 9)).

If the image pixels of a PhenoCam image starting from the bottom to the top were mapped to an actual position in the field, it would correspond to the positions starting from near to far from the camera. Hence, changing the object position can be thought of as selecting the sub-ROI in the vegetation field of view along the image's vertical direction. A set of three ROIs, which were horizontally at the same level, was assumed to be at the same distances from the vertical plane of the camera. However, as the camera was a point sensor, the horizontal sub-ROIs (as well as the vertical sub-ROIs) had different viewing angles, and this effect of the angle might contribute to the CVIs, which can be a potential parameter to analyze in the future but was not considered in this study. Therefore, to study the effect of object position from the camera on CVIs, the sub-ROIs within each row were considered as replications. The sub-ROIs 1, 2, and 3 served as a replication for testing "far"; similarly, the sub-ROIs 4, 5, and 6 for "middle"; and the sub-ROIs 7, 8, and 9 for "near" (Figure 2). The total distance between the PhenoCam's position in the field and the field edge in the camera's direction were ≈200 m and ≈210 m for Mandan H5 and I2, respectively (Figure 1). However, the distance covered in the field by the ROI between the bottom and the top of the whole ROI was ≈120 m, with some margin around the ROI.

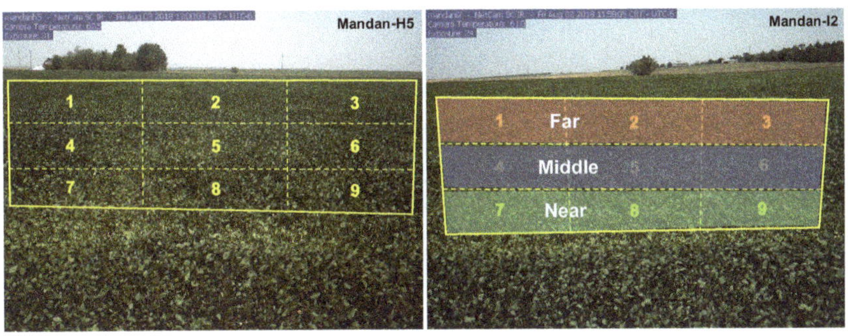

Figure 2. An example PhenoCam image of Mandan-H5 and Mandan-I2 soybean fields with ROI provided by the PhenoCam network (yellow solid line) and sub-ROIs (yellow dotted line) considered for analyzing the effect of image acquisition time and object position from the camera on various CVIs.

2.3. *Visual Assessment of Phenological Stages*

The visual assessment was performed weekly on each field to record the phenological stages. Additional information on the soybean phenological visual assessment is given in Supplementary Materials Section S4, Figures S9 and S10, and Table S1. Ten random locations were identified in each field and marked with flags for consistently recording phenological stages throughout the growing season (Figure 1; Supplementary Materials Figure S9). At each location, the phenological observation was performed, and five samples were considered as replications (Supplementary Materials Table S1). Soybean phenological stages were identified according to Endres and Kandel Endres and Kandel [44]. The phenological observation was started after the emergence of seedlings, and all the locations in both fields were assessed on the same day. The vegetative and reproductive stages were indicated with the letters V and R, respectively. The following phenological stages in soybean were considered (Supplementary Materials Figure S10): VC (cotyledon stage), V1 (first trifoliate), V2 (second trifoliate), V3 (third trifoliate), R1 (first flowering), R2 (full bloom), R3 and R4 (pod development), R5 and R6 (seed development), and R7 and R8 (plant maturation). These soybean phenological stages from the visual assessment were distinct and used in the crop phenological assessment, which was then compared to the CVI curves to determine whether any prominent curve features matched with soybean phenological stages.

2.4. CVIs for Phenological Analysis

The selection of the 15 CVIs (12 from RGB space, and 3 from other color spaces) was based on a comprehensive literature review focusing on three key criteria: (i) historical significance in vegetation monitoring—GCC and EXG were included as established benchmarks in PhenoCam networks [18,32]; (ii) proven performance in agricultural applications—NGRDI, GLI, and CIVE were selected based on their successful application in crop monitoring and precision agriculture [25,26]; and (iii) mathematical combinations of different color spaces—CVIs were chosen based on various applications in quantifying vegetation characteristics [27,28,45]. A summary of all the CVIs and their implication is presented in Table 1. We excluded indices requiring near-infrared bands or specific calibration procedures, focusing instead on RGB-based indices suitable for standard PhenoCam deployments.

Table 1. Description and applications of color vegetation indices (CVIs) used in this study.

CVI	Physiological Meaning	References
GCC	Represents the relative proportion of green reflectance in total visible reflectance, directly correlating with chlorophyll content and photosynthetic activity. It is widely used in PhenoCam networks for its simplicity and robustness across illumination conditions.	[18,32]
NGRDI	Normalizes the difference between green and red reflectance, making it sensitive to chlorophyll content and leaf structure changes. It is particularly effective in detecting early-season growth and senescence phases in crops.	[25,26]
NGBDI	Captures the relationship between green pigments and structural components represented by blue reflectance. It is used to distinguish vegetation from soil background, especially in the early growth stages.	[27]
MGRVI	A modified index that uses squared values to enhance sensitivity to green vegetation changes while reducing atmospheric effects. It is more responsive to subtle changes in canopy greenness than simple ratios.	[46]
RGBVI	Combines all three color channels to enhance vegetation detection, particularly effective in discriminating between vegetation and non-vegetation features under varying light conditions.	[47]
NDI	A scaled version of NGRDI that maintains the same physiological sensitivity while providing values in a more intuitive range (0–255), facilitating comparison across different imaging conditions.	[26]
GLI	Emphasizes green vegetation by incorporating all three channels with double weighting for green reflectance. It is effective in detecting subtle changes in canopy greenness during key growth stages.	[48]
RGRI	Simple ratio highlighting the inverse relationship between red and green reflectance, sensitive to chlorophyll degradation during senescence.	[49]

Table 1. *Cont.*

CVI	Physiological Meaning	References
EXG	Emphasizes excess green coloration by double weighting for the green component, making it particularly suitable for detecting healthy vegetation against soil backgrounds.	[50]
CIVE	A linear combination optimized for vegetation extraction, with coefficients derived from a statistical analysis of vegetation spectra. Effective in separating vegetation from the background under various illumination conditions.	[25,51]
VEG	A non-linear index developed to be less sensitive to illumination changes while maintaining sensitivity to vegetation changes. The parameter "a" helps optimize the index for specific vegetation types.	[52]
EXGR	Combines excess green and red information to improve vegetation detection, particularly useful in distinguishing vegetation during different phenological stages.	[26]
HSB_{hue}	Represents pure color information independent of brightness and saturation, making it robust against illumination changes. Particularly useful for tracking seasonal color changes.	[53]
$Lab_{a'}$	Utilizes the green–red opponent color axis from the Lab color space, providing a perceptually uniform measure of vegetation greenness that correlates well with human visual assessment.	[54]
DGCI	Combines hue, saturation, and brightness information to provide a comprehensive measure of "greenness" that correlates well with nitrogen status and overall plant health.	[45]

Note: The expansions and the formulas for calculating these CVIs are given in Sections 2.4.1 and 2.4.2.

2.4.1. CVIs from RGB Color Space

Twelve CVIs from the RGB color space, which are common in agricultural applications using other imaging platforms (e.g., handheld cameras, unmanned aerial systems) for estimating the vegetation fraction [25], identifying plant biomass [26], and weed monitoring [52], were considered [Equations (1)–(12)].

$$\mathrm{GCC} = \frac{G}{R+G+B} \tag{1}$$

$$\mathrm{NGRDI} = \frac{G-R}{G+R} \tag{2}$$

$$\mathrm{NGBDI} = \frac{G-B}{G+B} \tag{3}$$

$$\mathrm{MGRVI} = \frac{G^2-R^2}{G^2+R^2} \tag{4}$$

$$\mathrm{RGBVI} = \frac{G^2-B\times R}{G^2+B\times R} \tag{5}$$

$$\mathrm{NDI} = 128\left(\frac{G-R}{G+R}+1\right) \tag{6}$$

$$\mathrm{GLI} = \frac{2G-R-B}{2G+R+B} \tag{7}$$

$$\mathrm{RGRI} = \frac{R}{G} \tag{8}$$

$$\mathrm{EXG} = 2g-r-b \tag{9}$$

$$\mathrm{CIVE} = 0.441r-0.881g+0.385b+18.787{,}45 \tag{10}$$

$$\mathrm{VEG} = \frac{g}{r^a b^{(1-a)}} \tag{11}$$

$$\mathrm{EXGR} = \mathrm{EXG}-1.4r+g \tag{12}$$

In these equations, GCC is the green chromatic coordinate; R, G, and B are the red, green, and blue color intensity values of image pixels, respectively; NGRDI is the normalized green red difference index; NGBDI is the normalized green blue difference index; MGRVI is the modified green red vegetation index; RGBVI is the red green blue vegetation index; NDI is the normalized difference index; GLI is the green leaf index; RGRI is the red green ratio index; EXG is the excess green; g, r, b are the green/red/blue proportion; CIVE is the color index for vegetation; VEG is the vegitativen; $a = 0.667$ is a parameter in VEG [Equation (11)]; and EXGR is the excess green minus excess red.

2.4.2. CVIs from Other Color Spaces

Three CVIs from other color spaces, HSB (hue, saturation, and brightness) and Lab (Lightness, chroma1, chroma2), tested in the study were the HSB's hue ($\mathrm{HSB_{hue}}$), the Lab's inverted "a" ($\mathrm{Lab_{a'}}$), and the dark green color index (DGCI) derived from HSB. Of the three CVIs, the dark green color index (DGCI) is a common index used in identifying turfgrass quality [55] and nitrogen status in cotton [56]. The expressions for these CVIs are as follows:

$$\mathrm{HSB_{hue}} = H \tag{13}$$

$$\mathrm{Lab_{a'}} = 1/a \tag{14}$$

$$\mathrm{DGCI} = \frac{(H-60)/60) + (1-S) + (1-B')}{3} \tag{15}$$

where H, S, and B' are the hue, saturation, and brightness, respectively, of the HSB color space; and "a" is the chroma1 of the Lab color space.

Typically, the Lab "a" channel produced an inverted phenological trend; therefore, to follow the commonly associated phenological trend, the "a" channel value was inverted (as "$1/a$") and used for comparison.

2.5. *CVIs Analysis and Selection*

2.5.1. CVI Variation Comparison with Loess Smoothing

To evaluate the variations in a CVI observed trend, it is necessary to have a "reference curve" with which the variation can be compared. A smoothed curve from the respective PhenoCam data served as the reference curve in this study. The smoothed curve corresponding to the selected data was obtained using the `loess()` function [57], with a span value fixed at 0.3 [32]. This operation determined the smoothed curve parameters, and the R `predict()` function obtained the values of the smoothed curve.

It can be seen that the observed trend produced from normalized CVI values (raw points) was noisy (Figure 3) when compared to the loess-smoothed curve, and this variation may be due to the temporal variations in CVI values during the measurement period and not because of changes in the plant canopy among subsequent images. These patterns aligned with expected daily changes in solar position and intensity; the specific contribution of illumination conditions could not be quantified as direct light measurements were not collected. While the smoothed curve (red) had a gradual variation and was expected to reflect the actual crop phenological stages.

To determine the undulations in the trend, we used the ribbon or curve length. While other measures are also possible, this comprehensive curve length has physical meaning and is easy to interpret (Supplementary Materials Section S5; Figure S11). With the curve length, an observed trend with greater undulations is longer than a smooth trend [58]. The curve length was calculated as:

$$L_{\mathrm{curve}} = \sum_{i=1}^{n} \sqrt{1 + (\mathrm{CVI}_{i+1} - \mathrm{CVI}_i)^2} \tag{16}$$

where L_{curve} is the curve length, i is the day after planting, n is the total number of study days, and CVI_{i+1} and CVI_i are the normalized CVI values at a consistent time (e.g., 12:00 h; Figure 4) on the $i+1$ and ith day.

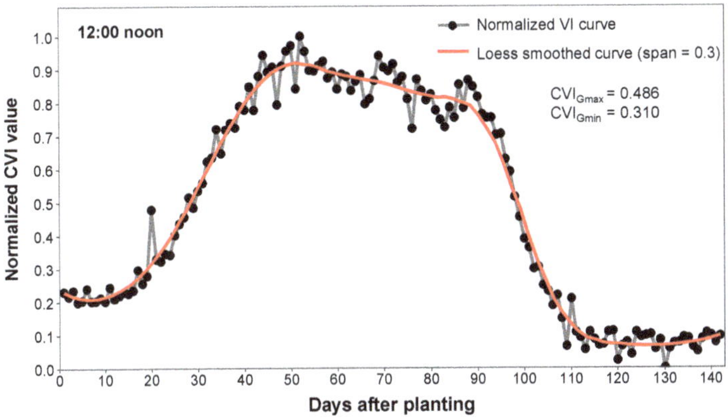

Figure 3. An example normalized GCC trend at 12:00 h (noon) and its loess-smoothed curve using a span of 0.3. Similar normalized CVI trends can be generated based on different times and ROIs.

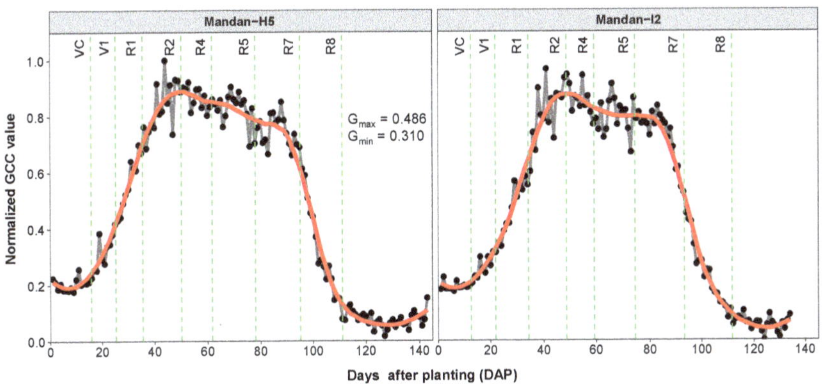

Figure 4. An example of a phenological curve using normalized GCC (gray) and its loess-smoothed values (red). Dashed lines indicate phenological stages from visual assessment. VC—cotyledon stage, V1—first trifoliate, R1—first flowering, R2—full bloom, R4—pod development, R5—seed development, R7—beginning maturity, and R8—matured plants.

It is practically challenging to minimize variations in the CVI trends caused by natural illumination in open atmospheric conditions. The deviation in length between the normalized raw CVI trend and the reference smooth curve provides a measure of variations caused by scene illuminations and other factors. Therefore, the deviation of the curve length is defined as:

$$L_{dev}\,(\%) = \frac{(L_{raw} - L_{smooth})}{L_{smooth}} \times 100 \qquad (17)$$

where L_{dev} is the deviation of the raw CVI with the smoothed trend (percent), L_{raw} is the length of the normalized raw CVI trend, and L_{smooth} is the length of the loess-smoothed curve (pixel). It represents the additional path length a curve takes compared to its smoothed curve. A few advantages of this L_{dev} are as follows: (i) scale independent—after normalization, L_{dev} provides a fair comparison between CVIs with different value ranges of other CVIs; (ii) physical interpretability—it directly represents curve roughness as excess path length; and (iii) sensitive to variations—it captures both magnitude and frequency of deviations. We have compared its functionality with the RMSE in (Supplementary Materials Section S5 and Figure S11).

2.5.2. Selection of Best Set of CVIs

The best set of CVIs was categorized as those that had a smooth variation across different times of image capture. In this study, the variation in a phenological curve caused by the illumination changes (time effect) was measured using the developed L_{dev} [Equation (17)]. This measure provided an idea of the extent of noise produced by each CVI with respect to its smoothed curve and was easy to physically interpret.

The L_{dev} produced by different CVIs was ranked in ascending order at the specific image acquisition time. The individual ranks at a specific time were summed across the time span (10:00 h–14:00 h), and a cumulative rank was evaluated. Based on the cumulative rank, a set of best CVIs was selected for further analysis.

2.6. Statistical Analysis and Visualizations

Statistical mean separation was performed using Duncan's multiple range test using the R package `agricolae` [59]. A mean comparison among L_{dev} values from different CVIs obtained from daily mean RGB value was performed to determine the CVIs that produced a similar variation. The daily mean RGB values are easily accessible, and they have been used to derive GCC and employed in various applications [60–62]. Therefore, it was desirable to compare among CVIs to determine those that produced a similar variation.

To study the effect of time and position (study variables) on the selected CVI's pattern, each CVI curve was divided into three major growth phases, vegetative, reproductive, and maturity, and the mean comparison was performed with the normalized CVI values to evaluate the separation of the curves. Dividing into three phases has been previously used in the estimation of biomass and other crop growth parameters [19]. The time span of 10:00 h–14:00 h at 30 min intervals, which produced nine variations, were grouped into three groups—start (10:00 h–11:00 h), midday (11:30 h–12:30 h), and end (13:00 h–14:00 h)—and mean comparisons among these groups were carried out. Similarly, the effect of position from the camera was statistically analyzed for three levels "far," "middle," and "near." The R package `ggplot2` [63] was used to visualize the results.

3. Results and Discussion

3.1. Phenology CVI Curves and Visual Assessment

3.1.1. Comparison of GCC Curve with Visual Assessment

The normalized GCC curve obtained from PhenoCam images of two fields, Mandan-H5 and Mandan-I2, followed a similar pattern (Figure 4). Even though the visual assessment was performed weekly, only the significant phenological stages are presented. In both fields, the GCC started to increase rapidly after the VC stage, which is ideally considered as the start of the growing season [44]. The VC stage in the field was characterized by the expansion of leaves, and this was precisely captured in the GCC trend with an increase in the GCC values. It was expected that the soybean would reach the VC stage within 10–15 days after planting (DAP), and this was consistent with our visual assessment and the GCC curve from PhenoCam images.

As the first vegetative stage (V1) was identified on ≈23 DAP, there was no prominent feature observed in the GCC trend except a consistent inclination in the GCC slope. The other vegetative stages (V2 and V3) were short and passed within a week, which was also not captured prominently in the GCC curve. The GCC values further increased and attained their maximum as the soybeans entered into the reproductive stage (R1). This was the stage when soybean started flowering; however, it happened only in the nodes and could not be seen at the canopy level (above) by the PhenoCam. To capture such subtle phenological stages, a close-up camera setup would be desirable, as employed in the PhenoCam experimental site "willamettepoplar" in Oregon [64].

As the soybeans entered the R2 stage, which was identified as "full bloom," the GCC curve attained the maximum peak on the 49th and 47th DAP in Mandan-H5 and Mandan-I2, respectively (Figure 4). This was a prominent phenological stage and was also characterized by canopy closure in the PhenoCam images. After R2, the GCC gradually decreased when the soybean "pod set" started. The decrease in GCC coincided with the end of flowering and the beginning of pod development that typically starts from the upper four nodes. Therefore, the canopy greenness declined as indicated by a downward slope after R2.

The gradual decrease continues up to 85–90 DAP, probably reaching the R6 stage called "full seed set." After R6, the seeds develop (R7) and turn from green to yellow. This phenomenon was observed by a sudden decrease in the GCC value at ≈95 DAP (Figure 4). At that stage, the crop is safe from major stresses [44]. As the crop matures and enters the R8 stage, it starts to shed leaves and transitions from yellow to brown. This was characterized by a further drop in GCC value to almost the minimum value, which indicated the end of the growing season. The crop is usually left in the field for 10–15 days after R8 to reduce the seed moisture content.

These results indicated that the GCC curve followed the actual soybean phenology as observed visually. With such interpretations, the GCC values can be used to obtain the phenological stages, such as VC, R2, and R8, without visiting the fields. The R6 stage might also be noticed with the start of a sharp decline at 85–90 DAP (Figure 4), which was missed in the visual assessment interval; however, this has to be validated in future studies. This could be a tool for organizations such as the USDA's National Agricultural Statistics Service and university research stations for preparing weekly crop progress reports during the growing season, which are now reported by manually visiting the fields. In future studies, the PhenoCam images can be potentially used to identify stress that causes discolorations in soybean leaves, such as iron deficiency chlorosis, as it has been proven successful with RGB images in aerial and ground sensing platforms [45,65–67].

3.1.2. Phenological Curve Pattern of Other CVIs

Soybean phenological curves for all CVIs showed a clear pattern highlighting different growth stages (Figure 5). Among the CVIs, RGRI and CIVE showed an inverse pattern because of their respective mathematical relationships. In the RGRI calculation [Equation (8)], the G value was considered inversely proportional to RGRI, whereas with CIVE [Equation (10)], it was a linear relationship. The increased weight and the negative sign assigned to g (0.881) produced the inverted pattern. Irrespective of the two different patterns (flipped and non-flipped), the phenological stages from visual assessment matched with the prominent features (e.g., peak or valley) of the CVI curve.

Ideally, the VC stage (Figure 5) is expected at 10–15 DAP in the CVI trend. In some of the CVIs, the VC stage was barely noticeable (e.g., NGBDI, RGRI, $Lab_{a'}$), making it difficult to identify the start of the growing season, while others (e.g., GCC, NGRDI, GLI, EXG, CIVE, EXGR, HSB_{hue}) prominently exhibited the VC stage. In previous PhenoCam studies, different threshold values, such as 10 %, 25 %, and 50 % of amplitude, were reported to identify the start of the growing season, which means the DAP at which the CVI value exceeds the threshold was identified as the VC stage [1,13]. In our study, since we normalized the CVI values, the threshold value could be directly seen with the graduations in the y-axis (Figure 5). Threshold values appeared to be specific to the CVI used. Based on the observations, a threshold value of 25% was good for many CVIs (e.g., GCC, RGBVI, EXG, VEG, HSB_{hue}) in representing the VC stage.

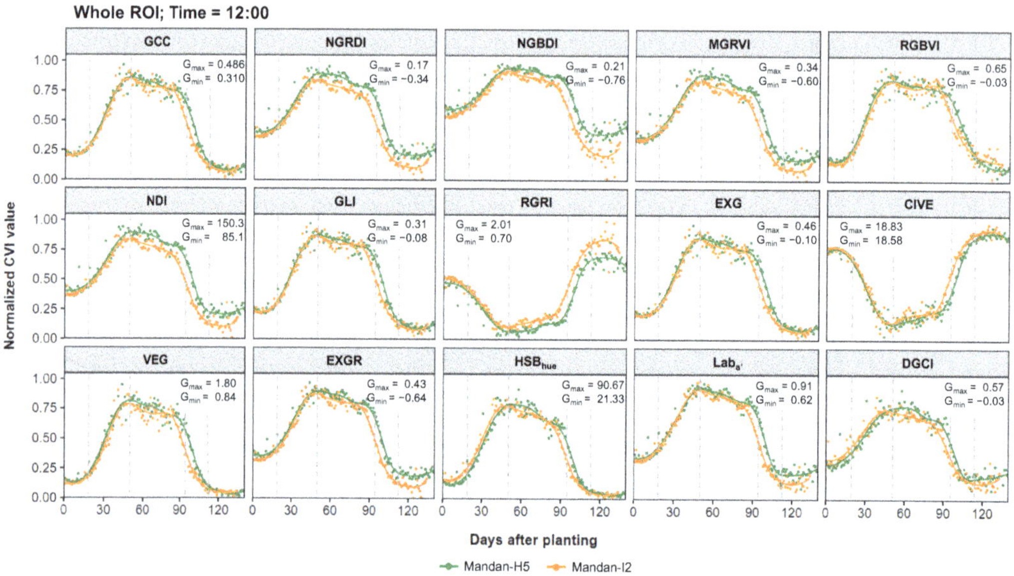

Figure 5. Phenological curves of other CVIs for Mandan-H5 and -I2 fields showing the different soybean phenological stages (points: normalized GCC; curve: loess-smoothed). The G_{max} and G_{min} are the global maximum and minimum values used in the normalization that can be used to infer the actual CVI values. The gray dashed lines indicate VC, R2, R7, and R8 phenological stages in soybean from left to right.

The peak bloom was identified in all the CVIs at 45–50 DAP, which matched the visual assessment. In a CVI curve, this was the point when it attained the maximum value except for RGRI and CIVE (minimum value). It was interesting to note that the crop growth was uniform in both fields until that point and displayed a minor separation afterward (Figure 5). This separation might be due to actual growth differences between fields or some of the agronomic factors such as field history, soil fertility, and nutrition.

The phenological transition into the maturity stage was faster in Mandan-I2, which matured faster than Mandan-H5 (Figure 5). This was observed with the separation in the CVI trend >85 DAP. However, this phenomenon was observed distinctly only with those CVIs in which the VC stage was not noticeable. Evaluating the precipitation data from both fields revealed that Mandan-H5 received comparatively more precipitation than Mandan-I2. This could be a possible reason for the delayed maturity in Mandan-H5, which caused a separation in the CVI curves. In addition, the inherent field conditions, such as topography and moisture availability, would have also produced some differences in crop growth between fields. Further investigations on the field history and other agronomic factors, such as soil type and quality and crop growth measures (e.g., biomass, leaf area index), may reveal the actual cause of separation at this phenological stage.

The end of the growing season was obvious in all CVI trends. This was characterized by the minimum value for all the CVIs, but the maximum value for RGRI and CIVE. After reaching the maturity stage, the CVIs attained a near-constant value without much variation. Some studies identified the end of the growing season using the same threshold value used to identify the start of season stage [13,68]. Similarly, in this study, the threshold value of 25 % was useful in identifying the end of the season with most of the CVIs tested. Although the manual selection of CVIs based on "user judgment" to screen the CVIs that matched the phenological stages is an important criterion, it is a subjective approach. Therefore, the selection of CVIs evaluated based on curve smoothness, which indicates the robustness of the image acquisition time and provides better insight into the nature of CVIs trends, is an objective and repeatable approach.

3.2. Statistical Comparison of L_{dev} and Selection of the Best CVIs

A statistical comparison of L_{dev} among various CVIs ($\alpha = 0.05$) helped in selecting a set of similar CVIs that produced the lowest L_{dev}. The GLI produced the minimum and DGCI produced the maximum deviations (Supplementary Materials Section S6; Figure S12. For all the CVIs from the RGB color space, Mandan-I2 consistently produced significantly higher L_{dev} ($\alpha = 0.05$) than Mandan-H5; while in HSB and Lab color spaces, Mandan-I2 produced lower deviations. This is an interesting observation and provides an insight that the mathematical relationships with the RGB color space and other color spaces are different. Also, the DGCI produced the minimum deviation between the two fields, but the phenological pattern was not as prominent as some of the CVIs, such as GCC, GLI, RGBVI, and VEG (Figure 5). Therefore, modifying the DGCI equation or deriving a new index similar to DGCI would possibly suit phenological applications, which could be an area for future research.

The selection of the best set of CVIs was based on the robustness of the curve at different times of image capture. The L_{dev} values of other CVIs between 10:00 h and 14:00 h at 30 min intervals are presented in Table 2. Among the CVIs, (i) GCC and EXG, and (ii) NGRDI and NDI, produced the same L_{dev}. This was because these two sets produced the same phenological pattern but on a different scale due to the similarity of the CVI expressions [Equations (1) and (9) and Equations (2) and (6)]. The normalized CVI curves were completely overlapping with each other as demonstrated with an example of GCC and EXG in (Supplementary Materials Figure S11). Therefore, for phenological comparisons, any one of the CVIs can be used (GCC/EXG and NGRDI/NDI).

The L_{dev} values for both fields ranged between 0.072 % and 0.412 %, indicating a wide span of noise (Table 2). Some of the CVIs were sensitive to illumination changes and produced high L_{dev} with low ranks. The CVIs from RGB produced lesser L_{dev} values and ranked higher than the CVIs from other color spaces. Out of all the CVIs, the EXGR produced the minimum L_{dev}, and DGCI produced the maximum L_{dev}.

Among the CVIs from other color spaces (HSB and Lab), HSB_{hue} ranked the best, followed by $\text{Lab}_{a'}$ and DGCI (Table 2). The reason for the low ranking (high L_{dev}) of DGCI was because the S and B' channels were influenced by color intensity variation [Equation (15)]. The DGCI was primarily developed for quantifying fine variations in green color in turfgrass [55]; hence, any small change in the intensity value produced more fluctuation, indicating sensitivity to illumination. When only HSB_{hue} [Equation (13)] was considered, ignoring S and B' channels, the L_{dev} was smaller compared to the DGCI (Table 2).

Based on the overall rank, the best four CVIs selected in the order of high to low ranking were EXGR, CIVE, GLI, and NGRDI/NDI. The L_{dev} from the daily mean RGB values of three of the selected CVIs, namely, EXGR, CIVE, and GLI, were statistically similar to those of GCC/EXG for both fields, while NGRDI/NDI was significantly different from GCC/EXG (Supplementary Materials S12). The L_{dev} produced by NGRDI/NDI was the highest among the selected CVIs, and this was evident from the cumulative rank as well (Table 2).

Table 2. Curve length deviations of 15 other CVIs from their respective loess-smoothed curve from the images captured between 10:00 h and 14:00 h during the 2018 growing season of soybean.

Vegetation Index	Curve Length Deviation from Smoothed Curve, L_{dev} (%)									Cumulative Rank	Overall Rank
	10:00 h	10:30 h	11:00 h	11:30 h	12:00 h	12:30 h	13:00 h	13:30 h	14:00 h		
Mandan-H5											
EXGR	0.119 ± 0.023 (2)	0.149 ± 0.014 (3)	0.127 ± 0.018 (3)	0.101 ± 0.014 (1)	0.124 ± 0.021 (1)	0.122 ± 0.014 (1)	0.094 ± 0.012 (1)	0.100 ± 0.015 (1)	0.098 ± 0.015 (3)	16	1
CIVE	0.115 ± 0.022 (1)	0.144 ± 0.013 (1)	0.142 ± 0.016 (5)	0.103 ± 0.017 (4)	0.132 ± 0.024 (3)	0.123 ± 0.019 (2)	0.113 ± 0.016 (5)	0.126 ± 0.024 (7)	0.143 ± 0.020 (8)	36	2
GLI	0.119 ± 0.021 (2)	0.144 ± 0.013 (1)	0.145 ± 0.015 (6)	0.102 ± 0.017 (3)	0.129 ± 0.023 (2)	0.125 ± 0.019 (3)	0.113 ± 0.015 (5)	0.129 ± 0.023 (8)	0.148 ± 0.019 (9)	39	3
NGRDI/NDI †	0.175 ± 0.025 (7)	0.177 ± 0.013 (7)	0.130 ± 0.019 (4)	0.123 ± 0.012 (6)	0.162 ± 0.021 (8)	0.172 ± 0.011 (8)	0.110 ± 0.009 (4)	0.113 ± 0.010 (4)	0.104 ± 0.014 (4)	52	4
RGRI	0.168 ± 0.027 (6)	0.211 ± 0.027 (13)	0.159 ± 0.032 (8)	0.130 ± 0.022 (8)	0.144 ± 0.027 (6)	0.166 ± 0.021 (6)	0.097 ± 0.015 (2)	0.113 ± 0.011 (4)	0.094 ± 0.013 (2)	55	5
GCC/EXG †	0.122 ± 0.022 (4)	0.150 ± 0.014 (4)	0.151 ± 0.017 (7)	0.109 ± 0.019 (5)	0.140 ± 0.026 (5)	0.132 ± 0.022 (4)	0.123 ± 0.017 (9)	0.139 ± 0.027 (9)	0.160 ± 0.023 (10)	57	6
NGBDI	0.189 ± 0.030 (10)	0.210 ± 0.033 (12)	0.171 ± 0.041 (10)	0.132 ± 0.028 (10)	0.135 ± 0.032 (4)	0.168 ± 0.031 (7)	0.098 ± 0.019 (3)	0.110 ± 0.015 (3)	0.090 ± 0.012 (1)	60	7
MGRVI	0.183 ± 0.025 (9)	0.172 ± 0.010 (6)	0.125 ± 0.016 (2)	0.125 ± 0.010 (7)	0.172 ± 0.020 (9)	0.178 ± 0.010 (9)	0.116 ± 0.009 (7)	0.117 ± 0.010 (6)	0.109 ± 0.015 (5)	60	7
VEG	0.140 ± 0.029 (5)	0.168 ± 0.020 (5)	0.171 ± 0.026 (10)	0.131 ± 0.028 (9)	0.185 ± 0.040 (11)	0.161 ± 0.035 (5)	0.165 ± 0.030 (11)	0.182 ± 0.047 (11)	0.215 ± 0.043 (11)	78	10
RGBVI	0.202 ± 0.029 (11)	0.186 ± 0.020 (8)	0.192 ± 0.026 (12)	0.159 ± 0.038 (12)	0.208 ± 0.041 (12)	0.225 ± 0.048 (11)	0.196 ± 0.029 (12)	0.229 ± 0.046 (12)	0.278 ± 0.042 (13)	103	12
HSB_{hue}	0.244 ± 0.028 (12)	0.196 ± 0.002 (9)	0.095 ± 0.006 (1)	0.101 ± 0.006 (1)	0.172 ± 0.010 (9)	0.237 ± 0.014 (12)	0.141 ± 0.008 (10)	0.107 ± 0.007 (2)	0.116 ± 0.010 (6)	62	9
$Lab_{a'}$	0.178 ± 0.006 (8)	0.198 ± 0.014 (10)	0.165 ± 0.006 (9)	0.136 ± 0.009 (11)	0.160 ± 0.004 (7)	0.179 ± 0.010 (10)	0.119 ± 0.007 (8)	0.158 ± 0.009 (10)	0.138 ± 0.010 (7)	80	11
DGCI	0.403 ± 0.009 (13)	0.206 ± 0.019 (11)	0.199 ± 0.017 (13)	0.176 ± 0.015 (13)	0.277 ± 0.009 (13)	0.342 ± 0.020 (13)	0.199 ± 0.012 (13)	0.251 ± 0.011 (13)	0.219 ± 0.006 (12)	114	13
Mandan-I2											
EXGR	0.113 ± 0.012 (1)	0.146 ± 0.007 (4)	0.106 ± 0.003 (4)	0.136 ± 0.009 (2)	0.116 ± 0.013 (4)	0.157 ± 0.006 (3)	0.136 ± 0.011 (4)	0.132 ± 0.011 (3)	0.173 ± 0.010 (3)	28	2
CIVE	0.119 ± 0.017 (4)	0.134 ± 0.014 (2)	0.115 ± 0.007 (6)	0.138 ± 0.006 (4)	0.143 ± 0.009 (8)	0.155 ± 0.015 (2)	0.162 ± 0.021 (7)	0.136 ± 0.022 (4)	0.169 ± 0.017 (2)	39	3
GLI	0.114 ± 0.016 (2)	0.133 ± 0.013 (1)	0.114 ± 0.007 (5)	0.136 ± 0.007 (2)	0.140 ± 0.008 (7)	0.146 ± 0.015 (1)	0.157 ± 0.020 (6)	0.13 ± 0.02 (2)	0.163 ± 0.016 (1)	27	1
NGRDI/NDI †	0.121 ± 0.010 (5)	0.166 ± 0.008 (6)	0.104 ± 0.011 (2)	0.145 ± 0.015 (7)	0.104 ± 0.018 (3)	0.172 ± 0.013 (6)	0.119 ± 0.010 (2)	0.137 ± 0.009 (5)	0.192 ± 0.014 (5)	41	4
RGRI	0.118 ± 0.012 (3)	0.178 ± 0.007 (8)	0.128 ± 0.018 (8)	0.153 ± 0.021 (8)	0.118 ± 0.024 (5)	0.176 ± 0.016 (8)	0.133 ± 0.017 (3)	0.146 ± 0.017 (8)	0.194 ± 0.022 (6)	57	8
GCC/EXG †	0.124 ± 0.019 (6)	0.138 ± 0.015 (3)	0.121 ± 0.008 (7)	0.144 ± 0.007 (6)	0.154 ± 0.008 (9)	0.160 ± 0.018 (4)	0.171 ± 0.023 (8)	0.141 ± 0.024 (7)	0.173 ± 0.019 (3)	53	6
NGBDI	0.133 ± 0.013 (8)	0.250 ± 0.008 (12)	0.174 ± 0.018 (11)	0.193 ± 0.020 (9)	0.159 ± 0.023 (10)	0.192 ± 0.014 (10)	0.149 ± 0.011 (5)	0.159 ± 0.012 (9)	0.208 ± 0.019 (8)	82	10
MGRVI	0.125 ± 0.010 (7)	0.158 ± 0.009 (5)	0.110 ± 0.010 (4)	0.143 ± 0.016 (5)	0.102 ± 0.018 (2)	0.173 ± 0.014 (7)	0.118 ± 0.010 (1)	0.138 ± 0.009 (6)	0.197 ± 0.014 (7)	44	5
VEG	0.180 ± 0.033 (10)	0.173 ± 0.028 (7)	0.164 ± 0.017 (10)	0.198 ± 0.016 (10)	0.238 ± 0.013 (12)	0.237 ± 0.038 (11)	0.258 ± 0.043 (13)	0.204 ± 0.049 (11)	0.234 ± 0.036 (11)	95	11
RGBVI	0.209 ± 0.034 (11)	0.236 ± 0.039 (11)	0.233 ± 0.015 (13)	0.246 ± 0.024 (13)	0.296 ± 0.020 (13)	0.269 ± 0.042 (12)	0.256 ± 0.032 (12)	0.224 ± 0.045 (12)	0.260 ± 0.049 (12)	109	12
HSB_{hue}	0.234 ± 0.009 (12)	0.180 ± 0.005 (9)	0.073 ± 0.005 (1)	0.127 ± 0.008 (1)	0.072 ± 0.012 (1)	0.178 ± 0.014 (9)	0.191 ± 0.008 (9)	0.116 ± 0.003 (1)	0.224 ± 0.012 (10)	53	6
$Lab_{a'}$	0.168 ± 0.012 (9)	0.228 ± 0.007 (10)	0.141 ± 0.018 (9)	0.199 ± 0.021 (11)	0.139 ± 0.024 (7)	0.166 ± 0.018 (5)	0.193 ± 0.012 (10)	0.164 ± 0.017 (10)	0.216 ± 0.022 (9)	80	9
DGCI	0.239 ± 0.011 (13)	0.272 ± 0.034 (13)	0.174 ± 0.032 (11)	0.245 ± 0.032 (12)	0.217 ± 0.030 (11)	0.307 ± 0.036 (13)	0.222 ± 0.021 (11)	0.282 ± 0.024 (13)	0.412 ± 0.025 (13)	110	13

Note: Values in parentheses indicate rank in ascending order of L_{dev} among CVIs at a specific time. † indicates both CVIs produced the same L_{dev} with the normalized CVI values (Supplementary Materials Section S5; Figure S11); hence, either of them can be used. EXGR — excess green minus excess red, CIVE—color index of vegetation, GLI—green leaf index, NGRDI —normalized green red difference index, NDI — normalized difference index, RGRI —red green ratio index, GCC —green chromatic coordinate, EXG — excess green, NGBDI —normalized green blue difference index, MGRVI —modified green red vegetation index, VEG —vegitativen, RGBVI —red green blue vegetation index, HSB_{hue}—hue channel value from HSB color space, $Lab_{a'}$ —inverted "a" channel value from Lab color space, and DGCI —dark green color index. Equations (1) and (15) define the CVIs used.

Overall, both fields produced similar ranks for the selected CVIs but in a different order based on L_{dev} values (Table 2). This difference may be due to the distance between the two fields (≈3.2 km apart) and possibly due to the difference in the cloud cover during image capture. Even though CIVE exhibited an inverted curve (Figure 5), it produced a low L_{dev} value among the tested CVIs and across different image acquisition times. This can also be corroborated by a previous study that reported CIVE was minimally affected by illumination conditions [69]. It is interesting to note that these CVIs produced lesser L_{dev} values and ranked higher than the commonly used GCC/EXG. These results support the hypothesis that some CVIs produce smaller variations than the existing CVIs in describing the soybean phenological stages.

Apart from the aforementioned CVIs, the GCC/EXG was also selected for further analysis and comparison as it has been widely used for phenological studies in the PhenoCam network [31,53]. The selected best four CVIs could be possibly considered in future PhenoCam related measurement and analysis. It will be also worthwhile to validate these CVIs with different crops and ecosystems (forest, rangeland, and agricultural).

3.3. Effect of Variables and Comparison of Selected CVIs

3.3.1. Effect of Image Acquisition Time on Selected CVIs

The GCC curves at different image acquisition times (10:00 h–14:00 h; nine trends) for the whole ROI are illustrated as an example (Figure 6). Irrespective of the image acquisition time, all GCC curves followed a similar trend (Figure 6A) but produced random fluctuations. The loess-smoothed curves (span = 0.3) of the three time groups (start: 10:00 h–11:00 h, midday: 11:30 h–12:30 h, and end: 13:00 h–14.00 h) are illustrated to demonstrate the separation among curves (Figure 6B). The time groups showed no separation throughout the growing season; however, the observed small separation occurred only during the reproductive phase. This phase is when many phenological changes occur in quick succession, influencing overall color (e.g., flowering, pod set, leaf yellowing).

In the normalized GCC curve with raw data (Figure 6A), all the curves were overlapping and steadily increased until 45–50 DAP (vegetative phase), beyond which more noise was observed until 40–85 DAP (reproductive phase). After 85 DAP (maturity phase), all the trends merged again and produced less variation among DAP. The variations among DAP were random and entirely due to ambient conditions. Ideally, the trends are expected to be smooth, but the fluctuations caused by image color differences due to cloud interference with the incoming sun's radiation are unavoidable. If desired, a color calibration procedure with a color calibration chart should be employed on captured images to minimize these variations [70].

Our uncertainty analysis revealed that temporal uncertainty was minimal during midday hours (11:30 h–12:30 h), with a coefficient of variation typically less than 9.6 % for normalized GCC. However, the coefficient of variation during the start (10:00 h–11:00 h) was 11.6 %, and it was 11.3 % during the end hours (12:00 h–14:00 h). Spatial uncertainty varied by growth phase, with 6.3 % during the vegetative phase as the canopy was developing, 5.2 % during the reproductive phase as the canopy was fully grown, and 13.9 % during the maturity phase as the leaves began to turn yellow and shed. The consistency of CVI calculations across both fields, combined with consistent image acquisition time and growth stages, supports the reliability of our methodology in minimizing measurement variability.

A statistical mean comparison analysis among image acquisition time groups (start, midday, and end) for the selected CVIs at each growth phase is presented in Table 3. The statistical analysis also revealed that the time groups did not have significant differences ($\alpha = 0.05$) in the CVI values for all three phases. The mean values at the reproductive phase were higher for all the CVIs except CIVE because of the inverted curve. The standard deviation (SD) values at the reproductive phase were less, as the distribution of CVI values was in a narrow range. More color changes occurred in the vegetative and maturity phases. The normalized CVI values among the selected CVIs for the overall CVI curve were significantly different within each time group (Supplementary Materials Section S7; Figure S13. Therefore, any time group can be used for comparing CVI curves with phenological stages. This result implies fewer images are required per day to build the phenological trend for a simple analysis, as any consistent time in the range from 10:00 h to 14:00 h can be suitable to capture the phenological stages.

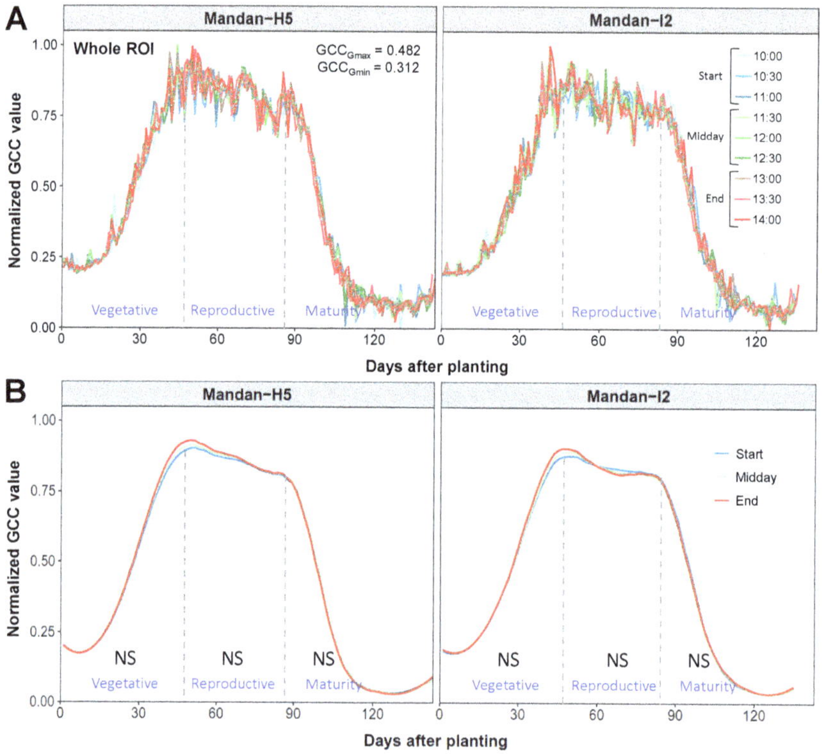

Figure 6. Effect of different image acquisition time (10:00 h–14:00 h at 30 min interval) on the commonly used GCC trend from the whole ROI. (**A**) Actual phenological curve of raw data showing random fluctuations obtained from images captured on each day; and (**B**) smoothed curves obtained after loess smoothing of the selected three time groups (start, midday, and end) that were statistically analyzed. Note: NS indicates no significant changes among the three time groups.

Table 3. Effect of image acquisition time and object position on selected CVIs within three growing phases and results of statistical comparisons.

Vegetation Index	Growth Phases	Image Acquisition Time: Start (10:00 h–11:00 h)	Image Acquisition Time: Midday (11:30 h–12:30 h)	Image Acquisition Time: End (13:00 h–14:00 h)	Object Position: Far (ROI-1–ROI-3)	Object Position: Middle (ROI-4–ROI-6)	Object Position: Near (ROI-7–ROI-9)
Mandan-H5							
EXGR	Vegetative	0.575 ± 0.228 a	0.574 ± 0.241 a	0.564 ± 0.263 a	0.565 ± 0.191 A	0.582 ± 0.195 A	0.544 ± 0.188 B
	Reproductive	0.887 ± 0.055 a	0.887 ± 0.061 a	0.898 ± 0.060 a	0.725 ± 0.085 C	0.789 ± 0.074 B	0.836 ± 0.054 A
	Maturity	0.390 ± 0.164 a	0.383 ± 0.161 a	0.374 ± 0.154 a	0.383 ± 0.139 C	0.427 ± 0.141 B	0.526 ± 0.139 A
	Overall	0.605 ± 0.264 a	0.602 ± 0.269 a	0.599 ± 0.281 a	0.549 ± 0.202 C	0.591 ± 0.208 B	0.631 ± 0.197 A
CIVE	Vegetative	0.494 ± 0.222 a	0.486 ± 0.224 a	0.486 ± 0.223 a	0.387 ± 0.176 C	0.428 ± 0.180 B	0.482 ± 0.152 A
	Reproductive	0.117 ± 0.059 a	0.109 ± 0.061 a	0.117 ± 0.059 a	0.146 ± 0.055 A	0.129 ± 0.053 B	0.142 ± 0.053 A
	Maturity	0.761 ± 0.258 a	0.759 ± 0.258 a	0.766 ± 0.245 a	0.590 ± 0.175 C	0.640 ± 0.195 B	0.685 ± 0.228 A
	Overall	0.474 ± 0.335 a	0.469 ± 0.337 a	0.474 ± 0.333 a	0.384 ± 0.236 B	0.410 ± 0.265 B	0.448 ± 0.280 A
GLI	Vegetative	0.459 ± 0.254 a	0.467 ± 0.260 a	0.469 ± 0.268 a	0.477 ± 0.221 A	0.457 ± 0.234 A	0.396 ± 0.208 B
	Reproductive	0.862 ± 0.057 a	0.871 ± 0.060 a	0.874 ± 0.057 a	0.737 ± 0.071 C	0.799 ± 0.063 B	0.830 ± 0.056 A
	Maturity	0.205 ± 0.240 a	0.206 ± 0.239 a	0.198 ± 0.232 a	0.240 ± 0.175 A	0.224 ± 0.193 A	0.232 ± 0.227 A
	Overall	0.492 ± 0.342 a	0.497 ± 0.345 a	0.496 ± 0.349 a	0.473 ± 0.266 A	0.481 ± 0.299 A	0.474 ± 0.312 A
NGRDI/NDI †	Vegetative	0.577 ± 0.188 a	0.581 ± 0.198 a	0.572 ± 0.194 a	0.593 ± 0.183 A	0.547 ± 0.187 B	0.498 ± 0.160 C
	Reproductive	0.906 ± 0.063 a	0.902 ± 0.066 a	0.893 ± 0.066 a	0.841 ± 0.066 B	0.831 ± 0.067 A	0.812 ± 0.059 A
	Maturity	0.286 ± 0.202 a	0.280 ± 0.201 a	0.268 ± 0.191 a	0.336 ± 0.159 B	0.270 ± 0.174 B	0.250 ± 0.202 A
	Overall	0.573 ± 0.305 a	0.570 ± 0.308 a	0.560 ± 0.306 a	0.578 ± 0.255 B	0.556 ± 0.279 B	0.607 ± 0.280 A
GCC/EXG †	Vegetative	0.441 ± 0.255 a	0.449 ± 0.262 a	0.452 ± 0.270 a	0.453 ± 0.222 A	0.434 ± 0.235 A	0.373 ± 0.208 B
	Reproductive	0.851 ± 0.061 a	0.860 ± 0.064 a	0.863 ± 0.062 a	0.714 ± 0.075 C	0.780 ± 0.067 B	0.813 ± 0.061 A
	Maturity	0.195 ± 0.233 a	0.196 ± 0.233 a	0.188 ± 0.226 a	0.222 ± 0.168 A	0.208 ± 0.186 A	0.217 ± 0.220 A
	Overall	0.479 ± 0.340 a	0.484 ± 0.344 a	0.484 ± 0.348 a	0.452 ± 0.263 A	0.462 ± 0.297 A	0.456 ± 0.311 A

Table 3. *Cont.*

Vegetation Index	Growth Phases	Image Acquisition Time			Object Position		
		Start (10:00 h–11:00 h)	Midday (11:30 h–12:30 h)	End (13:00 h–14:00 h)	Far (ROI-1–ROI-3)	Middle (ROI-4–ROI-6)	Near (ROI-7–ROI-9)
Mandan-I2							
EXGR	Vegetative	0.443 ± 0.281 a	0.438 ± 0.281 a	0.448 ± 0.310 a	0.498 ± 0.239 A	0.504 ± 0.228 A	0.478 ± 0.239 B
	Reproductive	0.797 ± 0.064 a	0.788 ± 0.087 a	0.796 ± 0.091 a	0.754 ± 0.092 B	0.757 ± 0.088 B	0.789 ± 0.086 A
	Maturity	0.307 ± 0.105 a	0.303 ± 0.107 a	0.312 ± 0.122 a	0.362 ± 0.142 C	0.406 ± 0.131 B	0.471 ± 0.133 A
	Overall	0.516 ± 0.272 a	0.510 ± 0.273 a	0.518 ± 0.284 a	0.536 ± 0.237 B	0.545 ± 0.227 B	0.591 ± 0.215 A
CIVE	Vegetative	0.443 ± 0.197 a	0.436 ± 0.201 a	0.438 ± 0.203 a	0.378 ± 0.157 B	0.399 ± 0.150 AB	0.421 ± 0.152 A
	Reproductive	0.112 ± 0.038 a	0.116 ± 0.051 a	0.120 ± 0.051 a	0.150 ± 0.048 A	0.133 ± 0.052 B	0.153 ± 0.051 A
	Maturity	0.763 ± 0.203 a	0.773 ± 0.195 a	0.768 ± 0.194 a	0.626 ± 0.146 B	0.654 ± 0.150 AB	0.687 ± 0.169 A
	Overall	0.439 ± 0.313 a	0.442 ± 0.315 a	0.442 ± 0.312 a	0.381 ± 0.233 B	0.391 ± 0.248 B	0.413 ± 0.260 A
GLI	Vegetative	0.459 ± 0.248 a	0.463 ± 0.251 a	0.469 ± 0.261 a	0.478 ± 0.222 A	0.455 ± 0.219 A	0.447 ± 0.219 A
	Reproductive	0.846 ± 0.043 a	0.842 ± 0.060 a	0.844 ± 0.064 a	0.790 ± 0.071 B	0.815 ± 0.070 A	0.829 ± 0.070 A
	Maturity	0.189 ± 0.175 a	0.172 ± 0.165 a	0.183 ± 0.171 a	0.223 ± 0.150 A	0.217 ± 0.153 A	0.213 ± 0.162 A
	Overall	0.495 ± 0.327 a	0.493 ± 0.328 a	0.498 ± 0.330 a	0.502 ± 0.283 A	0.501 ± 0.293 A	0.501 ± 0.302 A
NGRDI/NDI †	Vegetative	0.593 ± 0.184 a	0.595 ± 0.188 a	0.588 ± 0.195 a	0.645 ± 0.167 A	0.623 ± 0.158 A	0.617 ± 0.154 A
	Reproductive	0.854 ± 0.060 a	0.844 ± 0.077 a	0.838 ± 0.074 a	0.869 ± 0.064 A	0.851 ± 0.070 B	0.839 ± 0.068 B
	Maturity	0.205 ± 0.169 a	0.195 ± 0.157 a	0.202 ± 0.156 a	0.320 ± 0.131 A	0.292 ± 0.136 AB	0.278 ± 0.152 B
	Overall	0.551 ± 0.305 a	0.545 ± 0.306 a	0.543 ± 0.302 a	0.582 ± 0.266 B	0.593 ± 0.263 B	0.616 ± 0.260 A
GCC/EXG †	Vegetative	0.438 ± 0.249 a	0.443 ± 0.252 a	0.449 ± 0.263 a	0.453 ± 0.224 A	0.431 ± 0.220 A	0.422 ± 0.221 A
	Reproductive	0.831 ± 0.046 a	0.827 ± 0.065 a	0.829 ± 0.068 a	0.769 ± 0.075 B	0.795 ± 0.075 A	0.811 ± 0.076 A
	Maturity	0.167 ± 0.177 a	0.160 ± 0.167 a	0.167 ± 0.172 a	0.205 ± 0.143 A	0.199 ± 0.143 A	0.195 ± 0.154 A
	Overall	0.478 ± 0.326 a	0.476 ± 0.327 a	0.482 ± 0.329 a	0.481 ± 0.282 A	0.480 ± 0.293 A	0.481 ± 0.302 A

Note: † — Both CVIs produced the same mean and standard deviation values after normalization, therefore either of them can be used. EXGR —excess green minus excess red [Equation (12)], CIVE —color index of vegetation [Equation (10)], GLI —green leaf index [Equation (7)], NGRDI —normalized green red difference index [Equation (2)], NDI —normalized difference index [Equation (6)], GCC —green chromatic coordinate [Equation (1)], and EXG —excess green [Equation (9)]. Dissimilar letter groups denote significant differences ($\alpha = 0.05$) in the CVI values among the image acquisition time groups (a) and object position groups (A–C).

3.3.2. Effect of Object Position on Selected CVIs

The daily GCC during the growing season for each sub-ROI of the images, representing the object position, captured at 12:00 h for Mandan-H5 and Mandan-I2 are presented in Figure 7. In the vegetative phase (up to 45–50 DAP; Figure 7), the far region (sub-ROIs: 1–3) exhibited higher GCC values than the near region (sub-ROIs: 7–9). The GCC values from the middle region (sub-ROIs: 4–6) were between the observed far and near regions. The loess-smoothed curves are presented (Figure 7B) to observe the separation of the phenological curves from far, middle, and near positions. Even though the crop growth was uniform over the field, this phenomenon was observed because of the plants' position with respect to the PhenoCam and its oblique viewing angle.

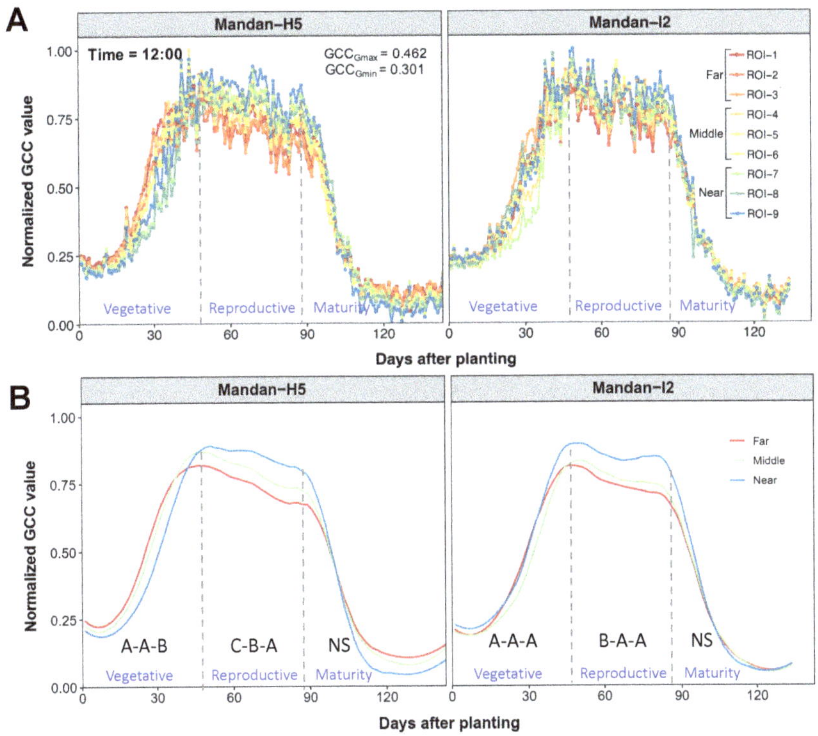

Figure 7. Effect of object position on the commonly used GCC trend from each sub-ROI of the images captured at 12:00 h. (**A**) Actual GCC phenological curve with raw data showing fluctuations with various sub-ROIs; and (**B**) loess-smoothed trend showing separation among curves in vegetative and reproductive phase but not in the maturity phase. Note: Uppercase letters (A through C) with different group labels indicate a statistically significant difference, and NS indicates non-significance ($\alpha = 0.05$). The order of letters A–A–B corresponds to far–middle–near ROIs.

In the vegetative phase, the far region appeared greener than the near region. Since the canopy was not closed, some portions of soil and residue were captured in the near region until 50 DAP. Hence, the GCC values were lower than in the far region, which actually represented a larger area than the area represented by the near ROI. The increased plant density (plants per ROI) in the far ROI presented a greener color even though the field plant density was uniform. Furthermore, the far region appeared blurry, due to the image depth and viewing angle, which eliminated the details of soil and residue in the vegetative phase. This pattern was more prominently expressed in Mandan-H5 than in Mandan-I2.

In the reproductive phase (after 50 DAP), the nature of trends reversed compared to the vegetative phase. The ROIs that were far from the camera showed lower GCC values than the ones that were near because of the image depth (which appeared blurry), which caused a color fading effect, while the regions near the camera showed sharp leaf features. Similar patterns were observed in a forest ecosystem with pine trees at high, mid, and low elevations, which indirectly represented distances [71].

In the maturity phase (after 85 DAP), all the curves converged as the leaves turned yellow and leaf drop ensued. As the crops passed on to the mid-maturity phase, all the curves merged and there was no visible difference among the object positions. This may be due to the appearance of only brown stems matching the topsoil color. The loess-smoothed GCC curves also displayed a similar pattern as that of raw GCC curves, but the separation was visually evident (Figure 7B).

A statistical mean comparison analysis indicated that there were significant differences among object positions on the curve pattern of selected CVIs (Table 3). Most of the CVIs showed mean groups of significant differences among object positions in at least two growth phases. In the vegetative phase, all the selected CVIs were significantly different in Mandan-H5, but a few of the CVIs (GLI, NGRDI/NDI, and GCC/EXG) were not different in Mandan-I2. In the reproductive phase, all the CVIs showed at least two mean groups among the object positions in both fields (Table 3). Therefore, for studies comparing the CVI curve with biomass content, it is necessary to consider choosing an appropriate ROI location as these two phases are important for such applications.

The maturity phase consistently showed significant differences among object positions for EXGR, CIVE, and NGRDI/NDI in both fields. Among the selected CVIs, the normalized CVI values were significantly different from each other within object position groups (Supplementary Materials Figure S13). This shows that if the ROIs are selected randomly across a landscape, as reported in [72,73], there is a possibility of having significantly different values among positions. Therefore, it is essential to use a fixed ROI at the center of the camera's field of view for an entire growing season for phenological comparisons.

As the object position significantly alters the CVI curves, these results are crucial to compare with crop growth parameters (e.g., biomass, leaf area index). It is therefore necessary to consider horizontally aligned smaller ROIs (consistent distance from the camera) or a larger ROI preferably in the middle (ROI #5) to address the variations. It can also be concluded that a smaller sub-ROI cannot be used to replace a larger ROI to capture good phenological variations based on the observed variations. Thus, the general conception that ROI placement in the image would have a negligible effect, in other words, the hypothesis of negligible variation among object positions, was rejected.

3.3.3. Comparison of L_{dev} Among Selected CVIs Within Time Groups and Position

A statistical comparison among the L_{dev} values of the four selected CVIs (EXGR, CIVE, GLI, and NGRDI/NDI) and GCC/EXG within time groups and object positions ($\alpha = 0.05$) offered insights on whether the CVIs were statistically similar (Supplementary Materials Section S8; Figure S14). Overall, within each time group and object position, the L_{dev} values followed a similar pattern, with the lowest from either GLI or EXGR and the highest from NGRDI/NDI.

Within each time group, the NGRDI/NDI was significantly different from other CVIs. The GCC/EXG produced a slightly higher L_{dev} compared to three of the selected CVIs (EXGR, CIVE, and GLI), but it was statistically similar (Supplementary Materials Figure S14). The selected CVIs produced less noise than the commonly used GCC/EXG even within each time group; therefore, the selected CVIs can be potentially employed in other phenological studies.

Within each object position, all the selected CVIs were statistically similar to each other. The GLI produced the minimum L_{dev}, and NGRDI/NDI produced the maximum L_{dev} (Supplementary Materials Figure S14). The near ROI position produced the highest L_{dev}, followed by the middle and far ROI positions. This can also be inferred from the phenological curve (Figure 7B) as the near ROI increased sharply and attained the maximum CVI value in the reproductive phase. In addition, the near ROI captured the sharp leaf features, thus producing more noise, while the middle and far ROI appeared blurry and produced lower noise. Therefore, while choosing an ROI for phenological analysis, the near region from the camera should be avoided, and the largest possible ROI in the middle region is recommended.

Our findings provide important advances in PhenoCam applications for agricultural environments. Recent studies have established quantitative relationships between developmental stages and phenotypic traits [5] but lack standardized protocols for image analysis. Our systematic evaluation of ROI position and its influence on color values provide insights for standardizing PhenoCam guidelines in agricultural environments, even though PhenoCam senses crop canopies homogeneously. The significant effects of object position on CVI values that we observed (Table 3) emphasize the importance of consistent ROI selection for reliable phenological comparisons.

Moreover, our temporal analysis demonstrates how image acquisition timing affects the detection accuracy of key phenological stages (green-up, peak greenness, and senescence), compared to manual field measurements. While previous studies used broad time windows [32,33], our finding shows that midday measurements (11:30 h–12:30 h) provide consistent results that help to optimize data collection protocols. This is particularly valuable for large-scale agricultural monitoring where resource use efficiency is important.

These methodological advances align with Richardson's vision [7] of PhenoCam's transformative potential, particularly in bridging the gap between high-frequency monitoring requirements and practical field applications in agricultural research. While our study focused on soybean, the geometric and optical principles underlying our guidelines for ROI position and image timing are applicable across different crops and seasons. Future work could validate these guidelines across diverse agricultural systems and multiple growing seasons to strengthen the standardization of PhenoCam protocols in agriculture.

4. Conclusions

This study generated valuable insights that can serve as guidelines in PhenoCam image processing for phenological measurement and analysis in an agricultural cropping environment. Specifically, applied to soybeans, the guidelines generated relate to reliable color vegetation indices (CVIs), suitable image acquisition time, and an appropriate object position through the image region of interest (ROI) for phenological analysis. Notably, certain features of the green chromatic coordinate (GCC) curve from PhenoCam images matched with the soybean phenological stages determined through visual assessment.

All the 15 CVIs tested followed a consistent pattern and captured the progression pattern of four significant growth stages: cotyledon (VC), peak bloom (R2), beginning maturity and seed development (R7), and end of maturity (R8). A developed measure, curve length deviation (L_{dev}), was compared to a loess-smoothed trend and adequately captured the variations in the phenological curve across different times. The best four CVIs selected in the RGB color space were (i) green leaf index (GLI), (ii) excess green minus excess red (EXGR), (iii) normalized green red difference index (NGRDI), and (iv) color index of vegetation (CIVE), based on the minimum L_{dev} across different times (10:00 h–14:00 h). On the other hand, the CVIs from other color spaces, such as HSB and Lab, were found less suitable.

The effect of image acquisition time groups (start: 10:00 h–11:00 h, midday: 11:30 h–12:30 h, and end: 13:00 h–14:00 h) on CVI values revealed no significant difference, suggesting that any consistent time with good illumination is sufficient. Object position (image ROI: far, middle, and near) exhibited a significant effect on CVI values, emphasizing the importance of selecting an appropriate region of the image for phenological comparisons. The normalized values among the selected CVIs exhibited significant differences within time and position groups, while the L_{dev} was significantly different only within time groups.

Overall, it is recommended to adopt a consistent image acquisition time that ensures sufficient light incidence, a larger ROI in the middle region of the field, and the application of the best-performing CVIs (GLI, EXGR, NGRDI, and CIVE). The GLI could be possibly considered in future phenological studies in the PhenoCam network due to its smooth phenological pattern and consistent performance comparable to GCC. These results have the potential to be incorporated as image processing guidelines within the standard protocol for an agricultural cropping environment in the PhenoCam network.

Supplementary Materials: The following supporting information can be downloaded at: https://www.mdpi.com/article/10.3390/rs17040724/s1. Section S1: Downloading Images From PhenoCam Network. Figure S1: Homepage of the PhenoCam network (Link: https://phenocam.nau.edu/webcam/, accessed on 8 February 2025). Figure S2: Login page for getting access to download images from PhenoCam network (Link: https://phenocam.nau.edu/webcam/accounts/login/, accessed on 8 February 2025). Figure S3: PhenoCam gallery page (top-portion) of the PhenoCam website (Source: https://phenocam.nau.edu/webcam/gallery/, accessed on 8 February 2025). Figure S4: PhenoCam image data download form (Link: https://phenocam.nau.edu/webcam/network/download/, accessed on 8 February 2025). Figure S5: Data request summary page showing entered details, number of images in the file, and download folder size (Link: https://phenocam.nau.edu/webcam/network/download/, accessed on 8 February 2025). Figure S6: The downloaded PhenoCam data is organized into folders by month. Section S2: Color Vegetation Indices (CVIs) Extraction From PhenoCam Provisional Data. Figure S7: An example page of downloading "Provisional Data" from PhenoCam network website (Link: https://phenocam.nau.edu/webcam/roi/mandanh5/AG1000/, accessed on 8 February 2025). Section S3: Image Color Value Extraction Through the Developed User-Coded ImageJ Plugin. Figure S8: Color values extraction process from PhenoCam images using the user-coded ImageJ plugin: (A) Process flowchart outlining the sequence of tasks, and (B) front panel of the developed plugin showing the options available for color values extraction and ROI specification. Section S4: Soybean Phenological Visual Assessment. Figure S9: Maps of Mandan-H5 and -I2 fields showing the paths of visual assessment and locations. Figure S10: Soybean grown stages information booklet and the URL information of this booklet and other soybean-related facts are given at the last panel. Table S1: LTAR common experiment at field scale in 2018 showing details of visual assessments performed. Section S5: Comparison Between GCC and EXG Before and After Normalization. Figure S11: Comparison of root mean square error (RMSE) and curve length deviation (L_{dev}) between GCC and EXG before and after normalization. Section S6: Statistical Comparison Among Studied CVIs. Figure S12: Curve length deviations for different CVIs with daily mean RGB values. Note: Dissimilar letter groups of the same color denote significant differences ($\alpha = 0.05$) among CVIs of Mandan-H5 (a–f) and Mandan-I2 (A–G). Section S7: Comparison of Normalized CVI Among Selected CVIs Within Time and Object Position Groups. Figure S13: Statistical comparison of normalized CVI values variation among selected CVIs on time groups (Start–End), and object position (Near–Far) on Mandan-H5 and -I2 fields. Dissimilar letter groups (a–c & A–D) denote the selected CVIs are significantly different ($\alpha = 0.05$). Section S8: Comparison of L_{dev} Among Selected CVIs Within Time and Object Position Groups. Figure S14: Statistical comparison of curve length deviations (L_{dev}) among selected CVIs on time groups (Start–End), and object position (Near–Far) on Mandan-H5 and -I2 fields. Dissimilar letter groups (a–b) denote the selected CVIs are significantly different ($\alpha = 0.05$).

Author Contributions: Conceptualization, S.S. and C.I.; methodology, S.S., N.S. and C.I.; formal analysis, S.S.; investigation, S.S.; resources, S.S., C.I., N.S., J.H. and D.A.; data curation, S.S.; writing—original draft preparation, S.S. and C.I.; writing—review and editing, S.S., C.I., N.S., J.H., D.A. and M.L.; visualization, S.S. and C.I.; supervision, C.I.; project administration, C.I.; funding acquisition, C.I., J.H. and D.A. All authors have read and agreed to the published version of the manuscript.

Funding: This work was supported in part by the Northern Great Plains Research Laboratory (NGPRL), USDA Agricultural Research Service (ARS), Mandan, ND, grant numbers: FAR0028541 and FAR0036174, the USDA National Institute of Food and Agriculture, Hatch Projects: ND01481 and ND01493. The development of PhenoCam has been supported by the Northeastern States Research Cooperative, NSF's Macrosystems Biology program (award EF-1065029), DOE's Regional and Global Climate Modeling program (award DE-SC0016011), and the US National Park Service Inventory and Monitoring Program and the USA National Phenology Network (grant number G10AP00129 from the United States Geological Survey). NGPRL research is funded by ARS project number 3064-21600-001-000D.

Data Availability Statement: The PhenoCam images are publicly available and can be downloaded at https://phenocam.nau.edu/webcam/gallery/ (accessed on 8 February 2025). The methodology outlined in the supplementary materials can be utilized to retrieve PhenoCam images. Details of visual assessment locations, dates, and soybean growth stages from field observation are included in the Supplementary Materials (S8).

Acknowledgments: The assistance provided by Justin Feld and Jessica Duttenhefner during the field crop visual assessment of soybean growth stages was highly appreciated. PhenoCam collaborators, including site PIs and technicians, are thanked for their efforts in support of PhenoCam. Site availability was facilitated by the Area 4 Soil Conservation Districts in North Dakota. This research was a contribution from the Long-Term Agroecosystem Research (LTAR) network. LTAR is supported by the United States Department of Agriculture.

Conflicts of Interest: The authors declare no conflicts of interest.

References

1. Richardson, A.D.; Hufkens, K.; Milliman, T.; Aubrecht, D.M.; Chen, M.; Gray, J.M.; Johnston, M.R.; Keenan, T.F.; Klosterman, S.T.; Kosmala, M.; et al. Tracking vegetation phenology across diverse North American biomes using PhenoCam imagery. *Sci. Data* **2018**, *5*, 1–24. [CrossRef] [PubMed]
2. Richardson, A.D. Tracking seasonal rhythms of plants in diverse ecosystems with digital camera imagery. *New Phytol.* **2019**, *222*, 1742–1750. [CrossRef] [PubMed]
3. Brown, T.B.; Hultine, K.R.; Steltzer, H.; Denny, E.G.; Denslow, M.W.; Granados, J.; Henderson, S.; Moore, D.; Nagai, S.; SanClements, M.; et al. Using phenocams to monitor our changing Earth: Toward a global phenocam network. *Front. Ecol. Environ.* **2016**, *14*, 84–93. [CrossRef]
4. Liu, Y.; Bachofen, C.; Wittwer, R.; Duarte, G.S.; Sun, Q.; Klaus, V.H.; Buchmann, N. Using PhenoCams to track crop phenology and explain the effects of different cropping systems on yield. *Agric. Syst.* **2022**, *195*, 103306. [CrossRef]
5. Aasen, H.; Kirchgessner, N.; Walter, A.; Liebisch, F. PhenoCams for field phenotyping: Using very high temporal resolution digital repeated photography to investigate interactions of growth, phenology, and harvest traits. *Front. Plant Sci.* **2020**, *11*, 593. [CrossRef]
6. Hill, A.C.; Vázquez-Lule, A.; Vargas, R. Linking vegetation spectral reflectance with ecosystem carbon phenology in a temperate salt marsh. *Agric. For. Meteorol.* **2021**, *307*, 108481. [CrossRef]
7. Richardson, A.D. PhenoCam: An evolving, open-source tool to study the temporal and spatial variability of ecosystem-scale phenology. *Agric. For. Meteorol.* **2023**, *342*, 109751. [CrossRef]
8. Guo, Y.; Chen, S.; Fu, Y.H.; Xiao, Y.; Wu, W.; Wang, H.; Beurs, K.d. Comparison of multi-methods for identifying maize phenology using phenocams. *Remote Sens.* **2022**, *14*, 244. [CrossRef]
9. Perez, M. Utilizing Phenocam Imagery and Convolutional Neural Network for Plant Phenological State Prediction. Master's Thesis, University of Twente, Enschede, The Netherlands, 2024.
10. Cao, M.; Sun, Y.; Jiang, X.; Li, Z.; Xin, Q. Identifying leaf phenology of deciduous broadleaf forests from phenocam images using a convolutional neural network regression method. *Remote Sens.* **2021**, *13*, 2331. [CrossRef]

11. Zhang, S.; Buttò, V.; Khare, S.; Deslauriers, A.; Morin, H.; Huang, J.G.; Ren, H.; Rossi, S. Calibrating PhenoCam data with phenological observations of a black spruce stand. *Can. J. Remote Sens.* **2020**, *46*, 154–165. [CrossRef]
12. Tran, K.H.; Zhang, X.; Ketchpaw, A.R.; Wang, J.; Ye, Y.; Shen, Y. A novel algorithm for the generation of gap-free time series by fusing harmonized Landsat 8 and Sentinel-2 observations with PhenoCam time series for detecting land surface phenology. *Remote Sens. Environ.* **2022**, *282*, 113275. [CrossRef]
13. Browning, D.; Karl, J.; Morin, D.; Richardson, A.; Tweedie, C. Phenocams bridge the gap between field and satellite observations in an arid grassland ecosystem. *Remote Sens.* **2017**, *9*, 1071. [CrossRef]
14. Watson, C.J.; Restrepo-Coupe, N.; Huete, A.R. Multi-scale phenology of temperate grasslands: Improving monitoring and management with near-surface phenocams. *Front. Environ. Sci.* **2019**, *7*, 14. [CrossRef]
15. Browning, D.M.; Snyder, K.A.; Herrick, J.E. Plant Phenology: Taking the Pulse of Rangelands. *Rangelands* **2019**, *41*, 129–134. [CrossRef]
16. Toda, M.; Richardson, A.D. Estimation of plant area index and phenological transition dates from digital repeat photography and radiometric approaches in a hardwood forest in the Northeastern United States. *Agric. For. Meteorol.* **2018**, *249*, 457–466. [CrossRef]
17. PhenoCam-Ag. PhenoCam Gallery—Agriculture Sites. 2024. Available online: https://phenocam.nau.edu/webcam/network/search/?sitename=&type=&primary_vegtype=AG&dominant_species=&active=unknown&fluxdata=unknown&group= (accessed on 8 February 2025).
18. Richardson, A.D.; Hufkens, K.; Milliman, T.; Frolking, S. Intercomparison of phenological transition dates derived from the PhenoCam Dataset V1. 0 and MODIS satellite remote sensing. *Sci. Rep.* **2018**, *8*, 5679. [CrossRef]
19. Zhang, X.; Jayavelu, S.; Liu, L.; Friedl, M.A.; Henebry, G.M.; Liu, Y.; Schaaf, C.B.; Richardson, A.D.; Gray, J. Evaluation of land surface phenology from VIIRS data using time series of PhenoCam imagery. *Agric. For. Meteorol.* **2018**, *256*, 137–149. [CrossRef]
20. Enquist, C.A.; Kellermann, J.L.; Gerst, K.L.; Miller-Rushing, A.J. Phenology research for natural resource management in the United States. *Int. J. Biometeorol.* **2014**, *58*, 579–589. [CrossRef]
21. Sun, Q. Plant Water Relations in Response to Drought and Different Cropping Systems. Ph.D. Thesis, ETH Zurich, Zurich, Switzerland, 2021.
22. Mendicino, R.; Tritini, S.; Mejia-Aguilar, A.; Monsorno, R. Development of a Smart Irrigation System for Apple Fields using a LoRaWAN Network. In Proceedings of the 9th International Conference on Sensors and Electronic Instrumentation Advances (SEIA' 2023), Funchal, Portugal, 20–22 September 2023; pp.70–73.
23. Richardson, A.D.; Jenkins, J.P.; Braswell, B.H.; Hollinger, D.Y.; Ollinger, S.V.; Smith, M.L. Use of digital webcam images to track spring green-up in a deciduous broadleaf forest. *Oecologia* **2007**, *152*, 323–334. [CrossRef]
24. Xue, J.; Su, B. Significant remote sensing vegetation indices: A review of developments and applications. *J. Sens.* **2017**, *2017*, 1353691. [CrossRef]
25. Torres-Sánchez, J.; Peña, J.M.; de Castro, A.I.; López-Granados, F. Multi-temporal mapping of the vegetation fraction in early-season wheat fields using images from UAV. *Comput. Electron. Agric.* **2014**, *103*, 104–113. [CrossRef]
26. Meyer, G.; Neto, J.C. Verification of color vegetation indices for automated crop imaging applications. *Comput. Electron. Agric.* **2008**, *63*, 282–293. [CrossRef]
27. Hatfield, J.L.; Prueger, J.H.; Sauer, T.J.; Dold, C.; O'brien, P.; Wacha, K. Applications of vegetative indices from remote sensing to agriculture: Past and future. *Inventions* **2019**, *4*, 71. [CrossRef]
28. Radočaj, D.; Šiljeg, A.; Marinović, R.; Jurišić, M. State of major vegetation indices in precision agriculture studies indexed in web of science: A review. *Agriculture* **2023**, *13*, 707. [CrossRef]
29. Barbedo, J.G.A. A review on the main challenges in automatic plant disease identification based on visible range images. *Biosyst. Eng.* **2016**, *144*, 52–60. [CrossRef]
30. Sunoj, S.; Igathinathane, C.; Hendrickson, J. *Monitoring Plant Phenology using Phenocam: A Review*; ASABE Paper No: 162461829; ASABE: St. Joseph, MI, USA, 2016. [CrossRef]
31. Andresen, C.G.; Tweedie, C.E.; Lougheed, V.L. Climate and nutrient effects on Arctic wetland plant phenology observed from phenocams. *Remote Sens. Environ.* **2018**, *205*, 46–55. [CrossRef]
32. Sonnentag, O.; Hufkens, K.; Teshera-Sterne, C.; Young, A.; Friedl, M.; Braswell, B.H.; Milliman, T.; O'Keefe, J.; Richardson, A.D. Digital repeat photography for phenological research in forest ecosystems. *Agric. For. Meteorol.* **2012**, *152*, 159–177. [CrossRef]
33. Linkosalmi, M.; Aurela, M.; Tuovinen, J.P.; Peltoniemi, M.; Tanis, C.M.; Arslan, A.N.; Kolari, P.; Böttcher, K.; Aalto, T.; Rainne, J.; et al. Digital photography for assessing the link between vegetation phenology and CO_2 exchange in two contrasting northern ecosystems. *Geosci. Instrum. Method. Data Syst.* **2016**, *5*, 417–426. [CrossRef]
34. Koebsch, F.; Sonnentag, O.; Järveoja, J.; Peltoniemi, M.; Alekseychik, P.; Aurela, M.; Arslan, A.N.; Dinsmore, K.; Gianelle, D.; Helfter, C.; et al. Refining the role of phenology in regulating gross ecosystem productivity across European peatlands. *Glob. Chang. Biol.* **2020**, *26*, 876–887. [CrossRef]

35. Richardson, A.D.; Braswell, B.H.; Hollinger, D.Y.; Jenkins, J.P.; Ollinger, S.V. Near-surface remote sensing of spatial and temporal variation in canopy phenology. *Ecol. Appl.* **2009**, *19*, 1417–1428. [CrossRef]
36. Richardson, A.D.; Seyednasrollah, B.; Hufkens, K.; Milliman, T. PhenoCam Installation Instructions (Revised August 2022). 2022. Available online: https://phenocam.nau.edu/pdf/PhenoCam_Install_Instructions.pdf (accessed on 8 February 2025).
37. Seyednasrollah, B.; Young, A.M.; Hufkens, K.; Milliman, T.; Friedl, M.A.; Frolking, S.; Richardson, A.D. Tracking vegetation phenology across diverse biomes using Version 2.0 of the PhenoCam Dataset. *Sci. Data* **2019**, *6*, 222. [CrossRef] [PubMed]
38. R Core Team. R: A Language and Environment for Statistical Computing. 2019. Available online: http://www.R-project.org/ (accessed on 8 February 2025).
39. RStudio Team. *RStudio: Integrated Development Environment for R. Ver. 1.2.1335;* RStudio, Inc.: Boston, MA, USA, 2018.
40. Seyednasrollah, B.; Young, A.M.; Hufkens, K.; Milliman, T.; Friedl, M.A.; Frolking, S.; Richardson, A.D.; Abraha, M.; Allen, D.W.; Apple, M.; et al. PhenoCam Dataset v2.0: Vegetation Phenology from Digital Camera Imagery, 2000–2018. 2019. Available online: https://daac.ornl.gov/cgi-bin/dsviewer.pl?ds_id=1674 (accessed on 8 February 2025).
41. Seyednasrollah, B.; Milliman, T.; Richardson, A.D. Data extraction from digital repeat photography using xROI: An interactive framework to facilitate the process. *ISPRS J. Photogramm. Remote Sens.* **2019**, *152*, 132–144. [CrossRef]
42. Rasband, W. ImageJ, US National Institutes of Health, Bethesda, MD, USA. 2018. Available online: https://imagej.net/ij/ (accessed on 8 February 2025).
43. Schindelin, J.; Arganda-Carreras, I.; Frise, E.; Kaynig, V.; Longair, M.; Pietzsch, T.; Preibisch, S.; Rueden, C.; Saalfeld, S.; Schmid, B.; et al. Fiji: An open-source platform for biological-image analysis. *Nat. Methods* **2012**, *9*, 676–682. [CrossRef] [PubMed]
44. Endres, G.; Kandel, H. Soybean Growth and Management Quick Guide. 2021. Available online: https://www.ndsu.edu/agriculture/sites/default/files/2021-11/a1174.pdf (accessed on 8 February 2025).
45. Hassanijalilian, O.; Igathinathane, C.; Doetkott, C.; Bajwa, S.; Nowatzki, J.; Esmaeili, S.A.H. Chlorophyll estimation in soybean leaves infield with smartphone digital imaging and machine learning. *Comput. Electron. Agric.* **2020**, *174*, 105433. [CrossRef]
46. Evstatiev, B.; Mladenova, T.; Valov, N.; Zhelyazkova, T.; Gerdzhikova, M.; Todorova, M.; Grozeva, N.; Sevov, A.; Stanchev, G. Fast Pasture Classification Method using Ground-based Camera and the Modified Green Red Vegetation Index (MGRVI). *Int. J. Adv. Comput. Sci. Appl.* **2023**, *14, 45–51*. . . [CrossRef]
47. Roth, R.T.; Chen, K.; Scott, J.R.; Jung, J.; Yang, Y.; Camberato, J.J.; Armstrong, S.D. Prediction of Cereal Rye Cover Crop Biomass and Nutrient Accumulation Using Multi-Temporal Unmanned Aerial Vehicle Based Visible-Spectrum Vegetation Indices. *Remote Sens.* **2023**, *15*, 580. [CrossRef]
48. Ribeiro, A.L.A.; Maciel, G.M.; Siquieroli, A.C.S.; Luz, J.M.Q.; Gallis, R.B.d.A.; Assis, P.H.d.S.; Catão, H.C.R.M.; Yada, R.Y. Vegetation Indices for Predicting the Growth and Harvest Rate of Lettuce. *Agriculture* **2023**, *13*, 1091. [CrossRef]
49. Sarkar, S.; Ramsey, A.F.; Cazenave, A.B.; Balota, M. Peanut leaf wilting estimation from RGB color indices and logistic models. *Front. Plant Sci.* **2021**, *12*, 658621. [CrossRef]
50. Woebbecke, D.M.; Meyer, G.E.; Von Bargen, K.; Mortensen, D. Color indices for weed identification under various soil, residue, and lighting conditions. *Trans. ASAE* **1995**, *38*, 259–269. [CrossRef]
51. Kataoka, T.; Kaneko, T.; Okamoto, H.; Hata, S. Crop growth estimation system using machine vision. In Proceedings of the 2003 IEEE/ASME International Conference on Advanced Intelligent Mechatronics (AIM 2003), Kobe, Japan, 20–24 July 2003; IEEE: Piscataway, NJ, USA, 2003; Volume 2; pp. b1079–b1083.
52. Hague, T.; Tillett, N.D.; Wheeler, H. Automated crop and weed monitoring in widely spaced cereals. *Precis. Agric.* **2006**, *7*, 21–32. [CrossRef]
53. Crimmins, M.A.; Crimmins, T.M. Monitoring plant phenology using digital repeat photography. *Environ. Manag.* **2008**, *41*, 949–958. [CrossRef] [PubMed]
54. Shajahan, S. Agricultural Field Applications of Digital Image Processing Using an Open Source ImageJ Platform. Ph.D. Thesis, North Dakota State University, Fargo, ND, USA, 2019.
55. Karcher, D.E.; Richardson, M.D. Quantifying turfgrass color using digital image analysis. *Crop Sci.* **2003**, *43*, 943–951. [CrossRef]
56. Raper, T.B.; Oosterhuis, D.M.; Siddons, U.; Purcell, L.C.; Mozaffari, M. Effectiveness of the dark green color index in determining cotton nitrogen status from multiple camera angles. *Int. J. Appl. Sci. Technol.* **2012**, *2*, 71–74. http://www.ijastnet.com/journals/Vol_2_No_1_January_2012/9.pdf.
57. Cleveland, W.S. Robust locally weighted regression and smoothing scatterplots. *J. Am. Stat. Assoc.* **1979**, *74*, 829–836. [CrossRef]
58. Sunoj, S.; Igathinathane, C.; Jenicka, S. Cashews whole and splits classification using a novel machine vision approach. *Postharvest Biol. Technol.* **2018**, *138*, 19–30. [CrossRef]
59. de Mendiburu, F. Agricolae: Statistical Procedures for Agricultural Research (V erion 1.3-1). 2019. Available online: http://CRAN.R-project.org/package=agricolae (accessed on 8 February 2025).
60. Hufkens, K.; Friedl, M.; Sonnentag, O.; Braswell, B.H.; Milliman, T.; Richardson, A.D. Linking near-surface and satellite remote sensing measurements of deciduous broadleaf forest phenology. *Remote Sens. Environ.* **2012**, *117*, 307–321. [CrossRef]

61. Toomey, M.; Friedl, M.A.; Frolking, S.; Hufkens, K.; Klosterman, S.; Sonnentag, O.; Baldocchi, D.D.; Bernacchi, C.J.; Biraud, S.C.; Bohrer, G.; et al. Greenness indices from digital cameras predict the timing and seasonal dynamics of canopy-scale photosynthesis. *Ecol. Appl.* **2015**, *25*, 99–115. [CrossRef]
62. Keenan, T.; Darby, B.; Felts, E.; Sonnentag, O.; Friedl, M.; Hufkens, K.; O'Keefe, J.; Klosterman, S.; Munger, J.W.; Toomey, M.; et al. Tracking forest phenology and seasonal physiology using digital repeat photography: A critical assessment. *Ecol. Appl.* **2014**, *24*, 1478–1489. [CrossRef]
63. Wickham, H. *ggplot2: Elegant Graphics for Data Analysis*; Springer: New York, NY, USA, 2016. Available online: https://ggplot2.tidyverse.org (accessed on 8 February 2025).
64. PhenoCam-Closeup. PhenoCam Close-Up Site—Willamettepoplar. 2019. Available online: https://phenocam.nau.edu/webcam/sites/willamettepoplar/ (accessed on 8 February 2025).
65. Kyaw, T.; Ferguson, R.B.; Adamchuk, V.I.; Marx, D.; Tarkalson, D.; McCallister, D. Delineating site-specific management zones for pH-induced iron chlorosis. *Precis. Agric.* **2008**, *9*, 71–84. [CrossRef]
66. Bai, G.; Jenkins, S.; Yuan, W.; Graef, G.L.; Ge, Y. Field-based scoring of soybean iron deficiency chlorosis using RGB imaging and statistical learning. *Front. Plant Sci.* **2018**, *9*, 1002. [CrossRef]
67. Dobbels, A.A.; Lorenz, A.J. Soybean iron deficiency chlorosis high throughput phenotyping using an unmanned aircraft system. *Plant Methods* **2019**, *15*, 97. [CrossRef] [PubMed]
68. Klosterman, S.; Hufkens, K.; Gray, J.; Melaas, E.; Sonnentag, O.; Lavine, I.; Mitchell, L.; Norman, R.; Friedl, M.; Richardson, A. Evaluating remote sensing of deciduous forest phenology at multiple spatial scales using PhenoCam imagery. *Biogeosci. Discuss.* **2014**, *11*, 4305–4320. [CrossRef]
69. Kazmi, W.; Garcia-Ruiz, F.J.; Nielsen, J.; Rasmussen, J.; Andersen, H.J. Detecting creeping thistle in sugar beet fields using vegetation indices. *Comput. Electron. Agric.* **2015**, *112*, 10–19. [CrossRef]
70. Sunoj, S.; Igathinathane, C.; Saliendra, N.; Hendrickson, J.; Archer, D. Color calibration of digital images for agriculture and other applications. *ISPRS J. Photogramm. Remote Sens.* **2018**, *146*, 221–234. [CrossRef]
71. Richardson, A.D.; Klosterman, S.; Toomey, M. Near-surface sensor-derived phenology. In *Phenology: An Integrative Environmental Science*; Springer: Berlin/Heidelberg, Germany, 2013; pp. 413–430. [CrossRef]
72. Yan, D.; Scott, R.L.; Moore, D.J.P.; Biederman, J.A.; Smith, W.K. Understanding the relationship between vegetation greenness and productivity across dryland ecosystems through the integration of PhenoCam, satellite, and eddy covariance data. *Remote Sens. Environ.* **2019**, *223*, 50–62. [CrossRef]
73. Snyder, K.A.; Huntington, J.L.; Wehan, B.L.; Morton, C.G.; Stringham, T.K. Comparison of Landsat and Land-Based Phenology Camera Normalized Difference Vegetation Index (NDVI) for Dominant Plant Communities in the Great Basin. *Sensors* **2019**, *19*, 1139. [CrossRef]

remote sensing

MDPI

Article

Sugarcane Mosaic Virus Detection in Maize Using UAS Multispectral Imagery

Noah Bevers [1], Erik W. Ohlson [2], Kushal KC [1], Mark W. Jones [2] and Sami Khanal [1,*]

[1] Department of Food, Agricultural, and Biological Engineering, The Ohio State University, Columbus, OH 43201, USA; bevers.5@osu.edu (N.B.); kc.7@osu.edu (K.K.)

[2] Corn, Soybean and Wheat Quality Research Unit, USDA-ARS, 1680 Madison Ave., Wooster, OH 44691, USA; erik.ohlson@usda.gov (E.W.O.)

* Correspondence: khanal.3@osu.edu

Abstract: One of the most important and widespread corn/maize virus diseases is maize dwarf mosaic (MDM), which can be induced by sugarcane mosaic virus (SCMV). This study explores a machine learning analysis of five-band multispectral imagery collected via an unmanned aerial system (UAS) during the 2021 and 2022 seasons for SCMV disease detection in corn fields. The three primary objectives are to (i) determine the spectral bands and vegetation indices that are most important or correlated with SCMV infection in corn, (ii) compare spectral signatures of mock-inoculated and SCMV-inoculated plants, and (iii) compare the performance of four machine learning algorithms, including ridge regression, support vector machine (SVM), random forest, and XGBoost, in predicting SCMV during early and late stages in corn. On average, SCMV-inoculated plants had higher reflectance values for blue, green, red, and red-edge bands and lower reflectance for near-infrared as compared to mock-inoculated samples. Across both years, the XGBoost regression model performed best for predicting disease incidence percentage ($R^2 = 0.29$, RMSE = 29.26), and SVM classification performed best for the binary prediction of SCMV-inoculated vs. mock-inoculated samples (72.9% accuracy). Generally, model performances appeared to increase as the season progressed into August and September. According to Shapley additive explanations (SHAP analysis) of the top performing models, the simplified canopy chlorophyll content index (SCCCI) and saturation index (SI) were the vegetation indices that consistently had the strongest impacts on model behavior for SCMV disease regression and classification prediction. The findings of this study demonstrate the potential for the development of UAS image-based tools for farmers, aiming to facilitate the precise identification and mapping of SCMV infection in corn.

Keywords: unmanned aerial systems; sugarcane mosaic virus; maize; multispectral sensor; machine learning

1. Introduction

Corn/maize (*Zea mays*) is one of the most important food crops globally and is the leading staple cereal in terms of annual production, exceeding 1 billion metric tons [1,2]. Ensuring the healthy and sustainable production of corn is, therefore, critical to maintaining food security and successful agricultural development on a global and local scale. One of the greatest challenges for farmers in maintaining profitable corn yields is the sustainable management of pests and pathogens. Annual global yield loss for corn due to pests and pathogens is approximately 22.5% of total yields [3]. One of the most important and widespread virus diseases in corn is maize dwarf mosaic (MDM), which is widely distributed in the USA and abroad.

MDM is induced by viruses in the family Potyviridae, most commonly by maize dwarf mosaic virus (MDMV) or by sugarcane mosaic virus (SCMV) [4]. SCMV infects various grain crops such as corn, sorghum, sugarcane, and other poaceous species, causing severe grain and forage yield losses among susceptible cultivars [4,5]. Monitoring the spread and emergence of new virus variants is critical for disease management [6,7]. Recently, a novel and highly virulent SCMV isolate was discovered in South Asia [8,9]. Furthermore, potyviruses are known to enhance other virus diseases. Notably, maize lethal necrosis (MLN) disease, caused by co-infection with maize chlorotic mottle virus and a potyvirus, devastated corn production in regions of East Africa, Southeast Asia, and South America [10–13]. The widespread distribution and diversity of MDM-inducing potyviruses, combined with their ability to cause severe disease individually or in co-infection with other viruses, highlight the long-term threat SCMV poses for the cultivation of corn and other grains [14,15].

Symptoms associated with MDM include mosaic, chlorosis, plant stunting, reduced biomass, and small ear size [16,17]. Early symptoms in young leaves appear as irregular, light, or dark green mosaicking or mottling patterns that may develop into greenish or yellowish streaks along the veins of the corn leaves [14]. Overall, visual symptoms alone are often insufficient for a positive identification of SCMV infection, and, thus, serological or molecular diagnostic tests are needed [14]. Currently, the preferred method for controlling SCMV is to grow resistant corn varieties. However, most commercial field corn hybrids commercially available in the USA are only partially resistant to SCMV and many sweet corn lines and varieties are highly susceptible to both MDMV and SCMV [18–21]. Furthermore, changes in climate, seasonal patterns, and extreme weather events may shift spatial and temporal SCMV population dynamics as well as the levels of SCMV virulence. For these reasons, the proactive development of better monitoring and identification techniques for virus detection is critical.

The early identification and continuous monitoring of pests and pathogens are key components in effective disease management and can help improve yields while also minimizing environmentally and economically costly control methods. Remote sensing leverages electromagnetic radiation as the information carrier [22], and so by capturing crop health information in the form of spectral signatures that cannot be detected by the human eye during traditional scouting approaches, multispectral remote sensing may provide an efficient and scalable approach to monitoring pathogen impacts [23]. UAS technology plays a vital role in precision agriculture by supporting the four pillars of farm input management: applying the right practice, at the right place, at the right time, and in the right quantity [24]. From pre-season planning to post-harvest, UAS technology can be integrated into every stage of production agriculture, enabling the collection of high-spatial- and -temporal-resolution images that can support efficient decision making, reductions in costs, and potential increases in yield and profit [24]. Since 2015, the use of UAS-based remote sensing has seen a sharp increase in agricultural applications [25], and there are now techniques that use spectral signatures, collected in the visible to near-infrared wavelengths by a multispectral sensor, to assess vegetation health. Specifically, vegetation indices derived from a combination of various spectral bands are useful in assessing the structural, physiological, or biochemical properties of vegetation [26–29]. Typically, healthy vegetation reflects large portions of near-infrared light and absorbs blue and red light for photosynthetic processes and chlorophyll production [30]. The reflectance spectra of vegetation can be analyzed via machine learning methods to identify anomalies or unique vegetation characteristics such as the presence of disease or nutrient stresses [25,31,32].

Several studies have investigated the spectral characteristics of mosaic viruses in crops, such as mosaic virus in sugarcane [33,34], mungbean yellow mosaic India virus in soybeans [35], wheat streak mosaic virus in wheat [36], and yellow mosaic disease in black beans [37]. In 2018, Moriya et al. [31] mapped the mosaic virus in sugarcane using in-field identification, spectroradiometer readings, and UAS-collected hyperspectral imagery. This involved creating a spectral reference library for healthy and infected leaves, weeds, and bare soil, followed by a spectral information divergence classification process of collected hyperspectral orthomosaics. The study successfully classified 74 out of 80 samples. Another similar study [34] used satellite imagery and the random forest algorithm to classify multi streak virus in corn, showing that the inclusion of vegetation indices as compared to images only improved classification accuracy by 3.4%, reaching 82.75%. However, hyperspectral imagery is expensive and complex to process, and satellite imagery has spatial resolution limitations.

Specific to corn, there have been a handful of studies that have investigated the spectral signatures of diseases, including foliar fungal diseases [29], maize dwarf mosaic virus [38,39], maize streak virus [40–42], northern leaf blight [43–45], grey leaf spot [46], and tar spot complex [47]. These studies utilize various spectral methodologies and data types, including multispectral satellite data, UAS-collected RGB/multispectral images, and proximal spectroradiometer reflectance measurements. However, to our knowledge, there have been no prior studies investigating the unique spectral characteristics of SCMV in corn and its corresponding relation to yields.

This study integrates a machine learning analysis of multispectral imagery collected via UAS for disease detection, focusing on three objectives: (i) identifying the most important/correlated spectral bands and vegetation indices with SCMV infection, (ii) comparing spectral signatures of mock-inoculated (noninfected) and SCMV-inoculated (infected) corn plants, and (iii) evaluating model prediction performances for four algorithms concerning early and late infections of SCMV.

2. Materials and Methods

2.1. Study Area and Experimental Setup

Experiments were conducted during corn growing seasons in 2021 and 2022 on research plots at Snyder and Schaffter Farms of the Ohio State University (OSU) located in Wooster, OH, USA (Figures 1 and 2). The Snyder Farm plots were approximately 0.64 acres in 2021 and 0.33 acres in 2022, containing Canfield silt loam soils. The 2022 Schaffter Farm plots were approximately 0.37 acres in size, containing Wooster-Riddles silt loams. Air temperature and relative humidity were similar for both years; however, the cumulative rainfall was greater for 2022 than for 2021 (Figure 3).

(**a**) (**b**) (**c**)

Figure 1. (**a**) Experimental layout in Synder field in 2021, overlaid on an aerial image captured on 14 July 2021. Red and blue boxes represent the randomized block design containing four single-row replicates of inoculated (red) and noninfected/mock-inoculated (blue) for each of the 51 hybrid varieties grown. (**b**) A corn crop that does not have SCMV infection or visual SCMV symptoms. (**c**) A corn crop that displays visual symptoms of SCMV infection.

Forty-eight commercial field corn hybrid varieties, provided by Rich Minyo, OSU, from the remnant seed of the Ohio Corn Performance Tests, were evaluated using a randomized block design in the 2021 field experiment. These varieties represent some of the highest-yielding varieties currently available from five major commercial seed companies and were found to be susceptible to SCMV from previous trials [20]. Thus, three in-house hybrid controls were used as SCMV-resistant (Oh28 × Pa405; Oh28 × Oh1VI) and -susceptible (Wf9 × Oh51A) controls. Seeds were planted with a Kinze 2100 planter (Kinze Manufacturing, Williamsburg, IA, USA) with Almaco seed distribution cones in single-row plots on May 20 in 2021.

Based on the 2021 field experiment, four susceptible varieties were selected from the 48 commercial hybrids for more in-depth characterization in the 2022 experiment, considering their disease incidence scores and yield penalty due to infection. Three hybrid controls were used in 2021 and Wf9 × Oh51A (hybrid 4)-susceptible control was used in 2022 to evaluate disease pressure and uniformity of inoculation. Seeds were planted in four-row plots on 23 May 2022. In each year, there were four replications per hybrid for each treatment (Figures 1 and 2). Row length was 6.7 m with 35 kernels per row. Rows were 0.76 m in width. For both years, the best agronomic practices achieved optimal yields without irrigation, as well as without the application of insecticides or fungicides.

The replicates included both SCMV-inoculated and mock-inoculated treatments (Figure 1). In 2021, Snyder Field had a total of 408 hybrid sample plots (rows) (51 hybrids with 4 replicates each). In 2022, each field had a total of 40 hybrid sample plots (blocks). Hybrids were inoculated as previously described in [19]. Briefly, SCMV-infected plant tissues were homogenized using a commercial blender in 0.01 M potassium phosphate buffer in a 1:10 wt/v ratio. The homogenate was filtered through fine cheesecloth. Then, 2 g of carborundum was mixed with 6 L of inoculum and applied using a gas-powered mist blower (Model 452 Solo, Newport News, VA, USA) at an approximate application rate of 1 L per 75 m. Mock inoculations were similarly applied using only 0.01 M potassium phosphate buffer and carborundum. Plants were inoculated between the V3 and V5 maturity stages. Inoculations were performed five times at five-day intervals between June and July (18 June, 24 June, 29 June, 1 July, and 6 July) in 2021 and four times (17 June, 22 June, 27 June, and 1 July) in 2022.

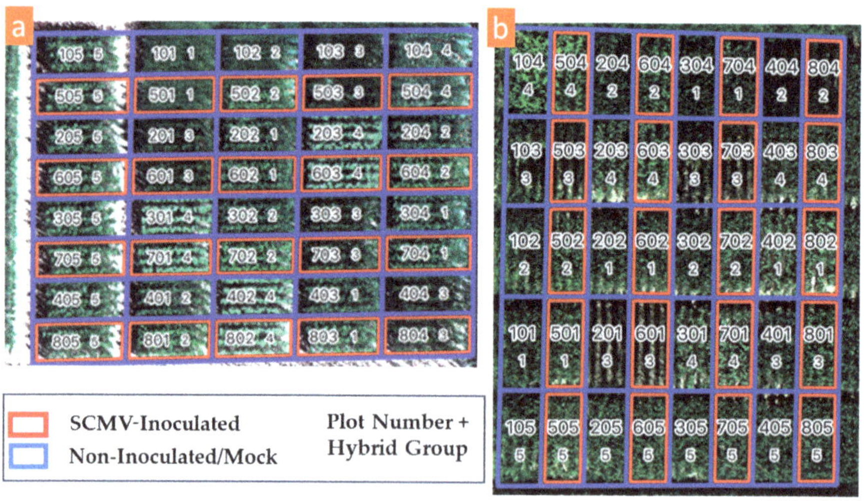

Figure 2. Experimental layout in Schaffter (**a**) and Synder (**b**) fields in 2022, overlaid on an aerial image captured on 1 September 2022, demonstrating randomized block designs with four 4-row replicates for each of the five hybrid varieties (including one control hybrid). Each box contains a unique plot number represented by a three-digit number, while the single-digit number corresponds to the hybrid group of that plot.

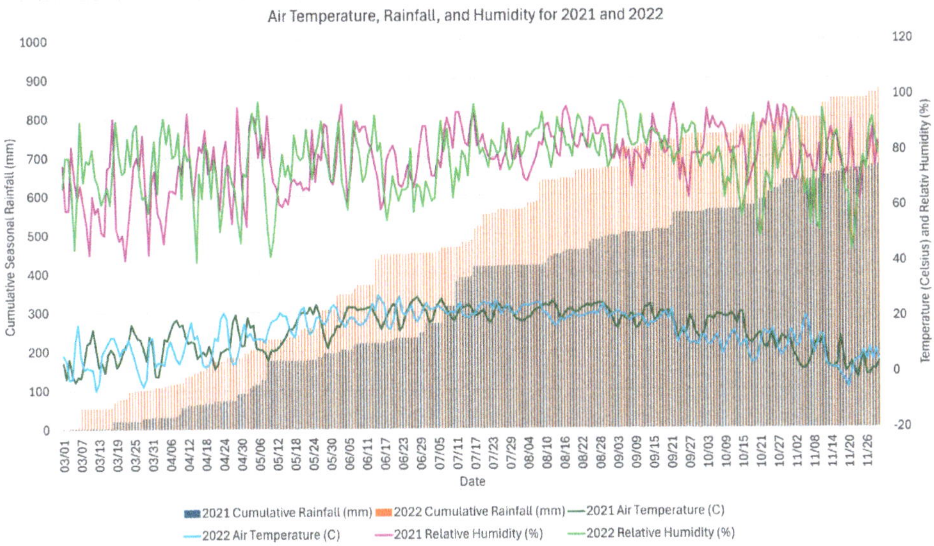

Figure 3. Cumulative rainfall (mm), average daily air temperature (Celsius), and average daily relative humidity (%) for 2021 and 2022 corn field seasons (March–November). Rainfall is displayed as bar plots that accumulate value as season progresses. Data collected from Ohio State University CFAES Weather System at the Wooster Station.

2.2. Data Collection

2.2.1. Imagery via UAS Flights

UAS flights were conducted using a DJI Matrice 200 drone (DJI, Shenzhen, China) on three dates each year for each field—(28 June, 14 July, and 28 July) in 2021 and (30 June, 28 June, and 1 September) in 2022—to capture trends in the spectral signatures of corn plants resulting from inoculations. During these flights, multispectral imagery with five spectral bands was collected using a MicaSense RedEdge-MX multispectral sensor (MicaSense, Seattle, WA, USA) at an altitude of 30m. The flights maintained a 75% front-overlap and 75% side-overlap, following a lawnmower pattern. To ensure the spatial accuracy of images respective to treatments on the field, four permanent ground control points (GCPs) were positioned at the corners of each corn field and remained visible and undisturbed throughout the growing season. Precise GPS coordinates for these GCPs were recorded using a Trimble real-time kinetic (RTK) GPS system (Trimble, Westminster, CO, USA).

Additional thermal imagery was collected with a DJI Zenmuse XT2 sensor (DJI, Shenzhen, China) using the same drone at an altitude of 20m with a 90% front-overlap and 90% side-overlap. Due to challenges in stitching the orthomosaic thermal images, usable images were obtained only for 28 June 2021 (Snyder), 14 July 2021 (Snyder), 30 June 2022 (Snyder), 30 June 2022 (Schaffter), 28 July 2022 (Snyder), and 1 September 2022 (Snyder).

LiDAR data were also collected using a Free Fly Alta X drone (Freefly Systems, Woodinville, WA, USA) at an altitude of 44m using a Velodyne VLP-16 Hi-Res sensor (Velodyne, San Jose, CA, USA). LiDAR data were collected for both fields only on 28 July 2022 to assess differences in canopy heights between treatments. For additional details on the data used in model development, refer to the subscript description in Table 1.

Table 1. Summary of available data collected for each field by date.

Dates [1,2]	Snyder Field	Schaffter Field
14 July 2021	Multispectral, Thermal IR	NA
28 July 2021	Multispectral	NA
28 July 2022	Multispectral, Thermal IR [3], LiDAR	Multispectral, LiDAR
1 September 2022	Multispectral, Thermal IR [3]	Multispectral

[1] Observations from plots (104, 204, 504, 604, 103, 203, 204, 604 in Figure 2b) across all dates in 2022 were excluded from the analyses due to flooding in the northwest corner of Snyder field. Additionally, observations from 'hybrid 4' were removed from analysis due to significant lodging. [2] All data collected during the June dates, i.e., 28 June 2021 and 30 June 2022, were excluded from machine learning model analyses as SCMV inoculation events had not yet been completed at this time, and, thus, the images do not reflect treatments implemented by the randomized block design. These portions of data were removed because any yield response or spectral response likely would not be due to the SCMV infection and would skew interpretation. The data from the June flights are only included in the model covered in Section 3.7, which includes data derived from additional sensors and also the models exploring the individual flight date performances. In total, 888 samples across both years were used for analysis across all flights and fields with 768 observations occurring in 2021 and 120 observations occurring in 2021. Additionally, in 2021, due to a data collection issue, multispectral imagery was only captured for 144 of the 408 sample plots on 28 June 2021. The in-house hybrid controls were also removed from model analysis. [3] Minor gaps were present in thermal orthomosaic representing 28 July 2022 and 1 September 2022 dates in Synder but were deemed useful to represent variations in canopy temperature due to SCMV inoculation. NA: not available.

2.2.2. Grain Yield

Yield values were recorded at the end of the season during harvest on 9 November 2021 and 21 November 2022 using a tractor-mounted yield monitor. Weight, moisture percentage, and bushels were used to calculate the standard 'bushels per acre' metric for each hybrid plot/block, and these values were interpreted as 'ground truth' yield values. Yields were normalized to a standard 15.5% moisture content and reported as kg/ha (or Bu/A).

2.2.3. Plant Disease Incidence

The percentage of plants that displayed mosaic symptoms for each hybrid plot was recorded using human scouting. Plant disease was evaluated based on the presence or absence of visible mosaic symptoms. Disease incidence for each hybrid variety was calculated based on the number of symptomatic plants divided by the final stand count. During 2021, symptoms were evaluated on 30 June 2021, 8 July 2021, and 12 July 2021. In 2022, the disease incidence scores were recorded on 15 July 2022 and 19 July 2022 for Schaffter Field and on 6 July 2022, 15 July 2022, and 18 July 2022 for Snyder Field. The percentage of plants symptomatic for SCMV within each hybrid sample plot (Figures 1 and 2) was recorded for each date. The value on the final scouting date was interpreted as the respective disease incidence rate and was used in subsequent analyses and modeling.

2.3. Data Processing

2.3.1. UAS Collected Imagery and Plant Reflectance

A flow diagram displaying the general methodological process implemented can be seen in Figure 4. The multispectral images collected by UAS were stitched into single-orthomosaic files of each field for each flight and for each of the five bands using Pix4D Mapper software Version 4.2.27 (Pix4D SA, Lausanne, Switzerland) [48]. To ensure that all images were aligned, the five-band composite files for each flight date were georeferenced based on the GCPs using ArcGIS Pro 3.2.0 software [49]. This process was also performed for the thermal images; however, some dates had stitching issues (see Table 1).

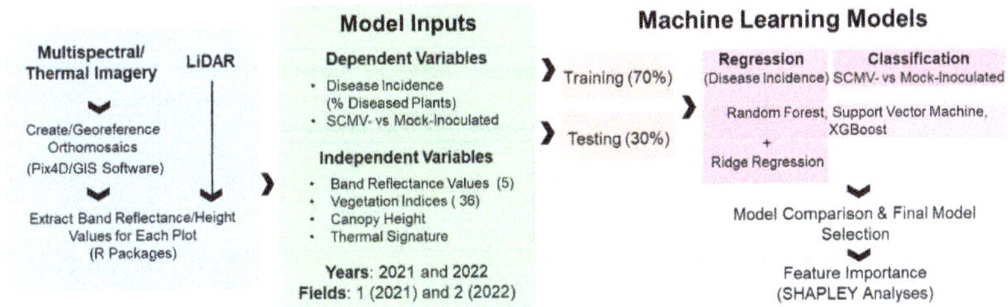

Figure 4. A flow diagram displaying the general methodological process implemented going from left to right. The best-performing regression models are identified for prediction of disease incidence percentage and binary classification prediction of disease presence. The best-performing models were further explored to identify the most important and most impactful features on the model's performance. The number in parentheses for independent variables defines the number of variables included as a model input.

Using the multispectral orthomosaics, mean pixel reflectance values were calculated and extracted for each of the five bands within each individual hybrid sample plot using the zonal statistics function in R v2022.07.0 Build 548 [50] for each field and flight date. To remove spectral noise from soil pixels, only pixels with corn vegetation were considered when summarizing reflectance values. For this, a threshold based on an excess green index, which amplifies green reflectance over red and blue [51,52], was used. In the threshold file, all vegetation pixels are assigned a value of one, and non-vegetation pixels are given a value of zero. For each flight, pixel-wise reflectance data for the five multispectral bands were extracted in R by multiplying each band file by the threshold file to remove non-vegetation pixels and then using an extract function to isolate reflectance values within each hybrid sample plot boundary. The same extraction process was applied to thermal orthomosaic pixels. A crop canopy height model was created from the LiDAR data collected using a .las file in CloudCompare v2.12.3 software [53], with more details reported in the Supplementary Materials.

2.3.2. Vegetation Indices

To further understand the effect of SCMV on corn vigor, 36 vegetation indices based on various combinations of spectral bands were calculated for each plot, alongside the individual spectral bands from multispectral (5) and thermal (1) sensors (Table 2). These vegetation indices include the following: BI, CI, CIG, CIRE, CVI, EVI, GARVI, GNDVI, gWDRVI 1, gWDRVI 2, HI, IRVI, lnRE, MCARI 1, MCARI 2, MCARIOSAVI, MSAVI, MSR, MTVI 1, MTVI 2, NDRE, NDVI, NGRDI, NIR/Green, NIR/Red, NIR/Red-Edge, OSAVI, RDVI, RI, SAVI, SCCCI, SI, TCARI, TCARIOSAVI, WDRVI 1, and WDRVI 2 (see Table 2).

Table 2. Vegetation indices (VIs) derived from the five-band multispectral images. VIs were used as features in modeling efforts.

Vegetation Index (VI)	Equation	Reference
Brightness Index (BI)	$(((R^2) + (G^2) + (B^2))/3)^{0.5}$	[54]
Coloration Index (CI)	$(R - G)/(R + G)$	[54]
Chlorophyll Index Green (CIG)	$(NIR/G) - 1$	[55]
Chlorophyll Index Red-Edge (CIRE)	$(NIR/Rdg) - 1$	[55]
Chlorophyll Vegetation Index (CVI)	$NIR-(R/(G^2))$	[56]
Enhanced Vegetation Index (EVI)	$(2.5 \times (NIR-R))/((NIR + 6 \times R-7.5 \times B) + 1)$	[57]
Green Atmospherically Resistant Vegetation Index (GARVI)	$(NIR - (G - (B - R)))/(NIR + (G - (B - R)))$	[58]
Green Normalized Vegetation Index (GNDVI)	$(NIR - G)/(NIR + G)$	[59]

Table 2. *Cont.*

Vegetation Index (VI)	Equation	Reference
Green Wide Dynamic Vegetation Index (α = 0.1; gWDRVI 1)	((0.1 × NIR − R)/(0.1 × NIR + R)) + ((1 − 0.1)/(1 + 0.1))	[60]
Green Wide Dynamic Vegetation Index (α = 0.2, gWDRVI 2)	((0.2 × NIR−R)/(0.2 × NIR + R)) + ((1 − 0.2)/(1 + 0.2))	[60]
Hue Index (HI)	(2 × R − G − B)/(G − B)	[54]
Inverse Ratio Index (IRVI)	R/NIR	[61]
Neparian Logarithm of the Red-Edge (lnRE)	100 × (ln_{NIR} − ln_R)	[62]
Modified Chlorophyll Absorption Ratio Index 1 (MCARI 1)	1.2 × (2.5 × (NIR − G) − 1.3 × (R − G))	[63,64]
MCARI 2	(3.75 × (NIR − R) − 1.95 × (NIR − G))/((((2 × NIR + 1)^2) − (6 × NIR − 5 × sqrtI) − 0.5))	[63,64]
Modified Chlorophyll Absorption Index/Optimized Soil-Adjusted Vegetation Index (MCARI/OSAVI)	(((Rdg − R) − 0.2 × (Rdg − G)) × (Rdg/R))/(1.16 × ((NIR − R)/(NIR + R + 0.16)))	[65]
Modified Soil-Adjusted Vegetation Index (MSAVI)	(2 × NIR + 1 − sqrt(((2 × NIR + 1)^2) − 8 × (NIR − R)))/2	[66]
Modified Simple Ratio (MSR)	((NIR/R) − 1)/(((NIR/R) + 1)$^{0.5}$))	[67]
Modified Triangular Vegetation Index 1 (MTVI 1)	(1.2 × (1.2 × (NIR − G) − 2.5 × (R − G)))	[64]
MTVI Index 2	(1.8 × (NIR − G) − 3.75 × (R − G))/(sqrt(((2 × NIR + 1)^2) − (6 × NIR − 5 × sI(R)) − 0.5))	[64]
Normalized Difference Red-Edge (NDRE)	(NIR − Rdg)/(NIR + Rdg)	[68]
Normalized Difference Vegetation Index (NDVI)	(NIR − R)/(NIR + R)	[69]
Normalized Green/REd Difference Index (NGRDI)	(G − R)/(G + R)	[70]
Ratio between NIR and Green bands (NIR/G)	NIR/G	[59]
Ratio between NIR and Red bands or Ratio Vegetation Index (NIR/R) (or RVI)	NIR/R	[71]
Ratio between NIR and Red-Edge bands (NIR/R-Edge)	NIR/Rdg	[71]
Optimized Soil-Adjusted Vegetation Index (OSAVI)	(1 + 0.16) × (NIR − R)/(NIR + R + 0.16)	[65]
Renormalized Difference Vegetation Index (RDVI) (broadband)	(NIR − R)/((NIR + R)$^{0.5}$)	[72]
Redness Index (RI)	(R^2)/(B × (G^3))	[54]
Soil-Adjusted Vegetation Index (L = 0.5, intermediate vegetation; SAVI)	1.5 × ((NIR − R)/(NIR + R + 0.5))	[73]
Simplified Canopy Chlorophyll Content Index (SCCCI)	NDRE/NDVI or $\frac{(NIR-Rdg)/(NIR+Rdg)}{(NIR-R)/(NIR+R)}$	[68,74]
Saturation Index or Normalized Pigment Chlorophyll Index (SI or NPCI)	(R − B)/(R + B)	[54,75]
Transformed Chlorophyll Absorption Reflectance Index (TCARI) (broadband)	(3 × ((Rdg − R) − 0.2 × (Rdg − G)) × (Rdg/R))	[76]
TCARI/Optimized Soil-Adjusted Vegetation Index (TCARI/OSAVI)	(3 × ((Rdg − R) − 0.2 × (Rdg − G)) × (Rdg/R))/(1.16 × ((NIR − R)/(NIR + R + 0.16)))	[76]
Wide Dynamic Vegetation Index 1 (α = 0.1; WDRVI 1)	(0.1 × NIR − R)/(0.1 × NIR + R)	[60]
WDRVI Index 2 (α = 0.2; WDRVI 2)	(0.2 × NIR − R)/(0.2 × NIR + R)	[60]

Note: The center wavelength and bandwidth of the five spectral bands included B—blue band (475 nm, 32 nm), G—green band (560 nm, 27 nm), R—red band (668 nm, 16 nm), Rdg—red-edge (717 nm, 12 nm), and NIR—near-infrared (842 nm, 57 nm).

2.4. Statistical Analysis and Machine Learning Models

2.4.1. Model Formation and Performance

A correlation scatterplot matrix was created using the psych library [77] in R [50] to investigate the linear correlation between each pair of available variables/features, including yield, canopy height, thermal value, disease incidence, the five spectral bands, and the 36 vegetation indices for each flight date. Specifically, a Pearson correlation (r) was analyzed for the SCMV-inoculated plots to explore the relationships between the spectral bands/vegetation indices and the 'disease incidence' scores. Correlation matrices were generated for all included variables and revealed the presence of multicollinearity, and this informed appropriate model selection. Kruskal–Wallis one-way analysis of variance [78] was also performed to test for statistically significant differences between SCMV-inoculated and mock-inoculated treatments for each of the spectral bands/vegetation indices.

To predict disease incidence and classify disease presence using multispectrally derived features, the performance of four commonly used machine learning models—ridge regression [79], support vector machine (SVM) [80], random forest (RF) [81], and XGBoost [82]—were evaluated. These models have robust capabilities for handling multicollinearity [79,83,84], which is often present among VIs derived using multispectral images. While XGBoost, SVM, and RF were used to predict the presence or absence of SCMV infection, XGBoost, SVM, RF, and ridge regression were used for predicting disease incidences.

For a detailed technical understanding of these algorithms, please see the original articles as cited. The modeling approach taken can be described primarily as a supervised machine learning-based strategy rather than a deep learning- or statistics-based approach [22]. Briefly outlining the selected models, ridge regression is a commonly used multiple regression technique for analyzing data that suffer from multicollinearity [79,85,86]. Ridge regression shrinks model coefficients and can help reduce overfitting and model complexity. The value for the penalty term or regularization parameter, lambda, was optimized for each ridge regression model using a grid search method. Support vector machine regression (SVM or SVR) tends to have very good generalization capability, is robust to outliers, and is effective even when the number of features/predictors is greater than the number of samples/observations [87–89]. For both ridge and SVR algorithms, all feature values were scaled to ensure features were contributing equally to the model. Random forest (RF) is a commonly used machine learning algorithm that averages the output of multiple individual regression trees to reach a single result [81]. RF models are an extension of bagging, which also randomly selects subsets of features [90]. RF models tend to be easily interpretable and efficient to train and display high generalization capabilities. XGBoost is a decision tree algorithm that implements regularized gradient boosting [82]. XGBoost training proceeds iteratively as new trees predict residuals of prior trees and then together yield a final result [82]. XGBoost is one of the leading machine learning algorithms, works well with non-scaled data, and is efficient and scalable. The SVM, rF, and XGboost algorithms were implemented in Python with scikit-learn libraries [91], while ridge regression was implemented in R using the caret and glmnet packages.

2.4.2. Model Performance Optimization (Using Multispectrally Derived Data)

The hyperparameters of all SVM, RF, and XGBoost models were optimized using Hyperopt in Python 3.7.16. Hyperopt is a library that provides sequential model-based optimization (also known as Bayesian optimization) for efficient function minimization, such as root mean squared error (RMSE), by exploring differing hyperparameter values [92]. While grid search (used for ridge regression) exhaustively examines all hyperparameter permutations, it becomes computationally demanding for algorithms with many parameters. Hyperopt navigates the hyperparameter space iteratively, leveraging prior evaluations for efficient exploration. Detailed hyperparameter spaces explored for each model are available in the Supplementary Materials (Table S1).

Initial models for predicting disease incidence used only multispectral-derived features, encompassing 5 bands and 36 indices, totaling 41. Models were trained for each individual year and combined years (all available data). Combined models included 888 samples across both years with 768 observations occurring in 2021 and 120 observations occurring in 2022. For this, five-fold cross-validation with a randomized split of 70% and 30% for training and testing sets, respectively, was used. Additionally, we also implemented the assigning of 2021 data for training and 2022 data for the testing of models, specifically for the XGBoost regression and SVM classification models due to their strong performances. Model performances were compared using RMSE and the coefficient of determination (R^2) as metrics.

For a more effective analysis of spectral bands and vegetation indices crucial for SCMV infection, a binary classification approach based on inoculation status was considered, rather than relying on a continuous 'disease incidence' score prone to human error and environmental factors. Classification models were built with two categories: SCMV-inoculated versus mock-inoculated. For the annotation of the classification models, the inoculated plots, regardless of disease incidence score, were set to values of 1, and non-inoculated plots were set to values of 0. Similar to regression models, all 41 features were incorporated into these classification models.

After identifying the best-performing models with multispectral-derived features, they were simplified using recursive feature elimination (RFE) as a feature selection tool to mitigate overfitting and enhance generalizability. RFE was applied to both years of data with the XGBoost model, maintaining optimized hyperparameters, and this was carried out using boostrfe and shap-hypetune in Python [93]. Following this simplification, feature importance and model behavior were analyzed using SHAP analysis. For the SVM classification model, a confusion matrix [94] was created to summarize SCMV infection status classification performance. Three metrics, including accuracy, precision, and recall, were estimated using predicted and actual observations (Equations (1)–(3)).

In the confusion matrix, true positive (*TP*) and true negative (*TN*) are represented by numbers in the diagonal and indicate the model's ability to accurately detect mock and SCMV inoculations, respectively. False positive (*FP*) refers to instances where the model predicts mock inoculations as SCMV inoculations. False negative (*FN*) indicates the model's failure to detect mock inoculations.

$$\text{Accuracy} = (TP + TN)/(TP + FP + TN + FN) \tag{1}$$

$$\text{Precision} = TP/(TP + FP) \tag{2}$$

$$\text{Recall} = TP/(TP + FN) \tag{3}$$

2.4.3. Modeling with Additional Available Data

Further models were investigated by incorporating additional features, such as LiDAR-derived canopy height and thermal imagery-based temperature values, to assess potential performance enhancement with data from extra sensors. These models utilized all available data from all sensors, extending beyond multispectral-derived information. XGBoost was selected for training these models due to its consistently superior performance with multispectral-only data. Hyperopt was employed once again to optimize the hyperparameters of these models. Additional variables such as thermal and canopy height that were available across both fields and years (Table 1) were incorporated into these models. This resulted in two additional features, totaling 43 features in the model. This modeling was only performed for all data in aggregate as the additional data occurred sporadically (i.e., both years were used and June dates were included so that all thermal data could be leveraged). Five-fold cross-validation was used with a total of 1059 observations and with a randomized split of 70% and 30% for training and testing sets, respectively.

2.4.4. Shapely Additive Explanations Analysis

SHAP analysis [95] was used to investigate model behavior and examine the influence of each feature on predicting the model output. This analysis was conducted for two best-performing models, one for disease incidence percentage and another for disease classification (inoculation vs. mock). TreeExplainer and KernelExplainer in Python were used for the SHAP analysis, which aids in interpreting the inner workings of machine learning models [96]. Feature importance and impact on model behavior were investigated using bar and summary plots. SHAP values quantify the contribution of each feature to model predictions. While SHAP values do not imply causation, they elucidate how the model behaves concerning predictions. In a SHAP plot, leftward dots along the x-axis indicate negative impacts, while rightward dots indicate a positive influence on the target variable (e.g., disease incidence percentage). The color gradient from red (high feature value) to blue (low value) illustrates feature significance. For instance, a red dot on the left side implies a negative contribution due to a high feature value. Features are ranked by average SHAP value, mirrored in the bar plots showing mean SHAP values along the x-axis, and offering insights into feature importance.

3. Results

3.1. Disease Incidence and Corn Yield

In 2021, disease incidence scores for commercial hybrids at Snyder Farm ranged from 5% to 100%, with median scores of 63% and mean scores of 60% (Figure 5). Susceptible (Wf9 × Oh51A) control had a mean of 96%, resistant (Oh28 × Pa405) control had 6%, and resistant (Oh28 × Oh1VI) control had 0%. Inoculated plot yields averaged 15,090 kg/ha (224 Bu/A), while non-inoculated plots averaged 16,821 kg/ha (250 Bu/A).

In 2022, disease incidence at Synder varied between 15% and 89%, with median and mean scores of 71% and 61%. Hybrid mean disease severities were 70%, 82%, 83%, 76%, and 21% for hybrids 1, 2, 3, 4, and 5, respectively. Mean yields for the inoculated plots were 12,987 kg/ha (193 Bu/A), while non-inoculated plots were 13,123 kg/ha (195 Bu/A).

At Schaffter Farm, disease incidences varied between 15% and 99%, with median and mean scores of 80% and 67%. Hybrid mean disease severities were 78%, 93%, 79%, 80%, and 18%, for hybrids 1, 2, 3, 4, and 5, respectively. The inoculated mean yield was 12,987 kg/ha (193 Bu/A), and the non-inoculated mean was 14,341 kg/ha (213 Bu/A).

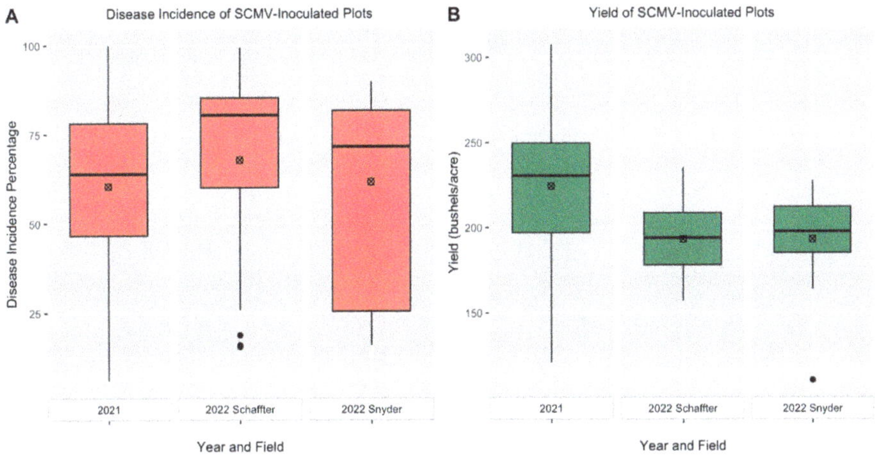

Figure 5. Boxplots displaying (**A**) the disease incidence percentage values and (**B**) average yield values (bushels/acre), for each of the fields and years involved in the study for the SCMV-inoculated field plots. The solid black lines inside the plots represent medians, and the X marks represent the mean for each plot.

3.2. Feature Correlation

Significant relationships between disease incidence and various vegetation indices (or spectral bands) were identified on 14 July 2021. SCCCI, TCARI/OSAVI, MACARI/OSAVI, and CVI showed the most significant correlations, with r values of −0.40, 0.30, 0.30, and 0.26, respectively. On 28 July 2021, SCCCI was found again as the most strongly correlated vegetation index with disease incidence, with an r value of −0.31, followed by NDRE, CIRE, and NIR/Red with r values of −0.24 each.

On 28 July 2022, no significant correlations were detected for disease incidence at Snyder. However, at Schaffter, disease incidence was significantly correlated with the NIR spectral band (r = −0.63), MCARI1 (r = −0.61), and MTVI1 (r = −0.60). On 1 September 2022, towards the later growth stage of corn, significant correlations were discovered at both locations. At Snyder, disease indices were most strongly correlated with NIR, MCARI1, and MTVI1 with r values of −0.63, −0.61, and −0.56, respectively. In Schaffter Field, MCARI2, NIR/Red, and MSR all had the highest r value of 0.56 (Table 3 and Supplementary Table S2). Compared to other VIs or spectral bands, SCCCI and MACRI were identified as consistently capturing variability in disease incidence at a higher percentage.

Table 3. Top vegetative indices and spectral bands with statistically significant correlation with disease incidence at each trial location.

2021	Snyder Farm	2022	Snyder Farm	Schaffter Farm
14 July 2021	SCCCI = −0.40	28 July 2022	Not Significant	NGRDI = 0.61
	TCARI/OSAVI = 0.3			
	MACARI/OSAVI = 0.3			CI = −0.61
	CVI = 0.26			HI = −0.60
28 July 2021	SCCCI = −0.31	1 September 2022	NIR = −0.63	MCARI2 = 0.56
	NDRE = −0.24			
	CIRE = −0.24		MCARI1 = −0.61	NIR/Red = 0.56
	NIR/Red = −0.24		MTVI1 = −0.56	MSR = 0.56

3.3. Analysis of Variance

On average for both years, SCMV-inoculated plots had a higher reflectance in the visible (red, green, blue) and red-edge portions of the electromagnetic spectrum and a lower average reflectance in the NIR portion as compared to mock-inoculated plots. On 14 July 2021, all bands showed significant differences, with significance in descending order: red, green, near-infrared, red-edge, and blue. On 28 July 2021, only red-edge and green had significant differences. Notably, no statistically significant differences were found in individual flight dates during 2022. Aggregated data for 2021 revealed significant differences in red-edge, red, and near-infrared bands, while no significant differences were found in the 2022 aggregated data (Figures 5, 6 and S1). Please see Figure S1 in the Supplementary Materials for boxplots displaying the comparison of the average reflectance values for each of the spectral bands for SCMV-inoculated and mock-inoculated corn plants.

On 14 July 2021, 37 out of 41 bands/indices exhibited statistically significant differences ($p < 0.05$) between SCMV-inoculated and mock-inoculated treatments. Of these variables, SCCCI exhibited the most difference with the highest chi-squared value, followed by NIR/Red-Edge, NDRE, and CIRE. On 28 July 2021, thirty of the variables showed significant differences, with SCCCI, NIR/Red-Edge, NDRE, and CIRE identified as being at the top.

In 2022, fewer significant differences in bands/indices between treatments were detected. On 28 July 2022, at Synder, no significant differences were found. However, at Schaffter, nearly significant differences were discovered between treatments for NIR/Green, CIG, and GNDVI ($p = 0.0596$). Similarly, on 1 September 2022 in Snyder Field, nearly significant differences were detected for NIR/Red-Edge, NDRE, and CIRE s ($p = 0.0596$). On 1 September 2022, at the Schaffter location, three variables were statistically significant: CVI, RI, and GARVI (ordered by descending chi-squared values).

Considering all 2021 data combined; thirty-seven variables (bands/indices) were found to be significantly different between treatments. The highest chi-squared values were found for SCCCI, followed by NIR/Red-Edge, NDRE, and CIRE. For all combined data (both fields) in 2022, no variables were found to be statistically significant. Thirteen variables were found to be significant for the aggregated Schaffter data in 2022. The highest chi-squared values were found for NIR/Red, IRVI, lnRE, gWDRVI_01, gWDRVI_02,

WDRVI_01, WDRVI_02, MSR, and NDVI. At Snyder in 2022, no variable was found to be statistically significant between SCMV-inoculated and mock-inoculated treatments.

Despite challenges in collecting and processing thermal and LiDAR canopy height data, useful insights were gained. On 28 July 2022, a positive and statistically significant correlation was found between canopy height and disease incidence for both Snyder (r = 0.34) and Schaffter (r = 0.53) fields. On the same date, Snyder showed no significant correlation between canopy height and corn yield (r = 0.00), while Schaffter displayed a statistically significant r value of 0.42. However, the relationship between thermal values and disease incidences was not statistically significant for any available flight dates (r = 0.14 for thermal on 28 June 2021, 0.02 for 14 July 2021, 0.07 for Schaffter on 30 June 2022, −0.26 for Snyder on 30 June 2022, −0.50 for Snyder on 28 July 2022, and 0.21 for Snyder on 1 September 2022). Thermal correlations with corn yield were significant on 28 June 2021 (r = −0.27) and 14 July 2021 (r = −0.17). All other thermal correlations with yield were not significant (r = −0.13, −0.32, −0.48, and −0.06 for 30 June 2022 at Snyder, 30 June 2022 at Schaffter, 28 July 2022 at Snyder, and 1 September 2022 at Snyder, respectively).

3.4. Regression Modeling of Disease Incidence

In 2021, both the random forest and XGBoost models for disease incidence achieved high performance, yielding R^2 values of 0.40 each, with RMSE values of 26.23 and 26.32, respectively (using a 70/30 random split). However, predictive ability declined notably in 2022, with R^2 values close to zero across all models. Combining data from both years, the XGBoost and random forest models remained the best performers, with R^2 values of 0.29 each and RMSE values of 29.35 and 29.26, respectively (Table 4). When using 2021 data to train and 2022 data to test XGBoost regression models, the model performed poorly, with an average RMSE of 45.67 and an R^2 value of −0.433.

Table 4. Model performances for prediction of disease incidence percentage by year.

Models	**2021**		**2022**		**2021 and 2022 Combined**	
	R^2	**RMSE**	**R^2**	**RMSE**	**R^2**	**RMSE**
Ridge Regression	0.30	28.78	0.02	39.50	0.21	30.99
Support Vector Regression	0.39	26.52	−0.06	39.34	0.25	30.43
Random Forest	0.40	26.23	−0.11	40.28	0.29	29.35
XGBoost	0.40	26.32	−0.07	39.66	0.29	29.26

Note: Models used 41 features as independent variables, and R^2 and RMSE are the average of three model repetitions using three unique 'seeds' for partitioning training and testing data.

Following feature selection via RFE, a simplified XGBoost disease incidence regression model incorporating data from both years with 14 features achieved an R^2 of 0.31 and an RMSE of 28.39. This closely matched the performance of the best XGBoost model utilizing all 41 features, with an R^2 of 0.29 and an RMSE of 29.26. The individual analysis of XGBoost models by UAS flight date revealed a stronger predictive ability for later (July) dates in 2021, particularly on 28 July 2021 with an R^2 of 0.43 and an RMSE of 26.19. This was closely followed by 14 July 2021 with an R^2 of 0.35 and an RMSE of 28.14. Despite the overall weak performance in 2022, a progressive improvement trend in model performances was observed throughout the season (Table 5).

Table 5. Performance of XGBoost model for prediction of disease incidence percentage by flight date.

Dates	R^2	RMSE
28 June 2021	0.03	33.98
14 July 2021	0.35	28.14
28 July 2021	0.43	26.19
30 June 2022	−0.32	41.94
28 July 2022	−0.20	38.85
1 September 2022	−0.10	37.16

Note: Models used 41 features as independent variables, and R^2 and RMSE are the average of three model repetitions using three unique 'seeds' for partitioning training and testing data.

3.5. Classification Modeling of SCMV Inoculation Status (Mock- vs. SCMV-Inoculated)

When evaluating the binary disease classification models for each year independently (using a 70/30 random split), SVM slightly outperformed XGBoost in 2021, with an accuracy of 0.759 (Table 6). Conversely, in 2022, random forest exhibited the highest performance, albeit poor, with an accuracy of 0.472. Combining data from both years, SVM remained the top performer with an accuracy of 0.729. SVM classification models trained on 2021 data and tested on 2022 data showed a slight improvement, with an average test accuracy of 57.2%. A confusion matrix is included alongside the SHAP analysis section for more details regarding the precision and recall of the best SVM performance (Figure 7).

Table 6. Accuracy of classification models for predicting SCMV-inoculated vs. mock-inoculated plots.

Models	2021	2022	2021 & 2022
Support Vector Machine	0.759	0.361	0.729
Random Forest	0.742	0.472	0.708
XGBoost	0.756	0.417	0.705

Note: Models included 41 spectral features. Accuracy reported is the average of three model runs with three unique 'seeds' for partitioning training and testing data.

3.6. XGBoost Regression Model for Disease Incidence

Based on the SHAP analysis of the best-performing XGBoost regression model for disease incidence prediction, the feature that had the overall strongest impact on model prediction behavior (i.e., most important) was the simplified canopy chlorophyll content index (SCCCI), followed by SI and TCARI/OSAVI (Figure 6). With 14 features determined by recursive feature elimination, the model achieved $R^2 = 0.312$ and RMSE = 28.39. Medium and low (purple and blue, respectively) values of SCCCI indicate a positive impact on the prediction value of disease incidence. The trend for the saturation index (SI) feature is less clear, but generally, low values have a negative impact, while medium to high values have a positive impact. Similarly, for TCARI/OSAVI, low to medium values have a negative impact, and mid–high to high values have a positive impact. The bar plot (5b) displays the average magnitude of the SHAP value for each feature, indicating the relative impact on XGBoost model predictions for disease incidence values.

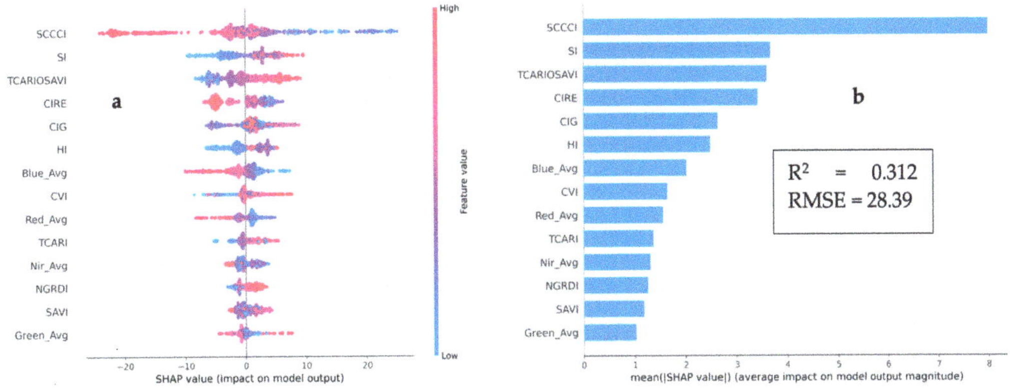

Figure 6. Summary plot ((**a**), left) and bar plot ((**b**), right) visualize SHAP analysis for the XGBoost regression model for the prediction of disease incidence.

3.7. Support Vector Machine Classification Model for SMCV Inoculation Status

Based on the SHAP analysis of the best-performing SVM classification model for binary SCMV infection prediction, the saturation index (SI) emerged as the feature with the strongest impact on model prediction behavior (Figure 7a). Following SI, TCARI/OSAVI, MCARI/OSAVI, and SCCCI were identified as the most impactful features. The confusion matrix provides more detailed metrics on the SVM model performance with the testing data (Figure 7b). For the test data, the model correctly recalled 100 out of 136 mock-inoculated samples (0.73) and 98 out of 131 SCMV-inoculated samples (0.75). The model misclassified 'mock-inoculated' and 'SCMV-inoculated' a total of 33 and 36 times, respectively. Precision scores for mock-inoculated and SCMV-inoculated samples were 0.75 and 0.73, respectively.

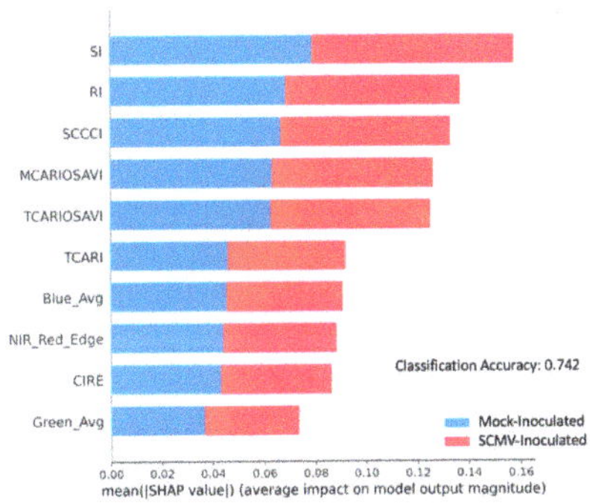

(**a**)

		Observed		
		Mock	SCMV	**Total (Predicted)**
Predicted	Mock	**100**	33	**133**
	SCMV	36	**98**	**134**
Total (Observed)		**136**	**131**	**267**

(**b**)

Figure 7. Bar plot (**a**) depicts SHAP analysis for SVM classification based on 41 features, using 'SCMV-inoculated' and 'mock-inoculated' as categories. The confusion matrix (**b**) displays model performance metrics on a total of 267 test data samples. Numbers in green and red-orange boxes were correctly and incorrectly classified, respectively.

3.8. Model Performance with Additional Features for Disease Incidence Prediction

The addition of thermal and canopy height features resulted in an R^2 of 0.225 and an RMSE of 30.73 for the XGBoost regression model trained on both years of data. This is a decreased performance in comparison to an R^2 of 0.29 and an RMSE of 29.26 for the equivalent XGBoost model that used 41 multispectrally derived features. Also, the equivalent RFE-simplified XGBoost model with just 14 multispectrally derived features outperformed both models with an R^2 of 0.312 and RMSE of 28.39. So, it appears that the additional sensor-based data did not improve model performance and that a reduction in features through RFE slightly improved overall performance.

3.9. Important Indices: SCCCI and SI

SHAP analysis of both the top-performing XGBoost regression and SVM classification models indicated that SCCCI and SI are two of the most important indices influencing model behavior in SCMV prediction. Further investigation of the SCCCI values across seasonal data indicates that, on average, mock-inoculated samples had consistently higher SCCCI values than SCMV-inoculated samples at all times post-inoculation (Figure 8a). Peak SCCCI values occurred during mid-to-late July and then declined as the season progressed toward harvest. Conversely, SI values exhibited an almost opposite trend (Figure 8b). Following inoculation, the SCMV-inoculated samples had higher SI values compared to mock-inoculated samples, with SI values generally increasing as the season progressed, peaking at the final data collection date.

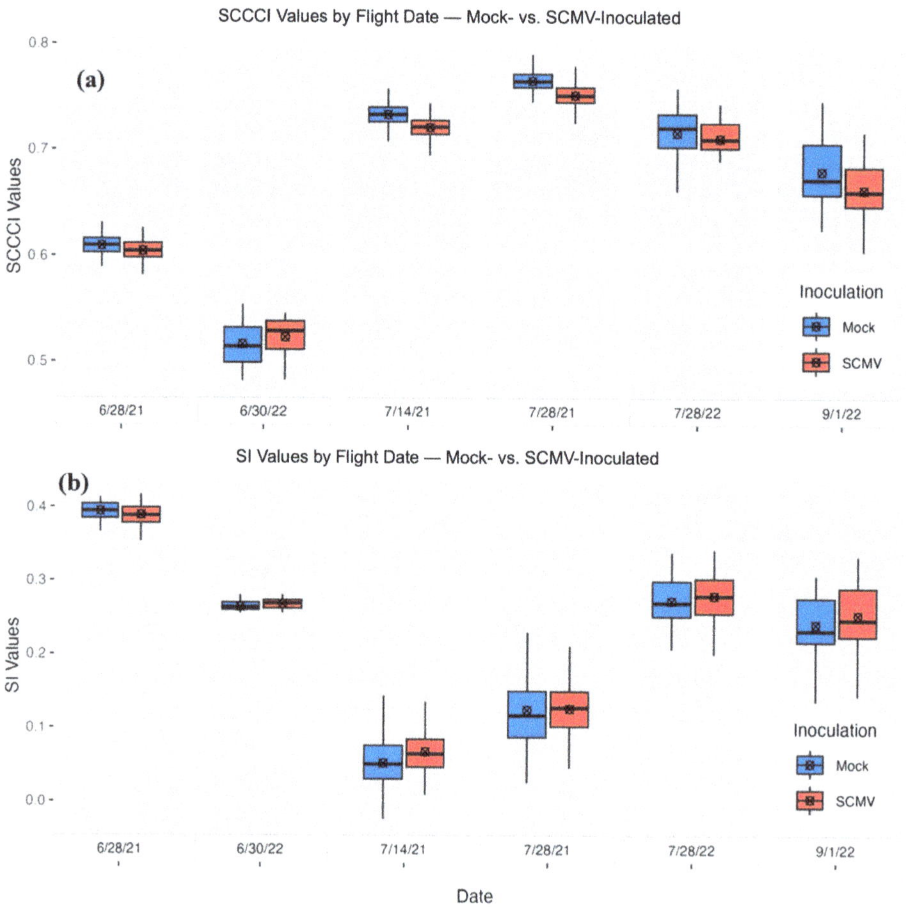

Figure 8. Boxplots displaying the average values of the (**a**) simplified canopy chlorophyll content index (SCCCI) and (**b**) saturation index (SI) for SCMV-inoculated (red bars) and mock-inoculated (blue bars) samples across the growing season dates for both 2021 and 2022.

4. Discussion

4.1. Regression and Classification Modeling

In this study, various modeling approaches were used to investigate the presence and absence of SCMV infection in corn. Given our primary metric for measuring disease incidence was the 'percentage of infected plants' per sample plot, our approach can be best viewed as exploratory, with the focus being on variable interpretation and impacts of features on model behavior, as opposed to inferential or predictive analysis.

In this study, treatment plots varied in size across years, making it challenging to interpret the 'percentage of symptomatic plants' into an interpretable pixel-wise prediction map. For disease mapping purposes, the classification of disease presence may be more appropriate. Using only multispectral imagery-derived features, the XGBoost regression model for disease incidence percentage had an RMSE value of 29.26, which was reduced to 28.39 after feature reduction via RFE from 41 to 14 features. In contrast, the binary classification SVM model for classifying SCMV inoculation vs. mock inoculation achieved an overall accuracy of 72.9%, with 0.75 recall and 0.73 precision for SCMV inoculation. While the classification model may not provide as much insight into the relationships between spectral features and SCMV disease severity, it could effectively highlight areas suspected to be infected with SCMV. Given that SCMV management largely relies on planting resistant corn varieties, the SVM classification model's performance may suffice for identifying disease presence across a field. Very precise field locations may not be as necessary as would be the case for a disease that may require the variable-rate precision spraying of pesticides.

Effective models for the prediction of disease incidence were obtained only in 2021 when examining models by year. The models did not capture the variability in the 2022 data very well, which may be attributed to several things. For instance, in 2021, the experimental design had a larger number of hybrid varieties in smaller spatial block sizes, resulting in a larger sample size compared to 2022 (768 observations vs. 120 observations for 2021 and 2022, respectively). This increased diversity of 48 unique hybrid varieties may have resulted in a wider variation in spectral responses due to SCMV inoculation, unlike the limited five hybrid samples in 2022. Generally, larger studies with more samples tend to produce more robust outcomes with reduced margins of error, which may partially explain the stronger fit in the XGBoost regression and SVM classification models for disease incidence in 2021. The lack of significant correlation between disease incidence and spectral variables at Snyder on 28 July 2022 may also be due to the small sample size. Additionally, disease progression at this timestamp might not have yet reached a spectrally significant stage since several significant correlations were noted at the next timestamp.

Furthermore, 2022 included an additional field location not used in the 2021 trials, potentially introducing variation in biotic and abiotic field parameters that could contribute to differences in disease response, thereby impacting model performances. Differences in environmental factors such as temperature, irradiance, and moisture between two years may have also contributed to differences in disease incidence and subsequent spectral responses. While it appeared that both years had similar air temperature and humidity, the 2022 season had higher cumulative rainfall than 2021. For mosaic disease in sugarcane fields, it has been observed that drought conditions and reduced rainfall environments favor the reproduction and activities of aphids, thereby facilitating the spread of mosaic disease [97]. Additionally, excessively hot climates constrain disease transmission, resulting in slower virus proliferation, fewer disease symptoms, and reduced severity of disease outbreaks [97]. This interplay between environmental parameters could contribute to different disease responses and subsequent spectral characteristics between the years.

Lastly, the timing of data collection differed between the years. In 2021, the last disease score date was 12 July 2021, with UAS flight dates on 14 July 2021 and 28 July 2021. In 2022, the last disease score date was 19 July 2022 for Schaffter and 18 July 2022 for Snyder, with UAS flight dates on 28 July 2022 and 1 September 2022. So, all data collection that was used for modeling occurred within 16 days during 2021 and within 46 days during 2022.

This discrepancy between plant disease evaluation and UAS flight dates could further explain the weaker model performance in 2022. There may have been continued disease progression that occurred throughout the season and was not comprehensively recorded by the disease incidence scores. Between the recording of disease incidence scores and the last UAS flights (a 16-day gap in 2021 and a 46-day gap in 2022), the SCMV infection may have progressed differently across plots, leading to shifts in disease scores that are not strongly represented in the reported disease scores. Perhaps the disease progressed strongly during this time period for plots that were initially recorded to have low disease scores (a smaller percentage of infected plants). This would render the disease scores less accurate and less informative, such that the ability of the regression model to capture a strong trend in disease scores was diminished. The seemingly better performance of the classification models indicates that the models were more effective at identifying infection presence rather than the severity of the infection (percentage of infected plants).

4.2. Selection of Sensors

Efforts were made to collect temperature data via thermal imagery and canopy height data via LiDAR sensors. However, the addition of these features did not notably enhance model performance for disease incidence prediction. Models built using all the multispectral-derived features (41 bands and indices) performed similarly to the simplified model (14 features). Given these findings, it may not be worth the resources and efforts to collect additional thermal and LiDAR data for SCMV detection purposes as it will require additional sensors and likely additional UAS flights. Moreover, the thermal data exhibited limited correlation with disease incidence, and the collection and processing of thermal data proved challenging, resulting in inconsistent data availability across fields and flight dates. Further investigation into thermal imagery may unveil previously undiscovered trends. However, it is essential to ensure that distinguishable physical characteristics are present in the field when using aerial thermal imagery for remote sensing applications in homogeneous agricultural fields. This will ensure smooth orthomosaic stitching processes.

4.3. Model Performance throughout the Season

This study aimed to compare the performance of models for SCMV detection across time periods and crop growth stages throughout the growing season using 2021 and 2022 data. Each year involved three data collection dates for UAS flights, with the first flight in June occurring before SCMV systemic infection. Thus, the June dates for both years exhibited poor predictive ability, leading to their exclusion from the final model building. Unfortunately, this resulted in less data available for modeling, suggesting potential benefits from additional flights post-inoculation.

In 2021, the performance of models for disease incidence prediction gradually improved throughout the season, whereas the effectiveness of disease prediction in 2022 generally lagged behind, although it did show improvement as the growing season progressed. This trend aligns with the progression of infection and increasing plant physiological symptoms. However, this trend may not hold true for MDM, as many hybrids lack visible symptoms at later growth stages. Based on these findings, aerial imagery scouting for SCMV infection during the latter half of July or potentially later into August may prove most effective and economical. Generating a predicted disease presence map could also facilitate targeted on-ground investigation, enabling effective disease management strategies for subsequent seasons, such as crop rotation or improved weed control, to mitigate virus overwintering and its aphid vector.

4.4. Spectral Features and Model Behavior

4.4.1. Spectral Bands

SCMV infection can alter plant pigment and canopy structure, influencing spectral reflectance values. In 2021, SCMV-inoculated plants displayed significant differences in red, red-edge, and NIR bands compared to mock-inoculated plants. SCMV-inoculated plants generally had higher reflectance in the visible and red-edge bands but lower reflectance in the NIR bands (Figure S1). These changes align with expected variations in spectral profiles in healthy and unhealthy vegetation [98,99].

Using chlorophyll content as a positive proxy for healthy vegetation, it is expected that healthier vegetation will have a higher absorption (lower reflectance) in the blue (peaks at 430 and 453 nm for chlorophyll *a*) and red (peaks at 642 and 662nm for chlorophyll *b*) bands and a lower absorption in green (in the 500–600 nm) bands as compared to unhealthy plants that lack the equivalent amount of chlorophyll [99–101]. SCMV infection causes initial chlorotic spotting, streaking, mosaicking, and the eventual progressive yellowing of leaves, which may explain the average differences found in the visible reflectance values. The higher reflectance values observed in SCMV-inoculated plots for both blue and red bands suggest an impact on chlorophyll content due to SCMV infection. This finding is supported by the significant differences observed in the red band, suggesting a stronger influence on chlorophyll *b* content by SCMV infection, leading to reduced red reflectance.

The red-edge portion (680–740 nm) of the spectrum serves as a transition zone between the red and NIR spectra where reflectance shifts from being primarily influenced by photosynthetic pigments (chlorophyll) to vegetation structural properties [102–104]. Red-edge radiation penetrates deeper into crop canopies than visible light, making red-edge-based indices less susceptible to chlorophyll saturations with increasing values of the leaf area index later in the growing season unlike NDVI [104,105]. Furthermore, the red-edge position typically shifts towards longer wavelengths in healthy plants, while it shifts towards shorter wavelengths in diseased plants [103,106]. In this study, SCMV-inoculated plants exhibited higher average red-edge reflectance across both fields and years, suggesting potential stress from disease infection. However, confirming this speculation is challenging due to the use of broadband multispectral images in the analyses.

4.4.2. Vegetation Indices and Feature Importance

SCCCI was identified as the most impactful feature for model behavior in both XGBoost Regression and SVM classification models (Figure 6a). SCCCI, derived from NDRE divided by NDVI, integrates NIR, red, and red-edge bands. Originally proposed as a red-edge index for cotton, SCCCI offers robustness compared to NDVI, particularly in avoiding early-season saturation [68,107,108]. Similarly, TCARI/OSAVI was ranked high in both regression and classification analyses. These two indices, along with MCARI/OSAVI, also show promise in explaining corn's nitrogen status using multispectral aerial imagery [76,108]. The prominence of SCCCI in disease incidence prediction suggests that SCMV-infected plants may exhibit reduced chlorophyll or photosynthetic pigment production across the canopy, while the leaf area index and vegetation cover remain relatively unaffected. This indicates that SCMV infection primarily impacts pigment performance rather than the structural integrity of the plant from an aerial imagery perspective. While SCMV infection may not drastically affect plant structure, it can lead to stunting or increased bushiness. LiDAR-derived canopy heights on 28 July 2022 did not show significant stunting between inoculated and mock-inoculated plants, but visible height differences in August/September suggest some stunting due to SCMV infection.

The saturation index (SI), also known as the normalized pigment chlorophyll index (NPCI), utilizes the red and blue regions and emerged as the most important feature for SCMV infection classification by SVM and the second most crucial feature for disease incidence prediction by XGBoost according to SHAP analyses. SI was originally designed to assess the carotenoid-to-chlorophyll ratio and is higher in nitrogen-limited leaves and inversely correlated with chlorophyll content [75]. The ratio between carotenoids and chlorophyll *a* typically decreases as plants grow and then increases as they senesce [109,110].

Carotenoids, with their photoprotective role in photosynthesis [111], contribute similarly to chlorophyll in visible absorption for wavelengths shorter than green [75]. However, they do not absorb as strongly in the red region as chlorophyll does. SI may, therefore, relate to the proportions of total photosynthetic pigments to chlorophyll proportions, influenced by nitrogen limitations or the protective effects of carotenoids [75,111]. The SHAP analysis of the disease incidence model indicates an almost inverse relationship of SI on model prediction compared to SCCCI. High SI values suggest increased disease incidence prediction, potentially indicating a discrepancy in carotenoid to chlorophyll content in heavily SCMV-infected plants [111]. However, these observations remain speculative without quantified estimations of photosynthetic pigment content levels in SCMV-inoculated and mock-inoculated plants. Notably, SI can be calculated using just an RGB camera, making it a potentially more cost-efficient method for predicting disease incidence in corn compared to using a multispectral camera.

4.5. Limitations and Steps Forward

The study revealed insights into the spectral characteristics of SCMV infection and proposed a methodology for its detection using multispectral data. However, variations in model performance between years suggested limited generalizability to new data. Specifically, the performance of the XGBoost regression model trained on 2021 data and tested with 2022 data highlighted this issue. This could be attributed to the limited dataset, particularly the lack of extensive 2022 data. This limitation can be addressed through the further collection of annotated multispectral imagery and ground truth data covering a range of field sites, environmental conditions, inoculation rates, corn hybrid varieties, crop growth stages, and management practices. This helps to build a larger, more diverse dataset, ultimately enhancing model performance and transferability to new fields and new seasons. Additionally, we had success using sequential model-based optimization (Hyperopt) to finalize model hyperparameters and then using recursive feature elimination to minimize the included features for XGBoost regression modeling. However, the additional fine-tuning of all the models with more data would confirm that XGBoost is indeed the best model and would also likely improve predictive accuracy.

The use of field scouting and disease incidence percentage as infection metrics to correlate with spectral properties can introduce human subjectivity and potential errors, which might have influenced the modeling process. Although these methods are practical and efficient, reflecting real-world SCMV scouting practices, obtaining quantitative estimates of virus titer through serological and molecular testing for individual plants could offer a deeper understanding of spectral relationships. However, these lab-based approaches are costly and time-consuming. Using the average disease incidence value and the average reflectance values of each hybrid plot has limitations regarding the precise mapping of disease incidence. For instance, some of the plants within a specific inoculated plot may be relatively healthy (e.g., they may have a high NDVI value) compared to other plants within the same plot, and, thus, the detection of disease presence or severity may be reduced or understated by averaging. Deriving a standard deviation from zonal statistics as a feature may also yield insightful results. However, as stated, given that SCMV treatment largely relies on planting resistant maize varieties during the next season, the identification of disease presence from means within a general area of a field can still be very informative for management decisions.

While limited observations of LiDAR and thermal imagery did not enhance model performance, exploring them further by incorporating data collected at various stages of SCMV infection may be valuable. To avoid concerns with the stitching of thermal imagery, it is suggested that special consideration be given to ground control points and georeferencing techniques specific to thermal imagery collection. Quantifying chlorophyll and pigment content through leaf tissues or plant sap analysis may elucidate the biochemical and physiological responses of corn to SCMV, which may then provide clarity on spectral indices' effectiveness for SCMV prediction. Additionally, hyperspectral imagery may offer a higher-resolution understanding of SCMV's impact on spectral characteristics and aid in disease identification.

5. Conclusions

This study primarily explores the potential of using UAS multispectral imagery for the detection of SCMV infection in corn fields using data collected over the 2021 and 2022 growing seasons at two farms in Ohio, USA. Field experiments with randomized block designs were implemented such that it was known which plots of each field were noninfected or infected with SCMV. Using the multispectral orthomosaic images, mean pixel reflectance values were calculated for each of the five spectral bands within each individual sample treatment plot. On average, SCMV-inoculated plants had higher reflectance values for blue, green, red, and red-edge bands and lower reflectance for near-infrared bands as compared to mock-inoculated samples. Machine learning algorithms were used for the exploratory analysis and predictive modeling of disease incidence percentage as regression and disease presence as a binary classification. For data aggregated from both years, an XGBoost model using just 14 multispectral image-derived features determined using recursive feature elimination performed best for 'disease incidence' prediction with an R^2 of 0.312 and an RMSE of 28.39. The SVM model was the best classification model for predicting SCMV disease presence with an accuracy of 0.729. A SHAP analysis demonstrated that the SCCCI, SI, and TCARI/OSAVI vegetation indices were generally the most impactful on model performance for the prediction and detection of SCMV in corn. A larger aerial multispectral and thermal image dataset is needed to build models that transfer the high performance of SCMV detection to additional unseen data and new fields. The methodology developed in this study demonstrates the potential for the development of a tool for farmers that may facilitate the precise identification and mapping of SCMV infection in corn.

Supplementary Materials: The following supporting information can be downloaded at https://www.mdpi.com/article/10.3390/rs16173296/s1: Figure S1: Boxplots displaying the comparison of the average reflectance values for each of the spectral bands for SCMV-inoculated and mock-inoculated corn plants; Table S1: Hyperparameters Spaces Explore; Table S2: Significant Pearson r Correlations.

Author Contributions: Conceptualization, S.K. and E.W.O.; methodology, N.B., E.W.O., M.W.J., K.K. and S.K.; software, N.B.; validation, N.B.; formal analysis, N.B.; investigation, N.B., E.W.O. and S.K.; resources, S.K., E.W.O. and M.W.J.; data curation, N.B. and K.K.; writing—original draft preparation, N.B.; writing—review and editing, N.B., E.W.O., S.K., M.W.J. and K.K.; visualization, Bever, N.; supervision, S.K.; project administration, S.K.; funding acquisition, S.K. and E.W.O. All authors have read and agreed to the published version of the manuscript.

Funding: This research was funded in part by USDA-ARS (58-5082-2-006), OSU Graduate School Fellowship, and NRT EmPowerment Fellowship.

Data Availability Statement: The data used in this study can be made available upon request.

Acknowledgments: We would like to thank Mark Bolin for collecting the UAS-based LiDAR data and Peter Thomison, Rich Minyo, and industry participants for providing hybrid seeds. We also thank Chris Nacci, Lynn Ault, Bob Napier, and Matthew Lowe for their assistance in the field.

Conflicts of Interest: The authors declare no conflicts of interest.

References

1. García-Lara, S.; Serna-Saldivar, S.O. Corn History and Culture (Third Edition). *Corn Chem. Technol. Third Ed.* **2019**, 1–18. [CrossRef]
2. Erenstein, O.; Jaleta, M.; Sonder, K.; Mottaleb, K.; Prasanna, B.M. Global maize production, consumption and trade: Trends and R&D implications. *Food Secur.* **2022**, *14*, 1295–1319. [CrossRef]
3. Savary, S.; Willocquet, L.; Pethybridge, S.J.; Esker, P.; McRoberts, N.; Nelson, A. The global burden of pathogens and pests on major food crops. *Nat. Ecol. Evol.* **2019**, *3*, 430–439. [CrossRef]
4. Zambrano, J.L.; Stewart, L.R.; Paul, P.A. Maize Dwarf Mosaic of Maize. Ohio State University Extension. 2016. Available online: https://ohioline.osu.edu/factsheet/plpath-cer-09 (accessed on 5 May 2023).
5. Shukla, D.D.; Tosic, M.; Jilka, J.; Ford, R.E.; Toler, R.W.; Langham, M.A.C. Taxonomy of potyviruses infecting maize, sorghum and sugarcane in Australia and the United States as determined by reactivities of polyclonal antibodies directed towards virus-specific N-termini of coat proteins. *Phytopathology* **1989**, *79*, 223–229. [CrossRef]

6. Tosic, M.; Ford, R.E.; Shukla, D.D.; Jilka, J. Differentiation of Sugarcane, Maize dwarf, Johnsongrass, and Sorghum mosaic viruses based on reactions of oat and some sorghum cultivars. *Plant Dis.* **1990**, *74*, 549–552. [CrossRef]
7. Frenkel, M.J.; Jilka, J.M.; McKern, N.M.; Strike, P.M.; Clark, J.M., Jr.; Shukla, D.D.; Ward, C.W. Unexpected sequence diversity in the amino-terminal ends of the coat proteins of strains of sugarcane mosaic virus. *J. Gen. Virol.* **1991**, *72*, 237–242. [CrossRef] [PubMed]
8. Gao, B.; Cui, X.-W.; Li, X.-D.; Zhang, C.-Q.; Miao, H.-Q. Complete genomic sequence analysis of a highly virulent isolate revealed a novel strain of Sugarcane mosaic virus. *Virus Genes* **2011**, *43*, 390–397. [CrossRef]
9. Viswanathan, R.; Karuppaiah, R.; Balamuralikrishnan, M. Identification of new variants of SCMV causing sugarcane mosaic in India and assessing their genetic diversity in relation to SCMV type strains. *Virus Genes* **2009**, *39*, 375–386. [CrossRef]
10. Niblett, C.; Claflin, L. Corn lethal necrosis—A new virus disease of corn in Kansas. *Plant Dis. Bull.* **1978**, *62*, 15.
11. Stewart, L.R.; Willie, K.; Wijeratne, S.; Redinbaugh, M.G.; Massawe, D.; Niblett, C.L.; Kiggundu, A.; Asiimwe, T. Johnsongrass mosaic virus contributes to maize lethal necrosis in East Africa. *Plant Dis.* **2017**, *101*, 1455–1462. [CrossRef] [PubMed]
12. Redinbaugh, M.G.; Stewart, L.R. Maize lethal necrosis: An emerging, synergistic viral disease. *Annu. Rev. Virol.* **2018**, *5*, 301–322. [CrossRef] [PubMed]
13. Ohlson, E.W.; Redinbaugh, M.G.; Jones, M.W. Mapping maize chlorotic mottle virus tolerance loci in the Maize 282 Association Panel. *Crop Sci.* **2022**, *62*, 1497–1510. [CrossRef]
14. Wu, L.; Zu, X.; Wang, S.; Chen, Y. Sugarcane mosaic virus—Long history but still a threat to industry. *Crop Prot.* **2012**, *42*, 74–78. [CrossRef]
15. Xu, D.L.; Park, J.W.; Mirkov, T.E.; Zhou, G.H. Viruses causing mosaic disease in sugarcane and their genetic diversity in southern China. *Arch. Virol.* **2008**, *153*, 1031–1039. [CrossRef]
16. Fuchs, E.; Grüntzig, M. Influence of sugarcane mosaic virus (SCMV) and maize dwarf mosaic virus (MDMV) on the growth and yield of two maize varieties. *J. Plant Dis. Prot.* **1995**, *102*, 44–50. Available online: http://www.jstor.org/stable/43386365 (accessed on 1 May 2023).
17. Janson, B.F.; Williams, L.E.; Findley, W.R.; Dollinger, E.J.; Ellett, C.W. Maize dwarf mosaic: New corn virus disease in Ohio. 1965. Available online: https://www.cabidigitallibrary.org/doi/full/10.5555/19641101624 (accessed on 21 April 2023).
18. Gustafson, T.J.; de Leon, N.; Kaeppler, S.M.; Tracy, W.F. Genetic analysis of sugarcane mosaic virus resistance in the wisconsin diversity panel of maize. *Crop Sci.* **2018**, *58*, 1853–1865. [CrossRef]
19. Meyer, M.D.; Pataky, J.K. Increased severity of foliar diseases of sweet corn infected with maize dwarf mosaic and sugarcane mosaic viruses. *Plant Dis.* **2010**, *94*, 1093–1099. [CrossRef]
20. Jones, M.W.; Ohlson, E.W. Susceptibility and yield response of commercial corn hybrids to maize dwarf mosaic disease. *Plant Dis.* **2024**, *108*, 1786–1792. [CrossRef]
21. Kerns, M.R.; Pataky, J.K. Reactions of Sweet Corn Hybrids with Resistance to Maize Dwarf Mosaic. *Plant Dis.* **1997**, *81*, 460–464. [CrossRef]
22. Shahi, T.B.; Xu, C.-Y.; Neupane, A.; Guo, W. Recent advances in crop disease detection using UAV and deep learning techniques. *Remote Sens.* **2023**, *15*, 2450. [CrossRef]
23. Zhang, J.; Huang, Y.; Pu, R.; Gonzalez-Moreno, P.; Yuan, L.; Wu, K.; Huang, W. Monitoring plant diseases and pests through remote sensing technology: A review. *Comput. Electron. Agric.* **2019**, *165*, 104943. Available online: https://www.sciencedirect.com/science/article/pii/S016816991930290X (accessed on 2 May 2023). [CrossRef]
24. Tsouros, D.C.; Bibi, S.; Sarigiannidis, P.G. A review on UAV-based applications for precision agriculture. *Information* **2019**, *10*, 349. [CrossRef]
25. Khanal, S.; KC, K.; Fulton, J.; Shearer, S.; Ozkan, E. Remote Sensing in Agriculture (Challenges and Opportunities). *Remote Sens.* **2020**, *12*, 3783. [CrossRef]
26. Chakravarthy, A.K. *Innovative Pest Management Approaches for the 21st Century: Harnessing Automated Unmanned Technologies*; Springer Nature: New York, NY, USA, 2020; pp. 255–272.
27. Bannari, A.; Morin, D.; Bonn, F.; Huete, A.R. A review of vegetation indices. *Remote Sens. Rev.* **1995**, *13*, 95–120. [CrossRef]
28. Huang, S.; Tang, L.; Hupy, J.P.; Wang, Y.; Shao, G. A commentary review on the use of normalized difference vegetation index (NDVI) in the era of popular remote sensing. *J. For. Res.* **2021**, *32*, 2719. [CrossRef]
29. Antolínez García, A.; Cáceres Campana, J.W. Identification of pathogens in corn using near-infrared UAV imagery and deep learning. *Precis. Agric.* **2023**, *24*, 783–806. [CrossRef]
30. Butcher, G. *Tour of the Electromagnetic Spectrum*; Government Printing Office: Washington, DC, USA, 2016.
31. Barbedo, J.G.A. A review on the main challenges in automatic plant disease identification based on visible range images. *Biosyst. Eng.* **2016**, *144*, 52–60. [CrossRef]
32. Mohanty, S.P.; Hughes, D.P.; Salathé, M. Using deep learning for image-based plant disease detection. *Front. Plant Sci.* **2016**, *7*, 1419. [CrossRef]
33. Moriya, E.A.S.; Imai, N.N.; Tommaselli, A.M.G.; Miyoshi, G.T. Mapping Mosaic Virus in Sugarcane Based on Hyperspectral Images. *IEEE J. Sel. Top. Appl. Earth Obs. Remote Sens.* **2017**, *10*, 740–748. [CrossRef]
34. Viswanathan, R.; Rao, G.P. Disease Scenario and Management of Major Sugarcane Diseases in India. *Sugar Tech* **2011**, *13*, 336–353. [CrossRef]

35. Gazala, I.F.S.; Sahoo, R.N.; Pandey, R.; Mandal, B.; Gupta, V.K.; Singh, R.; Sinha, P. Spectral reflectance pattern in soybean for assessing yellow mosaic disease. *Indian J. Virol.* **2013**, *24*, 242–249. [CrossRef]
36. Mirik, M.; Jones, D.C.; Price, J.A.; Workneh, F.; Ansley, R.J.; Rush, C.M. Satellite remote sensing of wheat infected by Wheat streak mosaic virus. *Plant Dis.* **2011**, *95*, 4–12. [CrossRef]
37. Prabhakar, M.; Prasad, Y.G.; Desai, S.; Thirupathi, M.; Gopika, K.; Rao, G.R.; Venkateswarlu, B. Hyperspectral remote sensing of yellow mosaic severity and associated pigment losses in Vigna mungo using multinomial logistic regression models. *Crop Prot.* **2013**, *45*, 132–140. [CrossRef]
38. Luo, L.; Chang, Q.; Wang, Q.; Huang, Y. Identification and severity monitoring of maize dwarf mosaic virus infection based on hyperspectral measurements. *Remote Sens.* **2021**, *13*, 4560. [CrossRef]
39. Ausmus, B.S.; Hilty, J.W. Reflectance studies of healthy, maize dwarf mosaic virus-infected, and Helminthosporium maydis-infected corn leaves. *Remote Sens. Environ.* **1971**, *2*, 77–81. [CrossRef]
40. Dhau, I.; Adam, E.; Ayisi, K.K.; Mutanga, O. Detection and mapping of maize streak virus using RapidEye satellite imagery. *Geocarto Int.* **2019**, *34*, 856–866. [CrossRef]
41. Dhau, I.; Dube, T.; Mushore, T.D. Examining the prospects of sentinel-2 multispectral data in detecting and mapping maize streak virus severity in smallholder Ofcolaco farms, South Africa. *Geocarto Int.* **2021**, *36*, 1873–1883. [CrossRef]
42. Dhau, I.; Adam, E.; Mutanga, O.; Ayisi, K.K. Detecting the severity of maize streak virus infestations in maize crop using in situ hyperspectral data. *Trans. R. Soc. S. Africa* **2018**, *73*, 8–15. [CrossRef]
43. Chen, S.; Zhang, K.; Wu, S.; Tang, Z.; Zhao, Y.; Sun, Y.; Shi, Z. A Weakly Supervised Approach for Disease Segmentation of Maize Northern Leaf Blight from UAV Images. *Drones* **2023**, *7*, 173. [CrossRef]
44. Garg, K.; Bhugra, S.; Lall, B. Automatic quantification of plant disease from field image data using deep learning. In Proceedings of the 2021 IEEE Winter Conference on Applications of Computer Vision, WACV 2021, Waikoloa, HI, USA, 3–8 January 2021; pp. 1964–1971. [CrossRef]
45. Wu, H.; Wiesner-Hanks, T.; Stewart, E.L.; DeChant, C.; Kaczmar, N.; Gore, M.A.; Nelson, R.J.; Lipson, H. Autonomous Detection of Plant Disease Symptoms Directly from Aerial Imagery. *Plant Phenome J.* **2019**, *2*, 1–9. [CrossRef]
46. Dhau, I.; Adam, E.; Mutanga, O.; Ayisi, K.; Abdel-Rahman, E.M.; Odindi, J.; Masocha, M. Testing the capability of spectral resolution of the new multispectral sensors on detecting the severity of grey leaf spot disease in maize crop. *Geocarto Int.* **2018**, *33*, 1223–1236. [CrossRef]
47. Loladze, A.; Rodrigues, F.A.; Toledo, F.; Vicente, F.S.; Gérard, B.; Boddupalli, M.P. Application of remote sensing for phenotyping tar spot complex resistance in maize. *Front. Plant Sci.* **2019**, *10*, 552. [CrossRef]
48. *Pix4D*, Version 4.2.27; Pix4D Mapper Photogrammetry Software: Lausanne, Switzerland.
49. ESRI. *ArcGIS Pro*; Environmental Systems Research Institute: Redlands, CA, USA, 2023.
50. R Core Team RF for SC. *R: A Language and Environment*; R Core Team: Vienna, Austria, 2019. Available online: https://www.r-project.org/ (accessed on 3 May 2023).
51. Parker, T.A.; Palkovic, A.; Gepts, P. Determining the genetic control of common bean early-growth rate using unmanned aerial vehicles. *Remote Sens.* **2020**, *12*, 1748. [CrossRef]
52. Woebbecke, D.M.; Meyer, G.E.; Von Bargen, K.; Mortensen, D.A. Color Indices for Weed Identification Under Various Soil, Residue, and Lighting Conditions. *Trans. ASAE* **1995**, *38*, 259–269. Available online: https://elibrary.asabe.org/azdez.asp?JID=&AID=27838&CID=t1995&v=38&i=1&T=1 (accessed on 3 May 2023). [CrossRef]
53. Girardeau-Montaut, D. *CloudCompare*; EDF R&D Telecom ParisTech: Paris, France, 2016. Available online: https://www.danielgm.net/cc/ (accessed on 4 January 2023).
54. Ray, S.S.; Singh, J.P.; Das, G.; Panigrahy, S.; Group, A.R.; Centre, S.A.; Potato, C. Use of high resolution remote sensing data for generating site-specific soil mangement plan. *Red* **2004**, *550*, 727.
55. Gitelson, A.A.; Vina, A.; Arkebauer, T.J.; Rundquist, D.C.; Keydan, G.; Leavitt, B. Remote estimation of leaf area index and green leaf biomass in maize canopies. *Geophys. Res. Lett.* **2003**, *30*, 4–7. [CrossRef]
56. Hunt, E.R.; Doraiswamy, P.C.; McMurtrey, J.E.; Daughtry, C.S.T.; Perry, E.M.; Akhmedov, B. A visible band index for remote sensing leaf chlorophyll content at the Canopy scale. *Int. J. Appl. Earth Obs. Geoinf.* **2012**, *21*, 103–112. [CrossRef]
57. Huete, A.; Didan, K.; Miura, T.; Rodriguez, E.P.; Gao, X.; Ferreira, L.G. Overview of the radiometric and biophysical performance of the MODIS vegetation indices. *Remote Sens. Environ.* **2002**, *82*, 195–213. [CrossRef]
58. Gitelson, A.A.; Kaufman, Y.J.; Merzlyak, M.N. Use of a green channel in remote sensing of global vegetation from EOS-MODIS. *Remote Sens. Environ.* **1996**, *58*, 289–298. [CrossRef]
59. Gitelson, A.A.; Merzlyak, M.N. Remote sensing of chlorophyll concentration in higher plant leaves. *Adv. Sp. Res.* **1998**, *22*, 689–692. [CrossRef]
60. Gitelson, A.A. Wide Dynamic Range Vegetation Index for Remote Quantification of Biophysical Characteristics of Vegetation. *J. Plant Physiol.* **2004**, *161*, 165–173. [CrossRef]
61. Junior, C.K.; Guimarães, A.M.; Caires, E.F. Use of active canopy sensors to discriminate wheat response to nitrogen fertilization under no-tillage. *Eng. Agric.* **2016**, *36*, 886. [CrossRef]
62. Portz, G.; Molin, J.P.; Jasper, J. Active crop sensor to detect variability of nitrogen supply and biomass on sugarcane fields. *Precis. Agric.* **2012**, *13*, 33–44. [CrossRef]

63. Banerjee, B.P.; Spangenberg, G.; Kant, S. Fusion of spectral and structural information from aerial images for improved biomass estimation. *Remote Sens.* **2020**, *12*, 3164. [CrossRef]
64. Haboudane, D.; Miller, J.R.; Pattey, E.; Zarco-Tejada, P.J.; Strachan, I.B. Hyperspectral vegetation indices and novel algorithms for predicting green LAI of crop canopies: Modeling and validation in the context of precision agriculture. *Remote Sens. Environ.* **2004**, *90*, 337–352. [CrossRef]
65. Rondeaux, G.; Steven, M.; Baret, F. Optimization of soil-adjusted vegetation indices. *Remote Sens. Environ.* **1996**, *55*, 95–107. [CrossRef]
66. Qi, J.; Chehbouni, A.; Huete, A.R.; Kerr, Y.H.; Sorooshian, S. A modified soil adjusted vegetation index. *Remote Sens. Environ.* **1994**, *48*, 119–126. [CrossRef]
67. Chen, J.M. Evaluation of vegetation indices and a modified simple ratio for boreal applications. *Can. J. Remote Sens.* **1996**, *22*, 229–242. [CrossRef]
68. Barnes, E.M.; Clarke, T.R.; Richards, S.E.; Colaizzi, P.D.; Haberland, J.; Kostrzewski, M.; Walker, P.; Choi, C.; Riley, E.; Thompson, T.; et al. Coincident detection of crop water stress, nitrogen status and canopy density using ground based multispectral data. In Proceedings of the Fifth International Conference on Precision Agriculture, Bloomington, MN, USA, 19 July 2000; Volume 1619.
69. Rouse, J.W., Jr.; Haas, R.H.; Deering, D.W.; Schell, J.A.; Harlan, J.C. *Monitoring the Vernal Advancement and Retrogradation (Green Wave Effect) of Natural Vegetation*; NASA/GSFCT Type III Final Report, 371; NASA: Washington, DC, USA, 1974.
70. Gitelson, A.A.; Kaufman, Y.J.; Stark, R.; Rundquist, D. Novel algorithms for remote estimation of vegetation fraction. *Remote Sens. Environ.* **2002**, *80*, 76–87. [CrossRef]
71. Ramoelo, A.; Skidmore, A.K.; Cho, M.A.; Schlerf, M.; Mathieu, R.; Heitkönig, I.M.A. Regional estimation of savanna grass nitrogen using the red-edge band of the spaceborne rapideye sensor. *Int. J. Appl. Earth Obs. Geoinf.* **2012**, *19*, 151–162. [CrossRef]
72. Roujean, J.L.; Breon, F.M. Estimating PAR absorbed by vegetation from bidirectional reflectance measurements. *Remote Sens. Environ.* **1995**, *51*, 375–384. [CrossRef]
73. Huete, A.R. A soil-adjusted vegetation index (SAVI). *Remote Sens. Environ.* **1998**, *25*, 295–309. [CrossRef]
74. Raper, T.B.; Varco, J.J. Canopy-scale wavelength and vegetative index sensitivities to cotton growth parameters and nitrogen status. *Precis. Agric.* **2015**, *16*, 62–76. [CrossRef]
75. Peñuelas, J.; Gamon, J.A.; Fredeen, A.L.; Merino, J.; Field, C.B. Reflectance indices associated with physiological changes in nitrogen- and water-limited sunflower leaves. *Remote Sens. Environ.* **1994**, *48*, 135–146. [CrossRef]
76. Haboudane, D.; Miller, J.R.; Tremblay, N.; Zarco-Tejada, P.J.; Dextraze, L. Integrated narrow-band vegetation indices for prediction of crop chlorophyll content for application to precision agriculture. *Remote Sens. Environ.* **2002**, *81*, 416–426. [CrossRef]
77. Revelle, W.; Revelle, M.W. Package 'psych'. *Compr. R Arch. Netw.* **2015**, *337*, 161–165.
78. Kruskal, W.H.; Wallis, W.A. Use of ranks in one-criterion variance analysis. *J. Am. Stat. Assoc.* **1952**, *47*, 583–621. [CrossRef]
79. McDonald, G.C. Ridge regression. *Wiley Interdiscip. Rev. Comput. Stat.* **2009**, *1*, 93–100. Available online: https://onlinelibrary.wiley.com/doi/full/10.1002/wics.14 (accessed on 3 May 2023). [CrossRef]
80. Hearst, M.A.; Dumais, S.T.; Osuna, E.; Platt, J.; Scholkopf, B. Support vector machines. *IEEE Intell. Syst. Their Appl.* **1998**, *13*, 18–28. [CrossRef]
81. Breiman, L. Random forests. *Mach. Learn.* **2001**, *45*, 5–32. Available online: https://link.springer.com/article/10.1023/A:1010933404324 (accessed on 3 May 2023). [CrossRef]
82. Chen, T.; Guestrin, C. XGBoost: A scalable tree boosting system. In Proceedings of the 22nd ACM SIGKDD International Conference on Knowledge Discovery and Data Mining, San Francisco, CA, USA, 13–17 August 2016; pp. 785–794. [CrossRef]
83. Ballabio, C.; Sterlacchini, S. Support vector machines for landslide susceptibility mapping: The Staffora River Basin case study, Italy. *Math. Geosci.* **2012**, *44*, 47–70. [CrossRef]
84. Triscowati, D.W.; Sartono, B.; Kurnia, A.; Domiri, D.D.; Wijayanto, A.W. Multitemporal remote sensing data for classification of food crops plant phase using supervised random forest. In Proceedings of the Sixth Geoinformation Science Symposium, Yogyakarta, Indonesia, 26–27 August 2019; Volume 11311, p. 1131102.
85. Fan, J.; Zhou, J.; Wang, B.; de Leon, N.; Kaeppler, S.M.; Lima, D.C.; Zhang, Z. Estimation of Maize Yield and Flowering Time Using Multi-Temporal UAV-Based Hyperspectral Data. *Remote Sens.* **2022**, *14*, 3052. [CrossRef]
86. Hoerl, A.E.; Kennard, R.W. Ridge Regression: Biased Estimation for Nonorthogonal Problems. *Technometrics* **1970**, *12*, 55–67. Available online: https://www.tandfonline.com/doi/abs/10.1080/00401706.1970.10488634 (accessed on 3 May 2023). [CrossRef]
87. Awad, M.; Khanna, R. Support Vector Regression. In *Efficient Learning Machines*; Apress: Berkeley, CA, USA, 2015; pp. 67–80. Available online: https://link.springer.com/chapter/10.1007/978-1-4302-5990-9_4 (accessed on 3 May 2023).
88. Smola, A.J.; Schölkopf, B. A tutorial on support vector regression. *Stat. Comput.* **2004**, *14*, 199–222. Available online: https://link.springer.com/article/10.1023/B:STCO.0000035301.49549.88 (accessed on 3 May 2023). [CrossRef]
89. Drucker, H.; Burges, C.J.C.; Kaufman, L.; Smola, A.; Vapoik, V. Support Vector Regression Machines. *Adv. Neural Inf. Process. Syst.* **1996**, *9*, 155–161.
90. Cutler, A.; Cutler, D.R.; Stevens, J.R. Random Forests. In *Ensemble Machine Learning*; Springer: New York, NY, USA, 2012; pp. 157–175. Available online: https://link.springer.com/chapter/10.1007/978-1-4419-9326-7_5 (accessed on 3 May 2023).
91. Pedregosa, F.; Varoquaux, G.; Gramfort, A.; Michel, V.; Thirion, B.; Grisel, O.; Blondel, M.; Prettenhofer, P.; Weiss, R.; Dubourg, V. Scikit-learn: Machine learning in Python. *J. Mach. Learn. Res.* **2011**, *12*, 2825–2830.

92. Bergstra, J.; Komer, B.; Eliasmith, C.; Yamins, D.; Cox, D.D. Hyperopt: A Python library for model selection and hyperparameter optimization. *Comput. Sci. Discov.* **2015**, *8*, 014008. Available online: https://iopscience.iop.org/article/10.1088/1749-4699/8/1/014008 (accessed on 3 May 2023). [CrossRef]
93. Cerilani, M. Shap-Hypetune. 2022. Available online: https://github.com/cerlymarco/shap-hypetune (accessed on 12 February 2023).
94. Cipriano, W. Pretty Print Confusion Matrix. 2018. Available online: https://github.com/wcipriano/pretty-print-confusion-matrix (accessed on 20 March 2023).
95. Lundberg, S.M.; Lee, S.I. A unified approach to interpreting model predictions. *Adv. Neural Inf. Process. Syst.* **2017**, *30*, 4766–4775.
96. Wei, H.E.; Grafton, M.; Bretherton, M.; Irwin, M.; Sandoval, E. Evaluation of the Use of UAV-Derived Vegetation Indices and Environmental Variables for Grapevine Water Status Monitoring Based on Machine Learning Algorithms and SHAP Analysis. *Remote Sens.* **2022**, *14*, 5918. [CrossRef]
97. Lu, G.; Wang, Z.; Xu, F.; Pan, Y.-B.; Grisham, M.P.; Xu, L. Sugarcane mosaic disease: Characteristics, identification and control. *Microorganisms* **2021**, *9*, 1984. [CrossRef]
98. Chang, J.; Clay, D.E.; Clay, S.A.; Reese, C. Using Field Scouting or Remote Sensing Technique to Assess Soybean Yield Limiting Factors Organic weed management View project Practical Agronomy and Mathematics for Precision Farming View project SEE PROFILE. 2013. Chapter 29. pp. 1–7. Available online: https://openprairie.sdstate.edu/cgi/viewcontent.cgi?filename=15&article=1001&context=plant_book&type=additional (accessed on 23 March 2023).
99. Slaton, M.R.; Hunt, E.R.; Smith, W.K. Estimating near-infrared leaf reflectance from leaf structural characteristics. *Am. J. Bot.* **2001**, *88*, 278–284. [CrossRef]
100. Broge, N.H.; Leblanc, E. Comparing prediction power and stability of broadband and hyperspectral vegetation indices for estimation of green leaf area index and canopy chlorophyll density. *Remote Sens. Environ.* **2001**, *76*, 156–172. [CrossRef]
101. Kim, M.S.; Daughtry CS, T.; Chappelle, E.W.; McMurtrey, J.E.; Walthall, C.L. The Use of the High Spectral Resolution Bands for Estimating Absorbed Photosynthetically Active Radiation. In Proceedings of the ISPRS'94, Val d'Isere, France; 1994; pp. 299–306.
102. Dawson, T.P.; Curran, P.J. A new technique for interpolating the reflectance red edge position. *Int. J. Remote Sens.* **1998**, *19*, 2133–2139. [CrossRef]
103. Shafri, H.Z.M.; Sall, M.A.M.; Ghiyamat, A. Hyperspectral Remote Sensing of Vegetation Using Red Edge Position Techniques. *Am. J. Appl. Sci.* **2006**, *3*, 1864–1871. [CrossRef]
104. Li, F.; Miao, Y.; Feng, G.; Yuan, F.; Yue, S.; Gao, X.; Liu, Y.; Liu, B.; Ustin, S.L.; Chen, X. Improving estimation of summer maize nitrogen status with red edge-based spectral vegetation indices. *F. Crop. Res.* **2014**, *157*, 111–123. [CrossRef]
105. Gu, Y.; Wylie, B.K.; Howard, D.M.; Phuyal, K.P.; Ji, L. NDVI saturation adjustment: A new approach for improving cropland performance estimates in the Greater Platte River Basin, USA. *Ecol. Indic.* **2013**, *30*, 1–6. Available online: https://www.sciencedirect.com/science/article/pii/S1470160X13000757 (accessed on 27 March 2023). [CrossRef]
106. Pu, R.; Gong, P.; Biging, G.S.; Larrieu, M.R. Extraction of red edge optical parameters from hyperion data for estimation of forest leaf area index. *IEEE Trans. Geosci. Remote Sens.* **2003**, *41*, 916–921. [CrossRef]
107. Sumner, Z.; Varco, J.J.; Dhillon, J.S.; Fox, A.A.A.; Czarnecki, J.; Henry, W.B. Ground versus aerial canopy reflectance of corn: Red-edge and non-red edge vegetation indices. *Agron. J.* **2021**, *113*, 2782–2797. [CrossRef]
108. Shaver, T.M.; Khosla, R.; Westfall, D.G. Evaluation of two crop canopy sensors for nitrogen variability determination in irrigated maize. *Precis. Agric.* **2011**, *12*, 892–904. [CrossRef]
109. Sanger, J. Quantitative investigations of leaf pigments from their inception in buds through autumn coloration to decomposition in falling leaves. *Ecology* **1971**, *52*, 1075–1089. [CrossRef]
110. Filella, I.; Serrano, L.; Serra, J.; Penuelas, J. Evaluating wheat nitrogen status with canopy reflectance indices and discriminant analysis. *Crop Sci.* **1995**, *35*, 1400–1405. [CrossRef]
111. Strzałka, K.; Kostecka-Gugała, A.; Latowski, D. Carotenoids and Environmental Stress in Plants: Significance of Carotenoid-Mediated Modulation of Membrane Physical Properties. *Russ. J. Plant Physiol.* **2003**, *50*, 168–173. [CrossRef]

remote sensing

MDPI

Article

Mapping Shrub Biomass at 10 m Resolution by Integrating Field Measurements, Unmanned Aerial Vehicles, and Multi-Source Satellite Observations

Wenchao Liu [1], Jie Wang [1,*], Yang Hu [2], Taiyong Ma [3], Munkhdulam Otgonbayar [4], Chunbo Li [1], You Li [1] and Jilin Yang [1]

[1] College of Grassland Science and Technology, China Agricultural University, Beijing 100193, China; sugarvinegar@outlook.com (W.L.); chunboli@cau.edu.cn (C.L.); 13194973225@163.com (Y.L.); jlyang@cau.edu.cn (J.Y.)

[2] School of Ecology and Environment, Ningxia University, Yinchuan 750021, China; huyang@nxu.edu.cn

[3] College of Forestry and Prataculture, Ningxia University, Yinchuan 750021, China; mataiyong506@stu.nxu.edu.cn

[4] Division of Physical Geography and Environmental Research, Institute of Geography and Geoecology, Mongolian Academy of Sciences, Ulaanbaatar 15170, Mongolia

* Correspondence: jiewang178@cau.edu.cn

Abstract: Accurately estimating shrub biomass in arid and semi-arid regions is critical for understanding ecosystem productivity and carbon stocks at both local and global scales. Due to the short and sparse features of shrubs, capturing the shrub biomass accurately by satellite observations is challenging. Previous studies mostly used ground samples and satellite observations to estimate shrub biomass by establishing a direct connection, which was often hindered by the limited number of ground samples and spatial scale mismatch between samples and observations. Unmanned aerial vehicles (UAVs) provide opportunities to obtain more samples that are in line with the aspects of satellite observations (i.e., scale) for regional-scale shrub biomass estimations accurately with low costs. However, few studies have been conducted based on the air-space-ground-scale connection assisted by UAVs. Here we developed a framework for estimating 10 m shrub biomass at a regional scale by integrating ground measurements, UAV, Landsat, and Sentinel-1/2 observations. First, the spatial distribution map of shrublands and non-shrublands was generated in 2023 in the Helan Mountains of Ningxia province, China. This map had an F1 score of 0.92. Subsequently, the UAV-based shrub biomass map was estimated using an empirical model between the biomass and the crown area of shrubs, which was aggregated at a 10 m × 10 m grid to match the spatial resolution of Sentinel-1/2 images. Then, a regional-scale estimation model of shrub biomass was developed with a random forest regression (RFR) approach driven by ground biomass measurements, UAV-based biomass, and the optimal satellite metrics. Finally, the developed model was used to produce the biomass map of shrublands over the study area in 2023. The uncertainty of the resultant biomass map was characterized by the pixel-level standard deviation (SD) using the leave-one-out cross-validation (LOOCV) method. The results suggested that the integration of multi-scale observations from the ground, UAVs, and satellites provided a promising approach to obtaining the regional shrub biomass accurately. Our developed model, which integrates satellite spectral bands and vegetation indices (R^2 = 0.62), outperformed models driven solely by spectral bands (R^2 = 0.33) or vegetation indices (R^2 = 0.55). In addition, our estimated biomass has an average uncertainty of less than 4%, with the lowest values (<2%) occurring in regions with high shrub coverage (>30%) and biomass production (>300 g/m^2). This study provides a methodology to accurately monitor the shrub biomass from satellite images assisted by near-ground UAV observations as well as ground measurements.

Keywords: remote sensing; vegetation mapping; environmental monitoring; UAV; land cover

1. Introduction

In recent years, more and more attention has been paid to environmental protection and ecosystem restoration due to the impacts of human activities and global climate change [1–4]. As an important vegetation type, shrubs play crucial roles in the conservation and restoration of ecosystems [5–7], such as soil conservation, sand fixation, habitat maintenance, and so on [8–11]. Shrub biomass is a critical measurement indicator to assess shrub growth status and productivity [12,13]. Therefore, accurately estimating shrub biomass is vital for monitoring the dynamics of vegetation growth, assessing ecological functions, and evaluating ecosystem carbon storage at the regional scale [14–16].

Field surveys, as a traditional approach, often have limitations in large-scale monitoring of shrub biomass considering the following factors [17]. Firstly, field surveys need substantial human resources and financial investment and are time-consuming, particularly in areas with intricate topography [18,19]. Secondly, field surveys can only obtain limited samples, making it challenging to fully capture the spatial distributions and temporal patterns of shrubs. Meanwhile, the destructive sampling techniques to quantify shrub biomass may potentially affect the surrounding environment [20]. Furthermore, the biomass of shrubs is influenced by various factors such as vegetation structure, environmental conditions, and climatic variations [21,22]. It is challenging to accurately represent the spatial distributions of shrub biomass at large scales by collecting shrub biomass samples in field surveys.

To address the limitations of conventional field surveys, remote sensing technology has emerged as an efficient tool for estimating shrub biomass [23–25]. For example, satellites can acquire extensive and continuous data with high spatial and temporal resolutions, facilitating large-scale estimation of shrub biomass such as Landsat and Sentinel series [26–28]. Previous studies have applied the combined Sentinel-1/2 satellite images to map the distribution of shrubland biomass from landscape to regional scales [27,28]. However, the performance of utilizing satellite images for shrub biomass estimation is related to selecting suitable satellite metrics [29]. For example, only utilizing the visible and near-infrared bands with wide wavelength ranges makes it challenging to capture the features related to shrub biomass, such as vegetation structure and leaf area index [30]. Therefore, more and more vegetation indices have been used as important indicators for biomass estimation [31]. The vegetation indices are numerical values calculated from spectral bands that can reflect key information about vegetation growth status like chlorophyll content, leaf area index, and other related features [32,33]. However, the sensitivity of existing vegetation indices to shrub biomass varies, and some indices may reach saturation in highly vegetated regions [34]. Therefore, it is still needed to understand the contributions of each vegetation indices or spectral bands to accurately estimate the shrub biomass.

In addition to the utilized satellite metrics, the simulation algorithms have been improved gradually in the last decades. Traditionally, statistical regression approaches were employed to estimate biomass by establishing empirical relationships between field-measured biomass and remote sensing variables [35]. However, these models had limitations in capturing complex nonlinear relationships, and the simulation accuracy was affected by the spatial heterogeneity [36]. Over time, several modeling approaches have been developed to enhance the precision of estimating shrub biomass [37–39]. In particular, machine learning methods have become the most used tools due to their superior performance compared to conventional modeling methods [40]. For example, machine learning methods like random forests, support vector machines, and boosted regression trees have been reported with great potential to handle complex and nonlinear relationships in biomass simulations [41,42]. Additionally, machine learning methods can effectively handle large volumes of remote sensing metrics, including multispectral and hyperspectral imagery, which provide rich remote sensing information for accurate biomass estimation across shrublands [40,43,44].

The accuracy of the satellite-based biomass estimation is often affected by two main features of training samples, including (1) the scale mismatch between the limited ground measurement area and the resolution of satellite images [27] and (2) the number of field samples [28]. Unmanned aerial vehicles (UAVs) have advantages to address these issues in shrub biomass estimation [37,45]. UAVs equipped with high-resolution sensors and advanced image processing techniques can capture high spatial resolution remote sensing imagery [46,47]. The high spatial resolutions allow for the capture of finer details of shrub structure and characteristics [48]. This capability is crucial for improving satellite-based biomass estimation by providing high-resolution near-ground observations and bridging the gaps of spatial scales between satellite observations and ground measurements [27,28]. Additionally, they can capture large areas of shrub coverage from a high-altitude perspective, which facilitates the rapid and accurate identification and extraction of shrub objects [37,49–51]. Thus, UAVs can improve workflow efficiency to gather more samples by lowering manpower, saving time, and reducing costs compared to traditional field measurements [52]. However, the studies are still limited to the estimation of shrub biomass at the regional scale by integrating UAV-derived shrub biomass and satellite observations.

This work aimed to propose a framework to estimate shrub biomass over a mountain region by integrating multi-scale data from field measurements, UAV-based biomass estimation, and satellite observations (Sentinel-1/2, Landsat). The Helan Mountains in Ningxia province, China, were selected as the study area. Firstly, the distribution of the shrublands and non-shrublands in the study area was identified by a land cover classification. Secondly, the shrub biomass was estimated based on the UAV images and the in situ measurements by establishing the allometric growth equation between shrub biomass and shrub structure parameters. Finally, the biomass distribution of shrubs was mapped based on the satellite observations using the optimal random forest regression (RFR) model across the Helan Mountains by utilizing the field measurements and UAV-based shrub biomass as samples. The optimal RFR was developed by selecting the sensitive satellite metrics to shrub biomass from the spectral bands, vegetation indices, and backscatter bands from the Landsat and Sentinel-1/2 images. The accuracy of the resultant map was evaluated over various ranges of shrub biomass or shrub coverage. This study provides a methodology to accurately estimate the biomass of shrubs at the satellite scale by incorporating near-ground UAV observations. This approach increases the training samples of shrub biomass and addresses the scale mismatch between ground measurements and satellite observations.

2. Material and Methods

2.1. Study Area

The Helan Mountains are a remnant range of the Kunlun Mountains and are situated at the intersection of the Ningxia Hui Autonomous Region and the Inner Mongolia Autonomous Region. The main peak of the Helan Mountains in Ningxia has an elevation of 3556 m and covers an area of approximately 2100 km^2. The mean annual temperature in the study area is around -0.8 °C, with a mean annual precipitation of about 430 mm (Figure 1).

The study area represents a typical temperate arid mountain region with diverse vegetation types. It is mainly covered by evergreen coniferous forests, deciduous broad-leaved forests, evergreen coniferous shrubs, deciduous broad-leaved shrubs, typical grasslands, and desert grasslands, with slope direction and elevation influencing distribution patterns. Furthermore, the region has several shrub species, including Amygdalus pedunculata, Caragana korshinskii, and Mongolian almond.

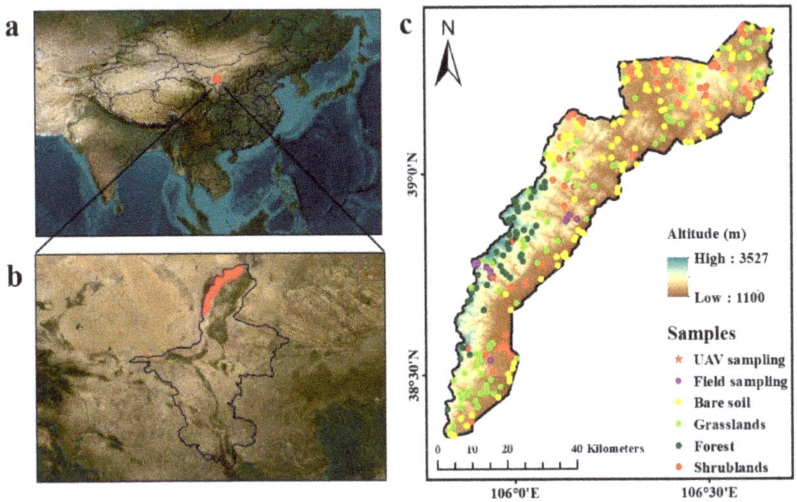

Figure 1. (**a**,**b**) Locations of the Helan Mountain in China and Ningxia province, and (**c**) the distribution of ground truth samples from field measurements, UAV, and visual interpretation.

2.2. Data

2.2.1. Landsat and Sentinel-1/2 Data and Pre-Processing

Our study utilized all available Landsat 8/9, Sentinel-2, and Sentinel-1 satellite images from late July to early August 2023. These satellite datasets were accessed through Google Earth Engine (GEE). Landsat-8 Operational Land Imager (OLI) and Landsat-9 OLI-2 data were utilized, which include visible, infrared, and shortwave infrared bands with a spatial resolution of 30 m. This study also used Sentinel-2 multispectral instrument (MSI) Level-2A products, which provide red-edge band data additionally at 10 m compared to Landsat. Sentinel-1 provides dual-polarization C-band Synthetic Aperture Radar (SAR) observations operating at 5.405 GHz (C band), including vertical receive (VV) and horizontal receive (VH) bands. Thus, our dataset includes three visible bands, three red-edge bands, one near-infrared band, two shortwave infrared bands, and two radar bands.

The spatial resolution of the Landsat 8/9 image was resampled to 10 m with the bilinear method in GEE to match the spatial resolution of the Sentinel-2. To identify the quality of the satellite images, the pixel-wise quality control operation images were processed by using the quality assurance (QA) band and CFMask algorithm [53]. Due to some differences in bandwidth between MSI, OLI, and ETM+ sensors, the band reflectance values were then matched among the three sensors to construct reliable time series by the ordinary least squares regression method proposed by Roy (2016) [54]. Thus, the harmonized Landsat/Sentinel-2 datasets were generated.

With the harmonized Landsat/Sentinel-2 datasets, various vegetation indices that are sensitive to shrub biomass were calculated [27] to select the optimal predictor variables for constructing the model. In total, 28 vegetation indices were selected and then computed on the GEE platform (Table 1).

Table 1. Vegetation indices (VIs) for the estimates of shrub biomass in this study. R = red; G = green; B = blue; RE = RedEdge1; NIR = near-infrared; SWIR1 = shortwave infrared 1; SWIR2 = shortwave infrared 2. ρ represents the band of NIR, $\sigma = 0.5 \times (\text{NIR} + \text{red})$.

No.	Index	Formula
1	Normalized difference vegetation index [55]	$\text{NDVI} = \frac{\text{NIR}-\text{R}}{\text{NIR}+\text{R}}$
2	Enhanced vegetation index 1 [56]	$\text{EVI} = \frac{2.5\times(\text{NIR}-\text{R})}{\text{NIR}+6\times\text{R}-7.5\times\text{B}+1}$
3	Land surface water index [57]	$\text{LSWI} = \frac{\text{NIR}-\text{SWIR1}}{\text{NIR}+\text{SWIR1}}$
4	Difference vegetation index [58]	$\text{DVI} = \text{NIR} - \text{R}$
5	Green normalized difference vegetation index [59]	$\text{GNDVI} = \frac{\text{NIR}-\text{G}}{\text{NIR}+\text{G}}$
6	Vegetation index green [60]	$\text{VIgreen} = \frac{\text{G}-\text{R}}{\text{G}+\text{R}}$
7	Infrared simple ratio [61]	$\text{ISR} = \frac{\text{NIR}}{\text{SWIR1}}$
8	Moisture stress index [62]	$\text{MSI} = \frac{\text{SWIR1}}{\text{NIR}}$
9	Ratio vegetation index [63]	$\text{RVI} = \frac{\text{R}}{\text{NIR}}$
10	Simple ratio [64]	$\text{SR} = \frac{\text{NIR}}{\text{R}}$
11	Enhanced vegetation index 2 [65]	$\text{EVI}_2 = \frac{2.5\times(\text{NIR}-\text{R})}{\text{NIR}+2.4\times\text{R}+1}$
12	Modified simple ratio [66]	$\text{MSR} = \frac{\frac{\text{NIR}}{\text{R}}-1}{\sqrt{\frac{\text{NIR}}{\text{R}}+1}}$
13	Optimized soil-adjusted vegetation index [67]	$\text{OSAVI} = (1+\text{L}) \times \frac{\text{NIR}-\text{R}}{\text{NIR}+\text{R}+\text{L}}$, L was set to 0.16
14	Renormalized difference vegetation index [68]	$\text{RDVI} = \frac{\text{NIR}-\text{R}}{\sqrt{\text{NIR}+\text{R}}}$
15	Soil adjusted vegetation index [69]	$\text{SAVI} = (1+\text{L}) \times \frac{\text{NIR}-\text{R}}{\text{NIR}+\text{R}+\text{L}}$, L = 0.5
16	Soil adjusted vegetation index 2 [70]	$\text{SAVI2} = \frac{\text{NIR}}{\text{NIR}+\frac{b}{a}}$, b was set to 0.025 and a to 1.25
17	Stress-related vegetation index 1 [71]	$\text{STVI}_1 = \frac{\text{SWIR1}\times\text{R}}{\text{NIR}}$
18	Stress-related vegetation index 2 [71]	$\text{STVI}_2 = \frac{\text{NIR}}{\text{SWIR2}\times\text{R}}$
19	Stress-related vegetation index 3 [71]	$\text{STVI}_3 = \frac{\text{NIR}}{\text{SWIR1}\times\text{R}}$
20	Red edge normalized difference vegetation index [72]	$\text{RENDVI} = \frac{\text{NIR}-\text{RE}}{\text{NIR}+\text{RE}}$
21	Anthocyanin reflectance index [73]	$\text{ARI} = \frac{1}{\text{G}} - \frac{1}{\text{RE}}$
22	Vogelmann red edge index [74]	$\text{VREI} = \frac{\text{NIR}}{\text{RE}}$
23	Radar ratio vegetation index	$\text{Ratio} = \frac{\text{VV}}{\text{VH}}$
24	Radar difference vegetation index	$\text{Difference} = \text{VV} - \text{VH}$
25	Radar normalized difference vegetation index (RNDVI)	$\text{RNDVI} = \frac{\text{VV}-\text{VH}}{\text{VV}+\text{VH}}$
26	Near-infrared reflectance of vegetation [75]	$\text{NIRv} = \left(\frac{\text{NIR}-\text{R}}{\text{NIR}+\text{R}} - \text{C}\right) \times \rho$ (C = 0.08)
27	kernel (NDVI) [76]	$\text{kNDVI} = \tanh((\frac{\text{NIR}-\text{R}}{2\sigma})^2)$
28	Normalized difference phenology index [77]	$\text{NDPI} = \frac{\text{NIR}-(0.74\times\text{R}+0.26\times\text{SWIR1})}{\text{NIR}+(0.74\times\text{R}+0.26\times\text{SWIR1})}$

2.2.2. Land Cover Data for Shrubland Mapping

The European Space Agency (ESA) WorldCover dataset provides global 10 m land cover types, which was based on Sentinel-1/2 images with 11 land cover types [78]. 500 samples were generated randomly, and the land cover type of each sample point was acquired. Then, the classification accuracy of each sample was examined by visual interpretation using high-resolution Google Earth images and field photographs taken in April, July, and August 2023. In this study, an area with more than 10% shrub cover was classified as shrubland based on the MODIS land cover dataset [79]. Finally, a total of 354 samples were obtained, including 97 shrublands, 86 grasslands, 112 barren lands, and

59 forests. These samples were randomly separated into training and validation points at a 7:3 ratio and used for shrubland mapping.

2.2.3. Field Measurement and UAV Images

Two comprehensive field surveys were conducted in the Helan Mountains in late April 2023 and late July to early August 2023. 13 individual shrubs were selected randomly from the study area for destructive sampling and measurement of structural parameters. The samples included the dominant species with different plant sizes to ensure the representativeness of the samples in the study area. Structural parameters include shrub length (CL), width (CW), height (CH), crown area (CA), and volume (CV). The shrub samples were then dried in an oven at 70 °C for 48 h and weighed to obtain the biomass data (Table 2). These measurements were used to develop the optimal allometric growth equation for estimating shrub biomass.

Table 2. Field samples, including structural parameters and biomass data of each shrub.

ID	CL (m)	CW (m)	CH (m)	CA (m^2)	CV (m^3)	Number of Branches	Single Branch Biomass (g)	Single Plant Biomass (g)
1	1.62	1.02	1.34	1.3	1.74	6	487.81	2926.88
2	1.74	1.55	1.83	2.12	3.88	13	243.91	3170.79
3	1.26	1.02	0.73	1.01	0.74	33	40.95	1351.46
4	0.72	0.58	0.5	0.33	0.16	1	183.43	183.43
5	0.51	0.48	0.45	0.19	0.09	1	513.86	513.86
6	1.15	0.96	0.84	0.87	0.73	10	134.43	1344.3
7	0.42	0.42	0.6	0.14	0.08	1	185.91	185.91
8	1.63	1.58	0.74	2.02	1.5	9	226.38	2037.45
9	0.88	0.99	1	0.68	0.68	9	121.95	1097.58
10	1.96	2.08	1.32	3.2	4.23	40	38.48	1539.07
11	1.32	1.47	1.11	1.52	1.69	15	144.95	2174.3
12	1.05	0.98	0.59	0.81	0.48	3	495.86	1487.58
13	0.89	0.68	0.75	0.48	0.36	5	51.95	259.77

In addition, 24 sampling plots with each size of 10 m × 10 m were chosen to measure the structural parameters of each shrub. The sampling plots were evenly distributed throughout the study area. Shrub cover within the plots ranged from low to high. By utilizing the previously fitted allometric growth equation, the total shrub biomass for each plot can be calculated.

In late April 2023, a UAV image of the Helan Mountain region was acquired using a DJI Phantom 4 Pro V2.0 (a UAV from DJI China). The image covered a large area of 588 m × 196 m, with a spatial resolution of 0.06 m. The Phantom 4 Pro V2.0 UAV provides six bands, including true color (RGB), blue, green, red, red edge, and near-infrared (NIR).

2.2.4. Auxiliary Datasets

Precipitation, temperature, aridity index, and elevation datasets were used in this study (Table 3). The precipitation data were obtained using the 1 km monthly precipitation dataset for China (1901–2022) [80], with a spatial resolution of 1 km. The temperature data were a 1 km monthly mean temperature dataset for China (1901–2022) [81]. Two datasets provide monthly average precipitation and temperature. Since the two datasets provided the most recent climate variables in 2022, the mean values of monthly precipitation and temperature from 2000 to 2022 were used to analyze the spatial distribution characteristics of shrub biomass. The 1 km aridity index data were the global aridity index and potential evapotranspiration (ET0), which calculated the aridity index based on the mean annual precipitation and evapotranspiration [82]. The lower the values, the more severe the drought conditions. The elevation data were obtained from the 30 m NASA SRTM Digital Elevation dataset [83]. These datasets were obtained through the GEE platform.

Table 3. Auxiliary datasets used in this study.

Influencing Factor	Dataset Name	Spatial Resolution
Precipitation	1 km monthly precipitation dataset for China (1901–2022)	1 km
Air temperature	1 km monthly mean temperature dataset for China (1901–2022)	1 km
Aridity index	Global aridity index and potential evapotranspiration (ET0) Climate Database v2	1 km
Elevation	NASA SRTM Digital Elevation	30 m

2.3. Methods

Figure 2 shows the workflow for shrub biomass estimation. It includes four main procedures. (1) A diverse ground sample dataset was constructed based on field surveys, ground measurements, UAVs, and Google Earth images. The dataset included samples of land cover types, shrub structure parameters, and shrub biomass. (2) The distribution map of shrublands was generated by land cover classification. (3) The shrub biomass map was generated for the Helan Mountains in 2023 by developing the optimal estimation model based on multi-scale data from ground, UAV, and satellite imagery. Here, the optimal model for shrub biomass estimation was built by comparing different RFR models driven by various combinations of predictor variables, including spectral bands, spectral vegetation indices, and backscatter bands (VV and VH). (4) The spatial distribution characteristics of the shrub biomass were analyzed comprehensively. Each of the steps is described in detail in the subsequent paragraphs.

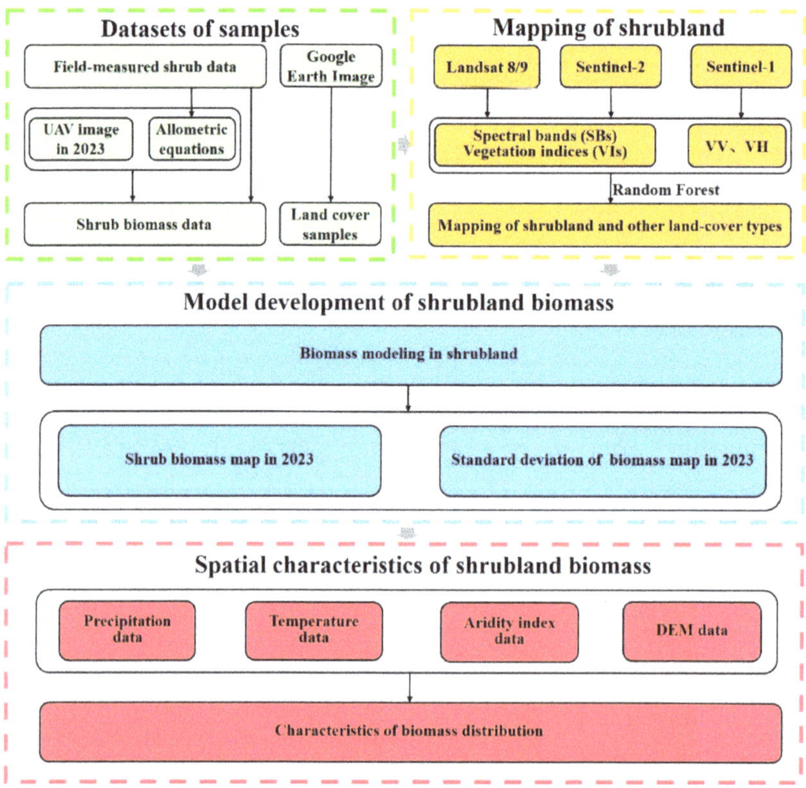

Figure 2. The workflow for estimating biomass of shrubland.

2.3.1. Shrub and Non-Shrub Mapping in the Study Area

The study area is mainly covered by four land cover types, including forest, bare soil, grassland, and shrubland, based on the field survey. For shrubland mapping, the time series of 11 spectral bands were acquired corresponding to our collected 354 samples (Section 2.2.2). The data from these 11 spectral bands were acquired from all available Landsat 8/9, Sentinel-2, and Sentinel-1 satellite imagery in late July to early August 2023 based on the GEE platform, and included three visible (RGB), three red-edge, one near-infrared (NIR), two shortwave infrared (SWIR), and two radar (VV, VH) bands. The spectral bands and the associated land cover type of each sample were then inputted into the random forest (RF) model.

RF is a widely used machine-learning method that combines the characteristics of decision trees and randomness [84,85]. It demonstrates good skills in handling large-scale and high-dimensional datasets and is also able to evaluate each feature's significance in terms of selecting important features. 70% of the samples were used to train the RF model, and the remaining samples (30%) were used to validate the performance of classification. The mapping result was evaluated by the metrics of accuracy, recall, and F1-score (Equations (1)–(3)). Python (version 3.11) and the scikit-learn library (version 1.5.1) were used for the model development and accuracy calculation. Thus, a detailed land cover map including shrublands and other land cover types has been generated over the study area based on the RF model and multi-source satellite images.

$$\text{Accuracy} = \frac{(\text{True Positives} + \text{True Negatives})}{(\text{True Positives} + \text{True Negatives} + \text{False Positives} + \text{False Negatives})} \tag{1}$$

$$\text{Recall} = \frac{\text{True Positives}}{(\text{True Positives} + \text{False Negatives})} \tag{2}$$

$$\text{F1 score} = \frac{2 \times (\text{Precision} \times \text{Recall})}{(\text{Precision} + \text{Recall})} \tag{3}$$

2.3.2. Estimation of Shrub Biomass at the Individual Level

The structural parameters of shrubs closely reflect their growth status and biomass accumulation, making them important indicators for estimating shrub biomass. By conducting field measurements of shrub length, width, and height, the shrub's crown area and volume can be calculated, both of which are positively correlated with shrub biomass [56]. Generally, shrub crown area (CA) and volume directly reflect biomass values. The crown area represents the horizontal structure of the shrub, while the volume represents its horizontal and vertical structural characteristics [56]. Studies have shown a linear link between the biomass of shrubs and CA [56].

Then allometric equations were established for shrub development. Seven samples were utilized in each iteration to fit the canopy area (CA) and crown volume (CV), with the remaining six samples being used for validation. A total of $5 \times C_{13}^{7}$ iterations were conducted, using the coefficient of determination (R^2), mean absolute error (MAE), and root mean square error (RMSE) as accuracy assessment criteria. The final selected allometric equations for shrub biomass estimation based on area and volume are as follows:

$$\text{AGB} = e^{(7.3+0.41 \times \ln(\text{CV}))} \tag{4}$$

$$\text{AGB} = 1060.93 \times \text{CA} \tag{5}$$

2.3.3. Estimation of Shrub Biomass at the UAV Plot Level

UAVs can provide a convenient approach to estimating shrub biomass at the plot level and alleviate the estimation uncertainty caused by the spatial mismatch between ground sampling and satellite observations [56]. In this study, we innovatively explored the feasibility of sample generation methods in remote sensing biomass inversion by combining ground survey and UAV rather than using only one of them. Specifically, as for the UAV-based sampling method, by leveraging the true color UAV image, we first identified the shrub based on an RF classification (RF) model (Figure 3a). The shrub (N = 106) and non-shrub (N = 229) samples were collected by visual interpretation of the UAV imagery using a random sampling approach. The RGB band values of each sample in the UAV imagery were extracted as input features for building the RF model. Thus, a binary image of shrub and non-shrub (i.e., the shrub and non-shrub pixels) in the UAV image was produced at the UAV level (spatial resolution about 0.06 m). The model fitness showed that all the accuracy, recall, and F1 score of the UAV-based shrub map exceeded 0.9.

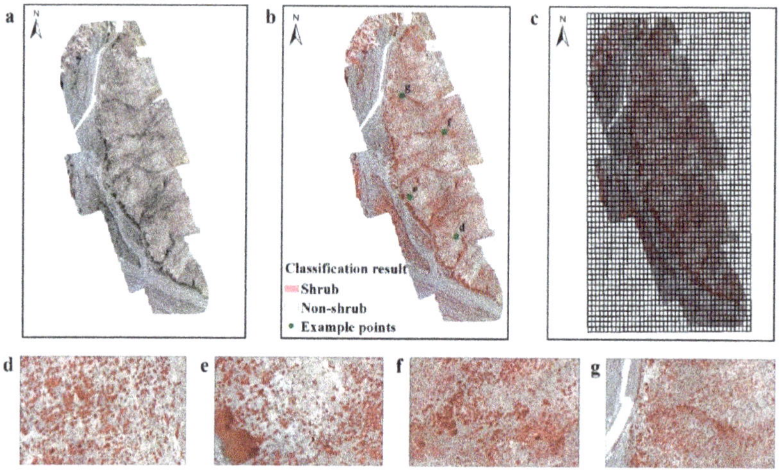

Figure 3. (**a**) The original unmanned aerial vehicle (UAV) image. (**b**) The classified map of shrublands. (**c**) The fishnet constructed based on the UAV imagery. (**d–g**) The zoomed-in views of four sample points in (**b**).

Then, the grids with a size of 10 m × 10 m were created by using the above UAV-based shrub and non-shrub map to match the spatial resolution of the satellite images we used in this study. This process was conducted using the "create fishnet" function in ArcMap 10.4. The grids were perfectly aligned with the pixel boundaries of the satellite image (Figure 3c). Within each grid, the crown area (CA) of shrubs was calculated by counting the shrub pixels in the UAV image. Third, the total shrub biomass within each grid was estimated by applying a CA-related biomass allometric equation (AGB = 1060.93 × CA (R^2 = 0.9); Equation (5)) [56]. In addition, to reduce the impacts caused by rugged terrain and clouds, the grids with black or bright backgrounds, steep slopes, or clouds were removed. At last, a total of 86 UAV-based samples were obtained for further biomass estimation.

2.3.4. Satellite-Based Modeling of Shrub Biomass at the Regional Level

Two methods for satellite-based biomass estimation at the regional level, the random forest regression (RFR) model and the leave-one-out cross-validation (LOOCV) method, were employed. The RFR model calculates the final score as the mean score of all decision trees rather than through voting, resulting in better predictive performance and higher accuracy than the RF model [86]. With the LOOCV method, the RFR model was trained N times (N = 89), where N-1 samples were used for training and the remaining sample was used for testing. The model with the highest accuracy was selected as the final output that was used to make full use of the samples [27].

It should be noted that the model's performance is not positively impacted by every feature. Certain characteristics may cause little variations in samples, which might reduce the classification accuracy. Thus, the combination of multiple feature variables may have the problem of data redundancy and complex computation. Accordingly, the optimization of multidimensional features is important for improving model performance [87–89]. To select the optimal features.

Three models were constructed with the RFR algorithm driven by three different satellite feature combinations: (1) the VV, VH, and spectral bands (SB), (2) the vegetation indices (VI), and (3) all available variables (SBVI). The accuracy of different combinations was calculated based on the R-squared (R^2) and root mean squared error (RMSE). Finally, the best model and its corresponding optimal feature combination could be identified for shrub biomass estimation. Two Python libraries (RFECV and KFold) based on scikit-learn library (version 1.5.1) were used in this study. The RFECV package was employed for constructing different combinations of features, and the KFold package was used for cross-validation.

2.3.5. Regional Implementation of the Satellite-Based Shrub Biomass Model

Through the LOOCV model, the distribution maps of shrubland biomass were obtained at 10 m spatial resolution with a total number of 89. The mean value map of the 89 shrubland biomass distribution maps was calculated as the final resultant map. Moreover, a map of standard deviation (SD) was generated for shrubland biomass to present the potential uncertainty based on the 89 shrubland biomass distribution maps. Subsequently, the uncertainty of the resultant biomass map was analyzed along with the magnitudes of shrubland biomass and coverage. To obtain the shrubland coverage map, the land cover map in Section 2.3.1 was classified and aggregated into the shrubland and non-shrubland maps. The shrubland and non-shrubland pixels were assigned into 1 and 0, respectively. Here, the non-shrubland included the forest, bare soil, and grassland types. This 10 m binary map was resampled into a shrubland coverage map at a spatial resolution of 1 km by the mean method. Then, the relationship between the SD map and the shrubland coverage map was examined at 1000 m by a spatial overlay analysis.

2.3.6. Spatial Characteristics of the Shrub Biomass

To examine the geographical patterns of shrub biomass, the resultant shrub biomass map was overlaid with different datasets of precipitation, temperature, aridity index, and elevation (Table 3). Then, the spatial distributions of shrub biomass along with the gradients of different environmental factors can be analyzed.

3. Results

3.1. Mapping of Shrubland and Other Land Cover Types

A land cover map for the Helan Mountains in 2023 was created by using an RF model constructed with 11 remote sensing features. Four land cover types have been identified, including forest, barren lands, grasslands, and shrublands. Figure 4 displays the result map, where only the shrublands were used for biomass estimation. It can be observed that shrublands are predominantly distributed in the southeastern plains and northeastern parts, while forests dominate the southwestern region. Additionally, there is relatively less grassland and more barren land in the study area, which can be attributed to the overall climate type being a typical temperate arid climate with low rainfall. The accuracy assessment on the shrubland and non-shrubland map was shown in Table 4, presenting an overall classification accuracy of 0.91, a recall of 0.92, and an F1 score of 0.92.

Table 4. Shrubland and non-shrubland classification accuracy assessment results.

Accuracy Index	Accuracy	Recall	F1 Score
Value	0.91	0.92	0.92

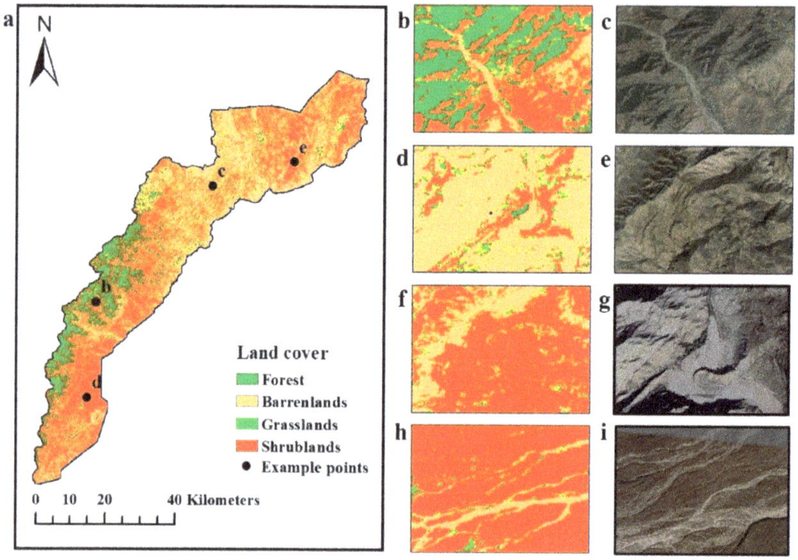

Figure 4. (**a**) The shrublands and other land over types of Helan Mountain, China, in 2023. (**b–i**) The zoom-in views of four example regions in the resultant map and the Google Earth images.

3.2. Selection of the Best Satellite-Based Biomass Model Based on Variable Importance Analysis

Three models driven by different features were conducted: VV, VH, and spectral bands (SB), another with only vegetation indices (VI), and a third with all available features (SBVI). The RFECV and KFold packages were used to determine the optimal variables for each model. The best variables for each model are shown in Table 5. The R^2 and RMSE for each of the three models are shown in Figure 5. The results suggested that the SBVI model achieved the best performance with R^2 of 0.62, which was higher than that of the SB model (R^2 = 0.33) and VI model (R^2 = 0.55). Therefore, the SBVI model was selected as the final model for estimating shrub biomass in the following studies. It also showed that a significant number of vegetation indices related to the near-infrared (NIR) band (e.g., DVI, SAVI2) are incorporated into the model, indicating the high correlation of NIR-based vegetation indices for shrub biomass.

Table 5. Model feature selection results.

Model	Features
SB	Blue, Red, NIR, SWIR2
VI	EVI, DVI, GNDVI, SAVI2, Ratio, RNDVI
SBVI	DVI, SAVI2, VH, VV

The accuracy was assessed based on shrub biomass samples across different groups of shrub coverages and biomass (Figure 6a). The ratio of RMSE to mean shrub coverage was calculated to present the error of one percent coverage of shrub (EOPC) [90]. EOPC decreases with the increase in shrub coverage. Similarly, the ratio of RMSE to mean biomass was calculated to present the error of one unit biomass (EOUB) (Figure 6b). EOUB decreases along with the increase in shrub biomass. The lowest values of R^2 were observed within the range of 20–30% shrub coverage (R^2 = 0.3) and 200–300 g/m^2 shrub biomass (R^2 = 0.19). These results indicate the estimation accuracy is relatively lower in the moderate shrub coverage and biomass ranges than that in the high and low levels.

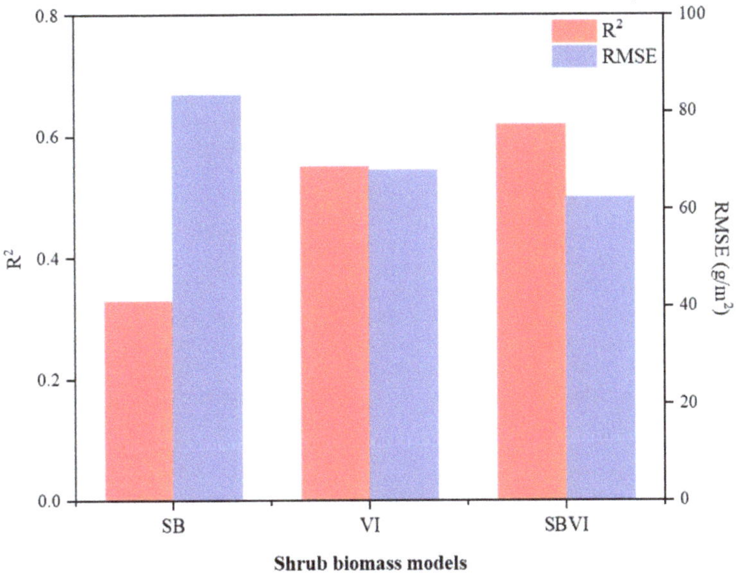

Figure 5. The comparison of accuracy among the three models. The x-axis represents three models driven by the basic bands (SB), the vegetation indices (VI), and the combination of the basic bands and vegetation indices (SBVI). Their performance is evaluated using R^2 and RMSE.

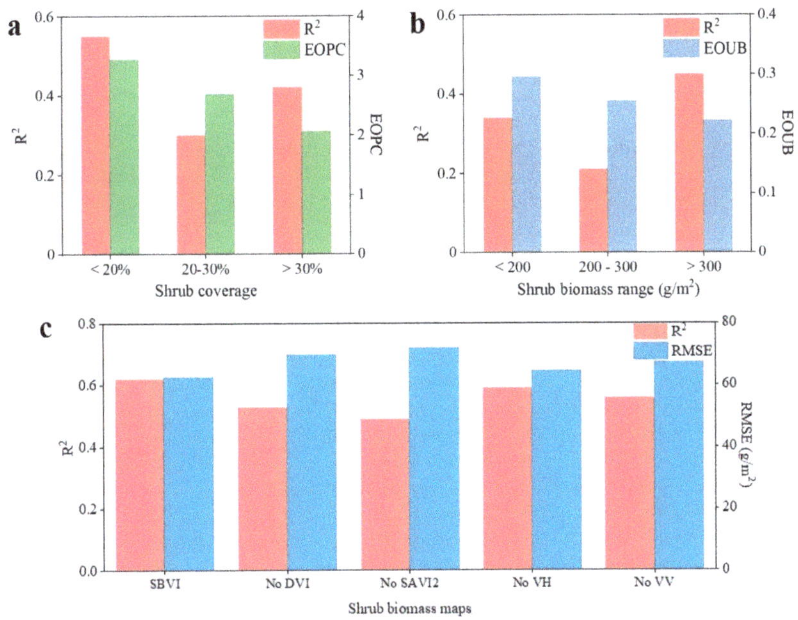

Figure 6. (**a**) The distribution of R^2 and EOPC within different ranges of shrub coverage. (**b**) The distribution of R^2 and EOUB within different ranges of shrub biomass. (**c**) The sensitivity of the biomass model to each variable examined by R^2 and RMSE. These analyses were conducted based on the ground samples. EOPC denotes the error of one percent coverage of shrub, calculated by RMSE/mean shrub coverage. EOUB denotes the error of one unit biomass, calculated by RMSE/mean shrub biomass.

To determine the contributions of four features to the model, a sensitivity analysis was conducted (Figure 6c). It can be observed that SAVI2 has the most significant impact on model accuracy. Removing SAVI2 resulted in a decrease in R^2 by 0.13 and an increase in RMSE by 9.53 g/m^2. VH had the least impact on the model, as removing VH only led to a decrease of 0.03 in R^2 and an increase of 1.93 g/m^2 in RMSE. The results showed that the four features are ranked from highest to lowest contribution as SAVI2, DVI, VV, and VH.

3.3. Shrub Biomass Mapping Based on Satellite Images in 2023

The 10 m shrub biomass distribution in the Helan Mountains was mapped by the SBVI model (Figure 7). Here, the shrub-covered regions were focused by excluding the non-shrub areas using the 10 m binary map (Section 2.3.5). The results showed the minimum value of shrub biomass was 104.99 g/m^2, while the maximum value was 564.54 g/m^2. Despite having more bushes than other vegetation, the biomass is often lower in the southeast. This is because the shrubs in this area are predominantly low-lying. By contrast, the central Helan Mountains have greater elevation and are mostly characterized by tall shrubs, which contribute to the high shrub biomass.

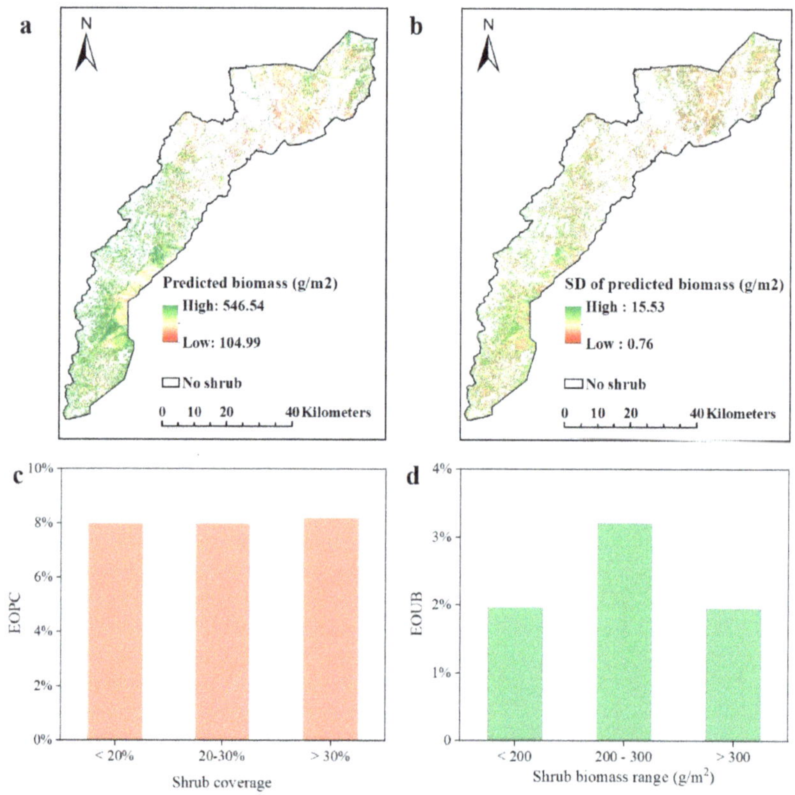

Figure 7. (**a**) The estimated distribution map of shrub biomass in the Helan Mountains. (**b**) The corresponding map of standard deviation (SD). (**c**) The distribution of EOPC within different ranges of shrub coverage. (**d**) The distribution of EOUB within different ranges of shrub biomass. These analyses were conducted based on the estimated distribution map of shrub biomass and the corresponding map of standard deviation. EOPC denotes the error of one percent coverage of shrub, calculated by SD/mean shrub coverage. EOUB denotes the error of one unit biomass, calculated by SD/mean shrub biomass.

Figure 7b depicts the spatial distribution of the standard deviation associated with the biomass predictions from the 89 models based on the LOOCV method. The result shows that higher standard deviations generally happened in the areas with higher shrub biomass, while the lower standard deviations were in the regions with lower shrub biomass. Figure 7c,d illustrate the distributions of the estimation errors under different shrub cover and biomass gradients, respectively. EOPC shows that the estimation errors vary, with the shrub cover with a mean value of around 8%. The EOUB indicates that the overall estimation error is less than 4%, and the estimation errors are even lower than 2% in both the high and low shrub biomass ranges.

3.4. Distribution of Shrub Biomass along with Different Environmental Factors

The distribution characteristics of shrub biomass were observed by temperature, precipitation, aridity index, and elevation. Figure 8a shows the shrub biomass mainly distributed in the regions with annual mean precipitation of 200–250 mm. Figure 8b displays the distribution of shrub biomass across different ranges of mean annual temperature. The shrub biomass mainly occurred in the regions with a mean annual temperature of 7–9 °C. As shown in Figure 8c, the shrub biomass is predominantly distributed within the aridity index range of 0.1–0.15, suggesting that shrubs may prefer to grow in more arid environments. Additionally, Figure 8d indicates that the shrub biomass is primarily distributed in areas with an elevation below 2500 m, implying that higher elevation regions may not be as suitable for shrub growth.

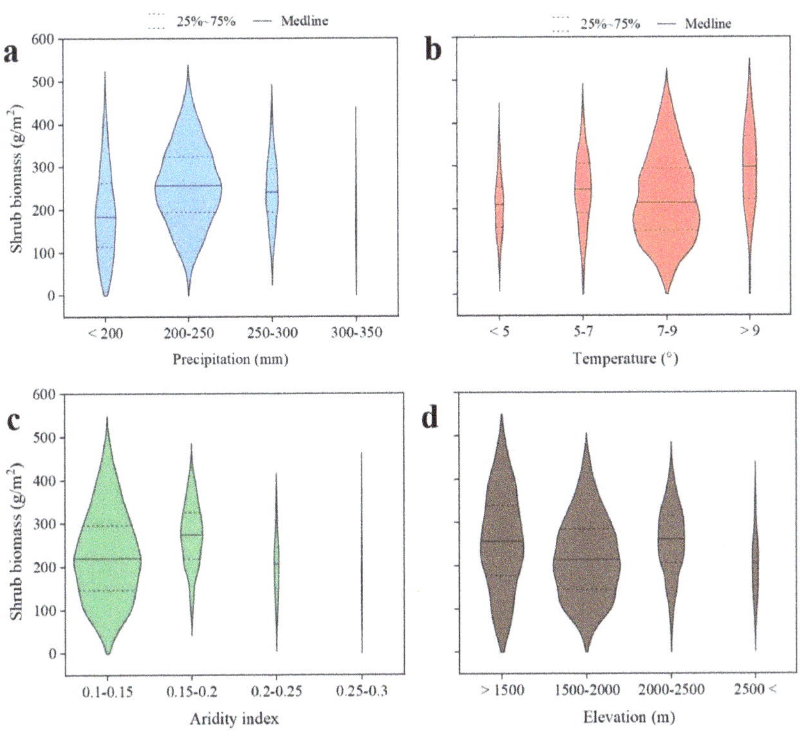

Figure 8. (**a**) The distribution of shrub biomass under precipitation gradients. (**b**) The distribution of shrub biomass under temperature gradients. (**c**) The distribution of shrub biomass within different ranges of aridity index. (**d**) The distribution of shrub biomass along elevation gradients.

4. Discussion

4.1. Advantages of Integrating UAV and Satellite Data for Shrub Biomass Estimations

The shrub biomass estimated based on the integration of UAVs and satellite imagery has significant implications for ecological research and environmental monitoring. Previous studies just relying on field-sampled shrub biomass and satellite imagery often have limited estimate accuracy ($R^2 < 0.6$) [91,92]. The accuracy is reported to be affected by the following factors, including (1) the scale mismatch between the limited ground measurement area and the resolution of satellite images [27] and (2) the number of field samples [28]. The UAVs can provide images to capture the canopy structure and coverage of shrubs at centimeter spatial resolution, which makes it possible to match the spatial scales of the ground measurements. The shrub, low stature, and patchiness of vegetation make UAVs well-suited for assisting in large-scale ground community surveys. Meanwhile, it is also possible to aggregate the UAV observations to match different satellite spatial resolutions. Thus, the scale effects on the shrub biomass estimations between ground measurements and satellite observations could be reduced by the utilization of UAVs. In this study, a grid was established with a consistent spatial resolution of the satellite images, which makes it possible to estimate shrub biomass within each grid cell. This approach provided an improved satellite-based shrub biomass estimation ($R^2 = 0.62$) with the integrated samples from ground measurements and UAV-based estimations (Figure 5). Furthermore, the utilization of UAV imagery provided a cost-effective and efficient method for estimating shrub biomass, which resulted in obtaining sufficient shrub biomass samples for developing robust models. However, the integration of UAV and satellite data also has limitations. For example, the temporal differences between the UAV and satellite imagery acquisitions may introduce uncertainties in capturing the shrub biomass. Therefore, further research should focus on exploring the potential of integrating multi-scale remote sensing data to achieve more comprehensive and accurate estimations of shrub biomass.

4.2. Uncertainties of Shrub Biomass Estimation

In this study, EOPC and EOUB were used to determine the estimation accuracy of the models in different shrub coverage or shrub biomass ranges. The spatial distribution maps generated using the leave-one-out cross-validation (LOOCV) and random forest regression (RFR) methods can also clearly characterize the accuracy of shrubland biomass estimation in different regions. The results show that in areas with shrub coverage exceeding 30% and biomass greater than 300 g/m^2, the estimation uncertainty is the lowest, with an average uncertainty below 4% and the lowest values even below 2%. These results suggested that the estimation accuracy of the models improved with the increases in shrub coverage or shrub biomass. Previous studies also found that the spectral signals of shrubs are weak in pixels with low shrub coverage, which makes it more difficult to identify the shrubs in these regions [93–96]. In the high biomass areas, the spectral signals of the shrubs are strong, which may allow the satellite-based models to better identify the distributions of shrub biomass [93–96]. This indicates that the integration approach of the multi-source data works well over most of the areas, especially for those with moderate or high shrub covers. For the regions with low shrub coverage, using more shrub biomass samples should be a possible way to improve the model's performance. This will be tested in our near future studies. The uncertainty analysis not only provided an assessment of the resultant biomass map but also directed our future improvements.

4.3. Spatial Characteristics of Biomass in Shrubland

The study found that shrub biomass was more extensively distributed within the precipitation range of 200–250 mm. As precipitation continued to increase beyond 250 mm, the distribution of shrub biomass correspondingly decreased. It is shown that precipitation below 250 mm may be an optimal range for shrub growth in the area. Excessive precipitation may impact the growth environment of shrubs, leading to a shift in dominant species from shrubs to other plant types in the Helan Mountains and resulting in a reduction in shrub biomass [97–99]. The phenomenon may be attributed to the fact that the increased availability of water under high precipitation regimes appears to favor the proliferation and competitive dominance of grass-like vegetation, thereby potentially suppressing the development and abundance of shrub communities [100]. Regarding temperature, the majority of shrub biomass is found in areas with an average annual temperature range of 7–9 °C. This suggests that shrubs are most suitable for growth within specific thermal conditions. Temperatures that are either too high or too low may exceed the adaptive range of the shrubs, affecting their physiological activities and growth rates [101]. The optimal temperature range facilitates the shrubs' ability to carry out photosynthesis and respiration, thereby sustaining their growth and biomass accumulation. Furthermore, the shrub biomass was mostly distributed within the aridity index range of 0.1–0.15. Aridity helps maintain the richness and diversity of species with different morpho-functional and life history traits in shrub canopies [99,102]. Regarding elevation, the distribution of shrub biomass did not exhibit a clear trend with changes in altitude. The study found that the majority of shrub biomass was concentrated in areas below 2500 m in elevation. At higher elevations, the lower air temperatures and harsher climate conditions may limit the growth cycles and rates of the shrub communities [103,104]. Additionally, the soil conditions and nutrient availability in the high-elevation areas may be less favorable compared to lower-elevation regions, further impacting the development of the shrub vegetation. Previous studies have suggested a correlation between shrub biomass distribution and the four environmental factors selected in this study in the southwestern region of China. Therefore, future research will be conducted to unravel the complex relationships between shrub biomass dynamics and environmental factors in further studies.

4.4. Limitations and Improvements

Although this research yielded some progress, certain limitations must be acknowledged. Firstly, shrub biomass may be impacted by factors such as soil moisture, topographical variances, and understory vegetation [8,105]. Future research could incorporate additional environmental factors to improve model accuracy. Second, applying our best model to other regions may result in additional errors. Due to spatial heterogeneity and differences in shrub species, the relationships between vegetation indices and shrub biomass may vary, leading to different optimal feature combinations. In the future, this research will be expanded to other regions to enhance the transferability of our model and apply it to regions with more diverse climatic environments. Additionally, a greater variety of data is aimed to be utilized, such as hyperspectral data, to develop more accurate shrub biomass estimation models. Finally, the results indicate that our model's precision is lower in areas with low shrub cover and biomass than in other regions; acquiring more shrub biomass data in these regions could be an improvement approach.

5. Conclusions

This study proposed a novel approach to improve the estimation of shrub biomass at a regional scale by increasing the shrub biomass samples and reducing the spatial scale mismatch between ground measurements and satellite observations by UAV. The workflow included the following parts: First, a shrubland distribution map was generated for the Helan Mountains, China. Subsequently, the shrub biomass was obtained at the UAV level based on the UAV images and the allometric growth equation for the shrub biomass fitted using field measurements. Then, the best shrub biomass estimation model at the satellite level was obtained by comparing three RFR models driven by different remote sensing features. This model was used to generate the distribution map of shrub biomass in the Helan Mountains in 2023. Based on the resultant map, the distribution of shrub biomass along with different environmental factors was assessed. The results indicate that using UAV imagery to calculate shrub biomass provides a more convenient and effective method to supplement shrub biomass samples, significantly reducing the workload and costs of fieldwork. Combining remote sensing imagery from different sensors will likely provide more assistance for future research on large-scale shrub biomass estimation. Although this study made some progress in the method, there are still limitations in sample size, data resolution, and incomplete environmental factors considered. In the following works, this work will be expanded to other regions to validate and improve the proposed methodology.

Author Contributions: Conceptualization, J.W.; methodology, W.L., J.W. and J.Y.; software, W.L.; validation, W.L.; formal analysis, W.L.; investigation, W.L., T.M., C.L. and Y.L.; resources, W.L.; data curation, W.L.; writing—original draft preparation, W.L.; writing—review and editing, W.L., J.W. and J.Y.; visualization, W.L. and M.O.; supervision, J.W.; project administration, J.W.; funding acquisition, J.W. and Y.H. All authors have read and agreed to the published version of the manuscript.

Funding: This study is supported by the Key Research and Development Projects of Ningxia Province, China (2022BEG03050, 2023BEG02049), the National Natural Science Foundation of China (42101355), and the Chinese Universities Scientific Fund (15053346, 10092004, 31051203). We express our gratitude to the anonymous reviewers for their valuable time and efforts in reviewing the manuscript.

Data Availability Statement: Data is contained within the article.

Acknowledgments: We acknowledge the data support from "Loess Plateau SubCenter, National Earth System Science Data Center, National Science & Technology Infrastructure of China. (http://loess.geodata.cn)".

Conflicts of Interest: The authors declare no conflict of interest.

References

1. Vezzoni, R.; Sijtsma, F.; Vihinen, H. Designing Effective Environmental Policy Mixes in the UN Decade on Ecosystem Restoration. *Ecosyst. Serv.* **2023**, *62*, 101541. [CrossRef]
2. Zhao, Y.; Wang, J.; Zhang, G.; Liu, L.; Yang, J.; Wu, X.; Biradar, C.; Dong, J.; Xiao, X. Divergent Trends in Grassland Degradation and Desertification under Land Use and Climate Change in Central Asia from 2000 to 2020. *Ecol. Indic.* **2023**, *154*, 110737. [CrossRef]
3. Trenberth, K.E. Climate Change Caused by Human Activities Is Happening and It Already Has Major Consequences. *J. Energy Nat. Resour. Law* **2018**, *36*, 463–481. [CrossRef]
4. Xiong, Y.; Mo, S.; Wu, H.; Qu, X.; Liu, Y.; Zhou, L. Influence of Human Activities and Climate Change on Wetland Landscape Pattern—A Review. *Sci. Total Environ.* **2023**, *879*, 163112. [CrossRef]
5. Parmenter, R.R.; MacMahon, J.A. Factors Determining the Abundance and Distribution of Rodents in a Shrub-Steppe Ecosystem: The Role of Shrubs. *Oecologia* **1983**, *59*, 145–156. [CrossRef]
6. Jankju, M. Role of Nurse Shrubs in Restoration of an Arid Rangeland: Effects of Microclimate on Grass Establishment. *J. Arid Environ.* **2013**, *89*, 103–109. [CrossRef]
7. Ballantyne, M.; Pickering, C.M. Shrub Facilitation Is an Important Driver of Alpine Plant Community Diversity and Functional Composition. *Biodivers. Conserv.* **2015**, *24*, 1859–1875. [CrossRef]
8. Li, J.; Zhao, C.Y.; Song, Y.J.; Sheng, Y.; Zhu, H. Spatial Patterns of Desert Annuals in Relation to Shrub Effects on Soil Moisture. *J. Veg. Sci.* **2010**, *21*, 221–232. [CrossRef]
9. Garcia-Estringana, P.; Alonso-Blázquez, N.; Marques, M.J.; Bienes, R.; González-Andrés, F.; Alegre, J. Use of Mediterranean Legume Shrubs to Control Soil Erosion and Runoff in Central Spain. A Large-Plot Assessment under Natural Rainfall Conducted during the Stages of Shrub Establishment and Subsequent Colonisation. *CATENA* **2013**, *102*, 3–12. [CrossRef]
10. Boelman, N.T.; Gough, L.; Wingfield, J.; Goetz, S.; Asmus, A.; Chmura, H.E.; Krause, J.S.; Perez, J.H.; Sweet, S.K.; Guay, K.C. Greater Shrub Dominance Alters Breeding Habitat and Food Resources for Migratory Songbirds in Alaskan Arctic Tundra. *Glob. Chang. Biol.* **2015**, *21*, 1508–1520. [CrossRef]
11. Návar, J.; Méndez, E.; Nájera, A.; Graciano, J.; Dale, V.; Parresol, B. Biomass Equations for Shrub Species of Tamaulipan Thornscrub of North-Eastern Mexico. *J. Arid Environ.* **2004**, *59*, 657–674. [CrossRef]
12. Chai, Y.; Zhong, J.; Zhao, J.; Guo, J.; Yue, M.; Guo, Y.; Wang, M.; Wan, P. Environment and Plant Traits Explain Shrub Biomass Allocation and Species Composition across Ecoregions in North China. *J. Veg. Sci.* **2021**, *32*, e13080. [CrossRef]
13. Gómez-Aparicio, L.; Zamora, R.; Gómez, J.M.; Hódar, J.A.; Castro, J.; Baraza, E. Applying Plant Facilitation to Forest Restoration: A Meta-Analysis of the Use of Shrubs as Nurse Plants. *Ecol. Appl.* **2004**, *14*, 1128–1138. [CrossRef]
14. Yu, G.; Li, X.; Wang, Q.; Li, S. Carbon Storage and Its Spatial Pattern of Terrestrial Ecosystem in China. *JORE* **2010**, *1*, 97–109. [CrossRef]
15. Fonseca, F.; de Figueiredo, T.; Bompastor Ramos, M.A. Carbon Storage in the Mediterranean Upland Shrub Communities of Montesinho Natural Park, Northeast of Portugal. *Agrofor. Syst.* **2012**, *86*, 463–475. [CrossRef]
16. Catchpole, W.R.; Wheeler, C.J. Estimating Plant Biomass: A Review of Techniques. *Aust. J. Ecol.* **1992**, *17*, 121–131. [CrossRef]

17. Fragaszy, D.M.; Boinski, S.; Whipple, J. Behavioral Sampling in the Field: Comparison of Individual and Group Sampling Methods. *Am. J. Primatol.* **1992**, *26*, 259–275. [CrossRef]
18. Chojnacky, D.C.; Milton, M. Measuring Carbon in Shrubs. In *Field Measurements for Forest Carbon Monitoring: A Landscape-Scale Approach*; Hoover, C.M., Ed.; Springer Netherlands: Dordrecht, The Netherlands, 2008; pp. 45–72. ISBN 978-1-4020-8506-2.
19. Rojo, V.; Arzamendia, Y.; Pérez, C.; Baldo, J.; Vilá, B. Double Sampling Methods in Biomass Estimates of Andean Shrubs and Tussocks. *Rangel. Ecol. Manag.* **2017**, *70*, 718–722. [CrossRef]
20. Zeng, H.-Q.; Liu, Q.-J.; Feng, Z.-W.; Ma, Z.-Q. Biomass Equations for Four Shrub Species in Subtropical China. *J. For. Res.* **2010**, *15*, 83–90. [CrossRef]
21. Huff, S.; Ritchie, M.; Temesgen, H. Allometric Equations for Estimating Aboveground Biomass for Common Shrubs in Northeastern California. *For. Ecol. Manag.* **2017**, *398*, 48–63. [CrossRef]
22. Laliberte, A.S.; Rango, A.; Havstad, K.M.; Paris, J.F.; Beck, R.F.; McNeely, R.; Gonzalez, A.L. Object-Oriented Image Analysis for Mapping Shrub Encroachment from 1937 to 2003 in Southern New Mexico. *Remote Sens. Environ.* **2004**, *93*, 198–210. [CrossRef]
23. Ramsey, R.D.; Wright, D.L.; McGinty, C. Evaluating the Use of Landsat 30m Enhanced Thematic Mapper to Monitor Vegetation Cover in Shrub-Steppe Environments. *Geocarto Int.* **2004**, *19*, 39–47. [CrossRef]
24. Serrano, L.; Peñuelas, J.; Ustin, S.L. Remote Sensing of Nitrogen and Lignin in Mediterranean Vegetation from AVIRIS Data: Decomposing Biochemical from Structural Signals. *Remote Sens. Environ.* **2002**, *81*, 355–364. [CrossRef]
25. Roy, P.S.; Ravan, S.A. Biomass Estimation Using Satellite Remote Sensing Data—An Investigation on Possible Approaches for Natural Forest. *J. Biosci.* **1996**, *21*, 535–561. [CrossRef]
26. Anderson, K.E.; Glenn, N.F.; Spaete, L.P.; Shinneman, D.J.; Pilliod, D.S.; Arkle, R.S.; McIlroy, S.K.; Derryberry, D.R. Estimating Vegetation Biomass and Cover across Large Plots in Shrub and Grass Dominated Drylands Using Terrestrial Lidar and Machine Learning. *Ecol. Indic.* **2018**, *84*, 793–802. [CrossRef]
27. Mao, P.; Ding, J.; Jiang, B.; Qin, L.; Qiu, G.Y. How Can UAV Bridge the Gap between Ground and Satellite Observations for Quantifying the Biomass of Desert Shrub Community? *ISPRS J. Photogramm. Remote Sens.* **2022**, *192*, 361–376. [CrossRef]
28. Chen, A.; Xu, C.; Zhang, M.; Guo, J.; Xing, X.; Yang, D.; Xu, B.; Yang, X. Cross-Scale Mapping of above-Ground Biomass and Shrub Dominance by Integrating UAV and Satellite Data in Temperate Grassland. *Remote Sens. Environ.* **2024**, *304*, 114024. [CrossRef]
29. Galidaki, G.; Zianis, D.; Gitas, I.; Radoglou, K.; Karathanassi, V.; Tsakiri–Strati, M.; Woodhouse, I.; Mallinis, G. Vegetation Biomass Estimation with Remote Sensing: Focus on Forest and Other Wooded Land over the Mediterranean Ecosystem. *Int. J. Remote Sens.* **2017**, *38*, 1940–1966. [CrossRef]
30. Lu, D. The Potential and Challenge of Remote Sensing-based Biomass Estimation. *Int. J. Remote Sens.* **2006**, *27*, 1297–1328. [CrossRef]
31. Kushida, K.; Kim, Y.; Tsuyuzaki, S.; Fukuda, M. Spectral Vegetation Indices for Estimating Shrub Cover, Green Phytomass and Leaf Turnover in a Sedge-shrub Tundra. *Int. J. Remote Sens.* **2009**, *30*, 1651–1658. [CrossRef]
32. Bannari, A.; Morin, D.; Bonn, F.; Huete, A.R. A Review of Vegetation Indices. *Remote Sens. Rev.* **1995**, *13*, 95–120. [CrossRef]
33. Chang, J.G.; Shoshany, M.; Oh, Y. Polarimetric Radar Vegetation Index for Biomass Estimation in Desert Fringe Ecosystems. *IEEE Trans. Geosci. Remote Sens.* **2018**, *56*, 7102–7108. [CrossRef]
34. Chang, G.J.; Oh, Y.; Goldshleger, N.; Shoshany, M. Biomass Estimation of Crops and Natural Shrubs by Combining Red-Edge Ratio with Normalized Difference Vegetation Index. *J. Appl. Remote Sens.* **2022**, *16*, 014501. [CrossRef]
35. Li, Y.; Andersen, H.-E.; McGaughey, R. A Comparison of Statistical Methods for Estimating Forest Biomass from Light Detection and Ranging Data. *West. J. Appl. For.* **2008**, *23*, 223–231. [CrossRef]
36. Wagner, H.H.; Fortin, M.-J. Spatial Analysis of Landscapes: Concepts and Statistics. *Ecology* **2005**, *86*, 1975–1987. [CrossRef]
37. Mao, P.; Qin, L.; Hao, M.; Zhao, W.; Luo, J.; Qiu, X.; Xu, L.; Xiong, Y.; Ran, Y.; Yan, C.; et al. An Improved Approach to Estimate Above-Ground Volume and Biomass of Desert Shrub Communities Based on UAV RGB Images. *Ecol. Indic.* **2021**, *125*, 107494. [CrossRef]
38. Tamiminia, H.; Salehi, B.; Mahdianpari, M.; Beier, C.M.; Klimkowski, D.J.; Volk, T.A. Comparison of Machine and Deep Learning Methods to Estimate Shrub Willow Biomass from UAS Imagery. *Can. J. Remote Sens.* **2021**, *47*, 209–227. [CrossRef]
39. Viana, H.; Aranha, J.; Lopes, D.; Cohen, W.B. Estimation of Crown Biomass of Pinus Pinaster Stands and Shrubland Above-Ground Biomass Using Forest Inventory Data, Remotely Sensed Imagery and Spatial Prediction Models. *Ecol. Model.* **2012**, *226*, 22–35. [CrossRef]
40. Wu, C.; Shen, H.; Shen, A.; Deng, J.; Gan, M.; Zhu, J.; Xu, H.; Wang, K. Comparison of Machine-Learning Methods for above-Ground Biomass Estimation Based on Landsat Imagery. *JARS* **2016**, *10*, 035010. [CrossRef]
41. Tang, Y.; Kurths, J.; Lin, W.; Ott, E.; Kocarev, L. Introduction to Focus Issue: When Machine Learning Meets Complex Systems: Networks, Chaos, and Nonlinear Dynamics. *Chaos Interdiscip. J. Nonlinear Sci.* **2020**, *30*, 063151. [CrossRef]
42. Cao, J.; Tao, T. Using Machine-Learning Models to Understand Nonlinear Relationships between Land Use and Travel. *Transp. Res. Part D Transp. Environ.* **2023**, *123*, 103930. [CrossRef]
43. Singh, N.; Singh, D.P.; Pant, B. A Comprehensive Study of Big Data Machine Learning Approaches and Challenges. In Proceedings of the 2017 International Conference on Next Generation Computing and Information Systems (ICNGCIS), Jammu, India, 11–12 December 2017; pp. 80–85.
44. Zhou, L.; Pan, S.; Wang, J.; Vasilakos, A.V. Machine Learning on Big Data: Opportunities and Challenges. *Neurocomputing* **2017**, *237*, 350–361. [CrossRef]

45. Abdullah, M.M.; Al-Ali, Z.M.; Srinivasan, S. The Use of UAV-Based Remote Sensing to Estimate Biomass and Carbon Stock for Native Desert Shrubs. *MethodsX* **2021**, *8*, 101399. [CrossRef] [PubMed]
46. Iizuka, K.; Itoh, M.; Shiodera, S.; Matsubara, T.; Dohar, M.; Watanabe, K. Advantages of Unmanned Aerial Vehicle (UAV) Photogrammetry for Landscape Analysis Compared with Satellite Data: A Case Study of Postmining Sites in Indonesia. *Cogent Geosci.* **2018**, *4*, 1498180. [CrossRef]
47. Li, Z.; Ding, J.; Zhang, H.; Feng, Y. Classifying Individual Shrub Species in UAV Images—A Case Study of the Gobi Region of Northwest China. *Remote Sens.* **2021**, *13*, 4995. [CrossRef]
48. Gonzalez Musso, R.F.; Oddi, F.J.; Goldenberg, M.G.; Garibaldi, L.A. Applying Unmanned Aerial Vehicles (UAVs) to Map Shrubland Structural Attributes in Northern Patagonia, Argentina. *Can. J. For. Res.* **2020**, *50*, 615–623. [CrossRef]
49. Abdullah, M.M.; Al-Ali, Z.M.; Abdullah, M.T.; Al-Anzi, B. The Use of Very-High-Resolution Aerial Imagery to Estimate the Structure and Distribution of the Rhanterium Epapposum Community for Long-Term Monitoring in Desert Ecosystems. *Plants* **2021**, *10*, 977. [CrossRef]
50. Ding, J.; Li, Z.; Zhang, H.; Zhang, P.; Cao, X.; Feng, Y. Quantifying the Aboveground Biomass (AGB) of Gobi Desert Shrub Communities in Northwestern China Based on Unmanned Aerial Vehicle (UAV) RGB Images. *Land* **2022**, *11*, 543. [CrossRef]
51. Poley, L.G.; Laskin, D.N.; McDermid, G.J. Quantifying Aboveground Biomass of Shrubs Using Spectral and Structural Metrics Derived from UAS Imagery. *Remote Sens.* **2020**, *12*, 2199. [CrossRef]
52. Shashkov, M.; Ivanova, N.; Shanin, V.; Grabarnik, P. Ground Surveys Versus UAV Photography: The Comparison of Two Tree Crown Mapping Techniques. In *Information Technologies in the Research of Biodiversity*; Bychkov, I., Voronin, V., Eds.; Springer International Publishing: Cham, Switzerland, 2019; pp. 48–56.
53. Zhu, Z.; Woodcock, C.E. Object-Based Cloud and Cloud Shadow Detection in Landsat Imagery. *Remote Sens. Environ.* **2012**, *118*, 83–94. [CrossRef]
54. Roy, D.P.; Kovalskyy, V.; Zhang, H.K.; Vermote, E.F.; Yan, L.; Kumar, S.S.; Egorov, A. Characterization of Landsat-7 to Landsat-8 Reflective Wavelength and Normalized Difference Vegetation Index Continuity. *Remote Sens. Environ.* **2016**, *185*, 57–70. [CrossRef] [PubMed]
55. Shoshany, M.; Karnibad, L. Mapping Shrubland Biomass along Mediterranean Climatic Gradients: The Synergy of Rainfall-Based and NDVI-Based Models. *Int. J. Remote Sens.* **2011**, *32*, 9497–9508. [CrossRef]
56. Huete, A.; Didan, K.; Miura, T.; Rodriguez, E.P.; Gao, X.; Ferreira, L.G. Overview of the Radiometric and Biophysical Performance of the MODIS Vegetation Indices. *Remote Sens. Environ.* **2002**, *83*, 195–213. [CrossRef]
57. Xiao, X.; Boles, S.; Liu, J.; Zhuang, D.; Frolking, S.; Li, C.; Salas, W.; Moore, B. Mapping Paddy Rice Agriculture in Southern China Using Multi-Temporal MODIS Images. *Remote Sens. Environ.* **2005**, *95*, 480–492. [CrossRef]
58. Richardson, A.J.; Wiegand, C.L. Distinguishing Vegetation from Soil Background Information. 1997. *Photogramm. Eng. Remote Sens.* **1977**, *43*, 1541–1552.
59. Gitelson, A.A.; Kaufman, Y.J.; Merzlyak, M.N. Use of a Green Channel in Remote Sensing of Global Vegetation from EOS-MODIS. *Remote Sens. Environ.* **1996**, *58*, 289–298. [CrossRef]
60. Gitelson, A.A.; Kaufman, Y.J.; Stark, R.; Rundquist, D. Novel Algorithms for Remote Estimation of Vegetation Fraction. *Remote Sens. Environ.* **2002**, *80*, 76–87. [CrossRef]
61. Fernandes, R.; Butson, C.; Leblanc, S.; Latifovic, R. Landsat-5 TM and Landsat-7 ETM+ Based Accuracy Assessment of Leaf Area Index Products for Canada Derived from SPOT-4 VEGETATION Data. *Can. J. Remote Sens.* **2003**, *29*, 241–258. [CrossRef]
62. Rock, B.N.; Vogelmann, J.E.; Williams, D.L.; Vogelmann, A.F.; Hoshizaki, T. Remote Detection of Forest Damage. *BioScience* **1986**, *36*, 439–445. [CrossRef]
63. Jordan, C.F. Derivation of Leaf-Area Index from Quality of Light on the Forest Floor. *Ecology* **1969**, *50*, 663–666. [CrossRef]
64. Birth, G.S.; McVey, G.R. Measuring the Color of Growing Turf with a Reflectance Spectrophotometer[1]. *Agron. J.* **1968**, *60*, 640–643. [CrossRef]
65. Jiang, Z.; Huete, A.R.; Didan, K.; Miura, T. Development of a Two-Band Enhanced Vegetation Index without a Blue Band. *Remote Sens. Environ.* **2008**, *112*, 3833–3845. [CrossRef]
66. Chen, J.M. Evaluation of Vegetation Indices and a Modified Simple Ratio for Boreal Applications. *Can. J. Remote Sens.* **1996**, *22*, 229–242. [CrossRef]
67. Rondeaux, G.; Steven, M.; Baret, F. Optimization of Soil-Adjusted Vegetation Indices. *Remote Sens. Environ.* **1996**, *55*, 95–107. [CrossRef]
68. Roujean, J.-L.; Breon, F.-M. Estimating PAR Absorbed by Vegetation from Bidirectional Reflectance Measurements. *Remote Sens. Environ.* **1995**, *51*, 375–384. [CrossRef]
69. Huete, A.R. A Soil-Adjusted Vegetation Index (SAVI). *Remote Sens. Environ.* **1988**, *25*, 295–309. [CrossRef]
70. Major, D.J.; Baret, F.; Guyot, G. A Ratio Vegetation Index Adjusted for Soil Brightness. *Int. J. Remote Sens.* **1990**, *11*, 727–740. [CrossRef]
71. Thenkabail, P.S.; Ward, A.D.; Lyon, J.; Merry, C.J. Thematic Mapper Vegetation Indices for Determining Soybean and Corn Growth Parameters. *Photogramm. Eng. Remote Sens.* **1994**, *60*, 437–442.
72. Gitelson, A.; Merzlyak, M.N. Quantitative Estimation of Chlorophyll-a Using Reflectance Spectra: Experiments with Autumn Chestnut and Maple Leaves. *J. Photochem. Photobiol. B Biol.* **1994**, *22*, 247–252. [CrossRef]

73. Gitelson, A.A.; Merzlyak, M.N.; Chivkunova, O.B. Optical Properties and Nondestructive Estimation of Anthocyanin Content in Plant Leaves. *Photochem. Photobiol.* **2001**, *74*, 38–45. [CrossRef]
74. Vogelmann, J.E.; Rock, B.N.; Moss, D.M. Red Edge Spectral Measurements from Sugar Maple Leaves. *Int. J. Remote Sens.* **1993**, *14*, 1563–1575. [CrossRef]
75. Badgley, G.; Field, C.B.; Berry, J.A. Canopy Near-Infrared Reflectance and Terrestrial Photosynthesis. *Sci. Adv.* **2017**, *3*, e1602244. [CrossRef] [PubMed]
76. Camps-Valls, G.; Campos-Taberner, M.; Moreno-Martínez, Á.; Walther, S.; Duveiller, G.; Cescatti, A.; Mahecha, M.D.; Muñoz-Marí, J.; García-Haro, F.J.; Guanter, L.; et al. A Unified Vegetation Index for Quantifying the Terrestrial Biosphere. *Sci. Adv.* **2021**, *7*, eabc7447. [CrossRef] [PubMed]
77. Wang, C.; Chen, J.; Wu, J.; Tang, Y.; Shi, P.; Black, T.A.; Zhu, K. A Snow-Free Vegetation Index for Improved Monitoring of Vegetation Spring Green-up Date in Deciduous Ecosystems. *Remote Sens. Environ.* **2017**, *196*, 1–12. [CrossRef]
78. Zanaga, D.; Van De Kerchove, R.; Daems, D.; De Keersmaecker, W.; Brockmann, C.; Kirches, G.; Wevers, J.; Cartus, O.; Santoro, M.; Fritz, S.; et al. ESA WorldCover 10 m 2021 V200. *Zenodo* **2022**. [CrossRef]
79. Friedl, M.; Sulla-Menashe, D. MODIS/Terra+Aqua Land Cover Type Yearly L3 Global 500m SIN Grid V061. *NASA EOSDIS Land Process. DAAC* **2022**, *10*. [CrossRef]
80. Peng, S. High-Spatial-Resolution Monthly Precipitation Dataset over China during 1901–2017. *Zenodo* **2019**. [CrossRef]
81. Peng, S. High-Spatial-Resolution Monthly Temperatures Dataset over China during 1901–2017. *Earth Syst. Sci. Data* **2019**, *11*, 1931–1946. [CrossRef]
82. Zomer, R.J.; Xu, J.; Trabucco, A. Version 3 of the Global Aridity Index and Potential Evapotranspiration Database. *Sci Data* **2022**, *9*, 409. [CrossRef]
83. Farr, T.G.; Rosen, P.A.; Caro, E.; Crippen, R.; Duren, R.; Hensley, S.; Kobrick, M.; Paller, M.; Rodriguez, E.; Roth, L.; et al. The Shuttle Radar Topography Mission. *Rev. Geophys.* **2007**, *45*. [CrossRef]
84. Belgiu, M.; Drăguţ, L. Random Forest in Remote Sensing: A Review of Applications and Future Directions. *ISPRS J. Photogramm. Remote Sens.* **2016**, *114*, 24–31. [CrossRef]
85. Biau, G.; Scornet, E. A Random Forest Guided Tour. *TEST* **2016**, *25*, 197–227. [CrossRef]
86. Cootes, T.F.; Ionita, M.C.; Lindner, C.; Sauer, P. Robust and Accurate Shape Model Fitting Using Random Forest Regression Voting. In *Computer Vision—ECCV 2012*; Fitzgibbon, A., Lazebnik, S., Perona, P., Sato, Y., Schmid, C., Eds.; Lecture Notes in Computer Science; Springer: Berlin/Heidelberg, Germany, 2012; Volume 7578, pp. 278–291. ISBN 978-3-642-33785-7.
87. Fu, B.; Liang, Y.; Lao, Z.; Sun, X.; Li, S.; He, H.; Sun, W.; Fan, D. Quantifying Scattering Characteristics of Mangrove Species from Optuna-Based Optimal Machine Learning Classification Using Multi-Scale Feature Selection and SAR Image Time Series. *Int. J. Appl. Earth Obs. Geoinf.* **2023**, *122*, 103446. [CrossRef]
88. Rogers, J.; Gunn, S. Identifying Feature Relevance Using a Random Forest. In *Subspace, Latent Structure and Feature Selection*; Saunders, C., Grobelnik, M., Gunn, S., Shawe-Taylor, J., Eds.; Springer: Berlin/Heidelberg, Germany, 2006; pp. 173–184.
89. Hasan, M.A.M.; Nasser, M.; Ahmad, S.; Molla, K.I. Feature Selection for Intrusion Detection Using Random Forest. *J. Inf. Secur.* **2016**, *7*, 129–140. [CrossRef]
90. Wang, J.; Xiao, X.; Basara, J.; Wu, X.; Bajgain, R.; Qin, Y.; Doughty, R.B.; Iii, B.M. Impacts of Juniper Woody Plant Encroachment into Grasslands on Local Climate. *Agric. For. Meteorol.* **2021**, *307*, 108508. [CrossRef]
91. Liang, T.; Yang, S.; Feng, Q.; Liu, B.; Zhang, R.; Huang, X.; Xie, H. Multi-Factor Modeling of above-Ground Biomass in Alpine Grassland: A Case Study in the Three-River Headwaters Region, China. *Remote Sens. Environ.* **2016**, *186*, 164–172. [CrossRef]
92. Jin, Y.; Yang, X.; Qiu, J.; Li, J.; Gao, T.; Wu, Q.; Zhao, F.; Ma, H.; Yu, H.; Xu, B. Remote Sensing-Based Biomass Estimation and Its Spatio-Temporal Variations in Temperate Grassland, Northern China. *Remote Sens.* **2014**, *6*, 1496–1513. [CrossRef]
93. Suess, S.; van der Linden, S.; Okujeni, A.; Griffiths, P.; Leitão, P.J.; Schwieder, M.; Hostert, P. Characterizing 32 years of Shrub Cover Dynamics in Southern Portugal Using Annual Landsat Composites and Machine Learning Regression Modeling. *Remote Sens. Environ.* **2018**, *219*, 353–364. [CrossRef]
94. Gan, L.; Cao, X.; Chen, X.; He, Q.; Cui, X.; Zhao, C. Mapping Shrub Coverage in Xilin Gol Grassland with Multi-Temporal Sentinel-2 Imagery. *Remote Sens.* **2022**, *14*, 3266. [CrossRef]
95. Macander, M.J.; Frost, G.V.; Nelson, P.R.; Swingley, C.S. Regional Quantitative Cover Mapping of Tundra Plant Functional Types in Arctic Alaska. *Remote Sens.* **2017**, *9*, 1024. [CrossRef]
96. Yang, X. Woody Plant Cover Estimation in Texas Savanna from MODIS Products. *Earth Interact.* **2019**, *23*, 1–14. [CrossRef]
97. Velasco, N.; Soto-Agurto, C.; Carbone, L.; Massi, C.; Bustamante, R.; Smit, C. Large-Scale Facilitative Effects for a Single Nurse Shrub: Impact of the Rainfall Gradient, Plant Community and Distribution across a Geographical Barrier. *J. Ecol.* **2024**, *112*, 233–245. [CrossRef]
98. Zhang, Y.; Huang, M.; Lian, J. Spatial Distributions of Optimal Plant Coverage for the Dominant Tree and Shrub Species along a Precipitation Gradient on the Central Loess Plateau. *Agric. For. Meteorol.* **2015**, *206*, 69–84. [CrossRef]
99. Holzapfel, C.; Tielbörger, K.; Parag, H.A.; Kigel, J.; Sternberg, M. Annual Plant–Shrub Interactions along an Aridity Gradient. *Basic Appl. Ecol.* **2006**, *7*, 268–279. [CrossRef]
100. Rinella, M.J.; Hammond, D.H.; Bryant, A.-E.M.; Kozar, B.J. High Precipitation and Seeded Species Competition Reduce Seeded Shrub Establishment during Dryland Restoration. *Ecol. Appl.* **2015**, *25*, 1044–1053. [CrossRef]

101. Andreu-Hayles, L.; Gaglioti, B.V.; Berner, L.T.; Levesque, M.; Anchukaitis, K.J.; Goetz, S.J.; D'Arrigo, R. A Narrow Window of Summer Temperatures Associated with Shrub Growth in Arctic Alaska. *Environ. Res. Lett.* **2020**, *15*, 105012. [CrossRef]
102. Laura, M.; Mónica, B.B.; Analía, L.C. Changes in Traits of Shrub Canopies across an Aridity Gradient in Northern Patagonia, Argentina. *Basic Appl. Ecol.* **2010**, *11*, 693–701. [CrossRef]
103. Zhao, Y.; Liu, X.; Li, G.; Wang, S.; Zhao, W.; Ma, J. Phenology of Five Shrub Communities along an Elevation Gradient in the Qilian Mountains, China. *Forests* **2018**, *9*, 58. [CrossRef]
104. Ensslin, A.; Rutten, G.; Pommer, U.; Zimmermann, R.; Hemp, A.; Fischer, M. Effects of Elevation and Land Use on the Biomass of Trees, Shrubs and Herbs at Mount Kilimanjaro. *Ecosphere* **2015**, *6*, art45. [CrossRef]
105. Bolstad, P.V.; Elliott, K.J.; Miniat, C.F. Forests, Shrubs, and Terrain: Top-down and Bottom-up Controls on Forest Structure. *Ecosphere* **2018**, *9*, e02185. [CrossRef]

remote sensing

MDPI

Article

New Insights on the Information Content of the Normalized Difference Vegetation Index Sentinel-2 Time Series for Assessing Vegetation Dynamics

César Sáenz [1,2,3], Víctor Cicuéndez [1,4], Gabriel García [5], Diego Madruga [1], Laura Recuero [3,6], Alfonso Bermejo-Saiz [1], Javier Litago [6], Ignacio de la Calle [2] and Alicia Palacios-Orueta [1,3,*]

1 Departamento de Ingeniería Agroforestal, ETSIAAB, Universidad Politécnica de Madrid, Av. Puerta de Hierro, nº 2—4, Ciudad Universitaria, 28040 Madrid, Spain; cesar.saenzf@alumnos.upm.es (C.S.); victcicu@ucm.es (V.C.); diego.madruga.ramos@alumnos.upm.es (D.M.); alfonso.bermejo@upm.es (A.B.-S.)

2 Quasar Science Resources, S.L., Camino de las Ceudas 2, Las Rozas de Madrid, 28232 Madrid, Spain; ignaciodelacalle@quasarsr.com

3 Centro de Estudios e Investigación para la Gestión de Riesgos Agrarios y Medioambientales (CEIGRAM), Universidad Politécnica de Madrid, C/Senda del Rey 13, 28040 Madrid, Spain; laura.recuero@upm.es

4 Departamento de Física de la Tierra y Astrofísica, Facultad de Ciencias Físicas, Universidad Complutense de Madrid (UCM), 28040 Madrid, Spain

5 Department of Computer Architecture and Automation, Complutense University of Madrid, 28040 Madrid, Spain; gabgar04@ucm.es

6 Departamento de Economía Agraria, Estadística y Gestión de Empresas, ETSIAAB, Universidad Politécnica de Madrid (UPM), Av. Puerta de Hierro, nº 2—4, Ciudad Universitaria, 28040 Madrid, Spain; javier.litago@upm.es

* Correspondence: alicia.palacios@upm.es

Abstract: The Sentinel-2 NDVI time series information content from 2017 to 2023 at a 10 m spatial resolution was evaluated based on the NDVI temporal dependency in five scenarios in central Spain. First, time series were interpolated and then filtered using the Savitzky–Golay, Fast Fourier Transform, Whittaker, and Maximum Value filters. Temporal dependency was assessed using the Q-Ljung-Box and Fisher's Kappa tests, and similarity between raw and filtered time series was assessed using Correlation Coefficient and Root Mean Square Error. An Interpolating Efficiency Indicator (IEI) was proposed to summarize the number and temporal distribution of low-quality observations. Type of climate, atmospheric disturbances, land cover dynamics, and management were the main sources of variability in five scenarios: (1) rainfed wheat and barley presented high short-term variability due to clouds (lower IEI in winter and spring) during the growing cycle and high interannual variability due to precipitation; (2) maize showed stable summer cycles (high IEI) and low interannual variability due to irrigation; (3) irrigated alfalfa was cut five to six times during summer, resulting in specific intra-annual variability; (4) beech forest showed a strong and stable summer cycle, despite the short-term variability due to clouds (low IEI); and (5) evergreen pine forest had a highly variable growing cycle due to fast responses to temperature and precipitation through the year and medium IEI values. Interpolation after removing non-valid observations resulted in an increase in temporal dependency (Q-test), particularly a short term in areas with low IEI values. The information improvement made it possible to identify hidden periodicities and trends using the Fisher's Kappa test. The SG filter showed high similarity values and weak influence on dynamics, while the MVF showed an overestimation of the NDVI values.

Keywords: Savitzky–Golay filter; Whittaker filter; fast Fourier transform filter; maximum value filter; interpolating efficiency indicator; vegetation dynamics

1. Introduction

Satellite time series constitutes one of the most extensive sources of information about the environment. The large spatial coverage acquired at regular time intervals results in enormous data sets that go beyond the concept of big data. This type of information provides spatial and temporal coherence to uncover patterns of environmental processes yielding, sometimes, more robust information than field data, especially in heterogeneous areas.

In this context, Sentinel-2 provides the most informative set of satellite information because of the combination of high spatial and temporal resolutions. On the one hand, the minimum pixel size of ten meters enables the assessment of heterogeneous landscapes such as crop mosaics [1–3] with high field size variability as well as areas with strong environmental gradients such as mountainous regions [4–6]. On the other hand, the high frequency of data acquisition (5 days) makes it possible to monitor vegetation seasonality in high detail [7]. Several studies have demonstrated its utility when identifying specific phenological events (e.g., start and end of growing season), which is highly relevant when evaluating the influence on the vegetation of agricultural management practices and the impact of extreme weather events [8–12]. This is especially interesting in regions with a Mediterranean type of climate, where the interannual variability of meteorological factors results in different vegetation adaptation strategies [13,14] and agricultural practices, such as crop rotation and sowing and harvest dates [15–17]. In these areas, rainfed crops, dependent on rainfall, have their main growing season in spring [18,19], whereas irrigated crops grow during summer in specific areas with infrastructure [20]. In addition, economic circumstances and political regulations increase cropping dynamics variability and, hence, make monitoring more complex [21].

The experience acquired from working with low-spatial-resolution time series, such as MODIS, highlights the potential of Sentinel-2 time series. To achieve Sentinel-2's full potential, there are two significant limitations: (1) the computational needs due to the increasingly longer time series with frequent observations and high spatial resolution, and (2) the presence of low-quality observations and gaps in time series [22] due to the presence of snow, sensor failure, and atmospheric disturbance (e.g., clouds) [23,24]. To overcome these issues, some authors used only cloud-free images, losing temporal resolution and limiting the dataset to some few images per year [25–28]. However, developing computationally efficient methods based on temporal analysis will make it possible to fully exploit the length of the Sentinel-2 time series while maintaining a high spatial resolution.

Compositing procedures such as Maximum Value Composite (MVC) [29] are frequently used before filtering time series from sensors with daily data acquisition (e.g., MODIS); although the effective temporal frequency is reduced, it does not significantly limit the information [30–32].

The impact of atmospheric disturbance becomes more limiting in sensors with longer revisiting periods such as Sentinel-2 and Landsat (5 and 16 days, respectively). In these cases, the decrease in temporal resolution due to compositing methods may [32] have a significant impact on the time series information contents, so other methods need to be explored.

The Copernicus System provides easily usable quality data that make it possible to accomplish the first assessment of the information content, including the number and length of gaps. The presence of long gaps in time series makes it more difficult to uncover vegetation dynamics patterns, and it seems appropriate to explore gap-filling and filtering strategies such as temporal filters and spectral domain procedures [33–36].

The methods used depend on the noise content in the series and the type of features to enhance. Usually, temporal filters are appropriate for preserving details and short-term patterns, while spectral domain procedures are better for highlighting long-term dynamics. In this sense, Zhou et al. [37] showed that different noise distributions require specific methodologies with temporal filters working better in the tropics and subtropics and spectral domain procedures in boreal latitudes.

Temporal filters such as the Savitzky–Golay (SG) [38] and Whittaker (WHI) [39,40] filters are based on time series smoothing algorithms using moving windows and are focused on the fact that the Normalized Difference Vegetation Index (NDVI) [41] time series are consistent with the vegetation growth curve [42]. The SG, based on a least square fitting, is the most widely used method for reducing noise in vegetation index time series [43]. The parameters used are the degree of the polynomial and the moving window size, which are defined based on the knowledge of the time series characteristics [37]. Htitiou et al. [44] and Stendardi et al. [45] applied the SG filter to the annual Sentinel-2 time series, preserving the NDVI profile and obtaining effective results in the identification of phenological metrics, especially the Start of Season (SOS). The basis of the WHI is to keep a balance between the fidelity of the data and the roughness of the filtered time series. Goldberg et al. [46] and Huang et al. [47] used it to smooth the Sentinel-2 NDVI time series to develop maps for different crops using classifiers, and a 90% accuracy was obtained. Nishanta Khanal et al. [48] compared the WHI, Fast Fourier Transform (FFT), and linear-fit averaging (LFA) smoothing methods to develop land cover classification maps in Nepal using Landsat 5, 7, and 8 images. They concluded that WHI worked better with stable classes while the FFT showed better results with the rest of the classes.

Conversely, procedures that work in the frequency domain, such as Fast Fourier Transform (FFT) [49] and Harmonic Analysis (HANTS) [37], are effective for assessing periodic time series [43].

Despite the potential of Sentinel-2 vegetation index time series for assessing vegetation dynamics with high spatial detail, only few studies testing filters and reconstruction strategies have been conducted [22,50]. Selecting the appropriate process is important to fill the gaps in time series due to the lack of images or the presence of low-quality values while maintaining the dynamics [51,52].

Evaluating the behavior of the filters and comparing their fit with the original Sentinel-2 time series data and dynamics is essential [53,54] for extracting relevant and reliable information, such as phenological metrics.

The main objective of this study was to assess the information content based on Sentinel-2 NDVI time series in terms of temporal dependency in an area with a significant climatic gradient and high variation in land cover dynamics. Specifically, five scenarios of NDVI time series patterns with different land covers and environmental conditions were assessed: (1) wheat and barley, (2) maize, (3) alfalfa, (4) beech forests, and (5) pine forests.

The specific objectives were as follows:

(1) To assess the information content in terms of temporal dependency in the raw NDVI time series and evaluate the interactions between land cover dynamics and other sources of variability.
(2) To assess the impact of interpolation and filtering procedures on the information content, specifically changes in the dynamics and the significance of the main periodicities.
(3) To calculate a concise indicator to evaluate the number and distribution of invalid observations due to meteorological perturbances based on the image classification scene provided by Copernicus.

2. Materials and Methods

2.1. Study Area

The study area covered the Sentinel-2 tile 30TUN (212,056.05 km^2), with the upper left corner at 43°19′34″N–5°28′0″W and lower right corner at 42°21′31″N–4°5′43″W. This tile includes the provinces of León, Palencia, Cantabria, Asturias, and Burgos (Figure 1). The study period was from January 2017 to August 2023.

This area has a temperate climate with dry summer in the southern area (Koppen Csb), and it is fully humid in the northern area (Koppen Cfb) [55,56]. Precipitation is usually concentrated in autumn, winter, and spring, with an accumulated yearly rainfall of 940 mm [55]. Soils are Fluvisols and Cambisols according to the FAO-UNESCO system [57]. Forested areas constitute approximately 35% of the Sentinel-2 tile, and they are distributed at higher altitudes than croplands. The area shows a strong topographical gradient that results in high variability in short distances.

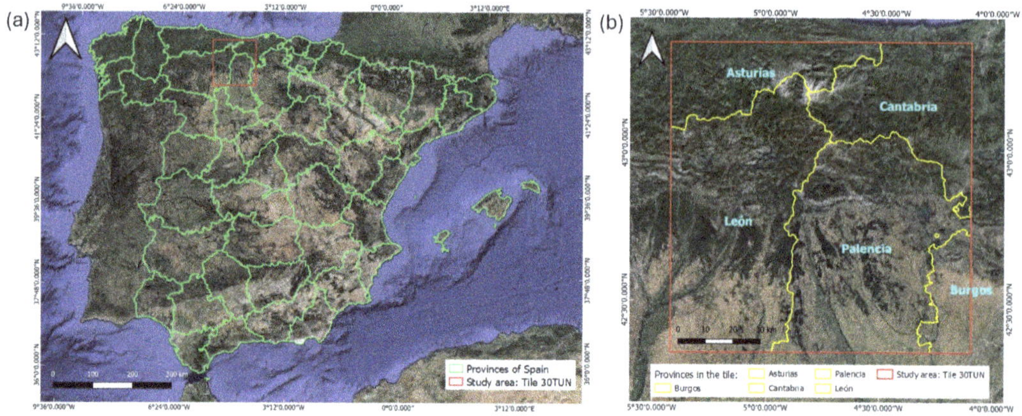

Figure 1. Image in true color from © Google Earth (Landsat/Copernicus). (**a**) Map of Spain and (**b**) study area.

2.2. *Experimental Design and Working Scheme*

The work was focused on assessing the information content of Sentinel-2's complete NDVI time series and how several pre-processing methods can enhance the information content. To accomplish a comprehensive analysis, the following criteria were proposed: (1) the presence of distinct sources of variability, (2) the use of different pre-processing methods, and (3) the use of several evaluation methodologies.

To assure variability, a complete Sentinel-2 tile with a significant climatic gradient, representative land cover types, and vegetation dynamics was chosen; data were assessed and pre-processing levels were from 2017 to 2023. In the agricultural area, the crops selected were wheat, barley, and irrigated crops alfalfa and maize. For forest species, beech and pine forests were selected.

Figure 2 shows the flowchart from the building of the time series to the filtering process. The Jupyter Notebook application was used in Python 3.9 language.

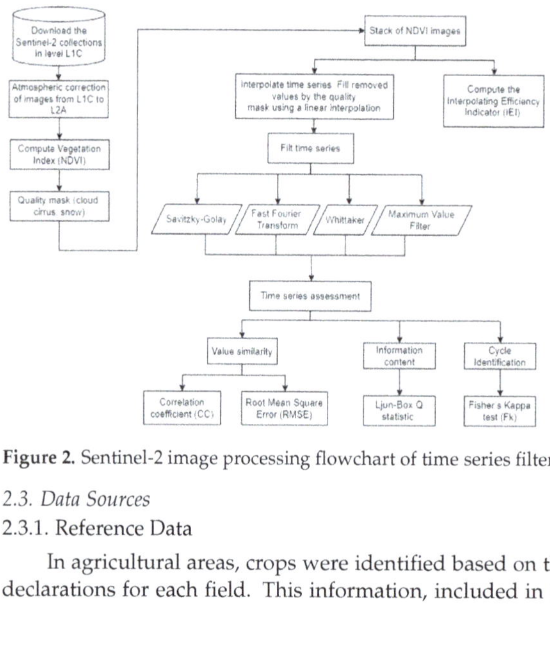

Figure 2. Sentinel-2 image processing flowchart of time series filtering.

2.3. *Data Sources*

2.3.1. Reference Data

In agricultural areas, crops were identified based on the 2018 to 2023 PAC farmers' declarations for each field. This information, included in the "Sistema de Información

Geográfica de Parcelas Agrícolas (SIGPAC)", was downloaded from the "Fondo Español de Garantía Agraria (FEGA)" webpage (https://www.fega.gob.es/atom/es.fega.sigpac.xml, accessed on 14 June 2023). The surface evaluated was 455.74 km^2, including rainfed and irrigated crops. Forest information was obtained from the forest map of Spain at a scale of 1:50,000 [58]; the area evaluated was 872.63 and 563.82 km^2 for beech and pine forests, respectively. Figure 3 shows the spatial distribution of the land covers studied. Only fields with the same vegetation type throughout the study period were evaluated.

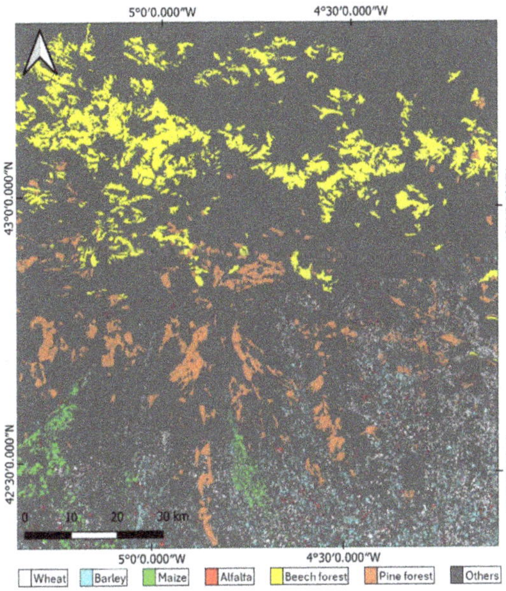

Figure 3. Spatial distribution of the vegetation species in the study area.

Non-irrigated wheat and barley evolution depends on meteorological conditions while irrigated maize shows a more stable cycle during summer. Alfalfa shows a distinct variability due to cuts during the growing period. Table 1 shows the growing period for different crops in our study area. Rainfed crops such as wheat and barley have a growing period, which is highly dependent on rainfall, during spring. Alfalfa, cultivated under irrigated conditions, is cut two to six times per year, and irrigated maize grows in more stable conditions during summer [59].

Table 1. Growing calendar of different crops in northern Spain.

Crop	Months											
	Jan.	Feb.	Mar.	Apr.	May	Jun.	Jul.	Aug.	Sep.	Oct.	Nov.	Dec.
Wheat												
Barley												
Maize												
Alfalfa					Several cuts							
Beech forest										leaf fall		
Pine forest												

Light green represents the growing period for the different vegetation types, the dark green represents the several cuts in Alfalfa, and the orange represents the leaf fall in Beech Forest.

Beech and pine forests were selected for this study due to their distinct dynamics, being deciduous and evergreen, respectively. This choice was aimed to enhance our understanding of the impact of interpolation and filtering in mountainous areas with a higher cloud cover.

2.3.2. Sentinel-2 Data

The Sentinel-2 Multispectral Instrument (MSI) on board the satellites, 2A and 2B, from the Copernicus program provides information in 13 bands in the optical range at spatial resolutions of 10, 20, and 60 m and every five days [60].

In this research, we used the highest resolution images of 10 m to be able to work with pure pixels. A total of 485 images from January 2017 to August 2023 for the 30TUN tile were downloaded at Level-1C: Top-Of-Atmosphere (TOA) from https://dataspace.copernicus.eu/browser/ (accessed on 5 September 2023), and then they were atmospherically corrected using the Sen2Cor V.02.09 tool (https://step.esa.int/main/snap-supported-plugins/sen2cor/, accessed on 15 October 2023) to convert them to Level-2A: Bottom-Of-Atmosphere (BOA). This atmospheric correction process generated the Scene Classification Map (SCL) used to assess data quality; thus, an observation was considered invalid when the pixel's class was "no data", "saturated or defective", "dark area pixels", "cloud shadows", "cloud medium probability", "cloud high probability", and "snow". In this study, we did not use "thin cirrus" clouds because they are partially transparent and still allow for the capture of useful information [61]. Using this classification could significantly reduce the available data, especially in regions with frequent cloud cover, like our study area in the northern region.

2.4. *Compilation of NDVI and Quality Data Time Series*

The NDVI was calculated from the Level-2A images (b4-red: 646–685 nm and b8-nir: 767–908 nm), and the complete time series were generated. The SCL quality band of each date was used to identify the low-quality observations and assess their frequency and distribution in the time series. In the context of time series analysis, a "gap" refers to a period of time when data are scarce, missing, or incomplete. These gaps may arise due to various factors such as technical issues in data collection or simply a lack of data availability for certain periods. Identifying and understanding these gaps is crucial to ensure the accuracy and integrity of time series analysis.

The Interpolating Efficiency Indicator (IEI) was proposed for assessing the potential information content of a pixel time series based on the number of invalid observations and the presence of long gaps in terms of length and distribution within the year. While short gaps (i.e., 1 to 3 invalid consecutive observations) may be interpolated with high reliability, the interpolation process may be limited when gaps are longer, resulting in inaccurate information, especially when they occur during significant periods within the growing cycle.

The IEI (Equation (1)) values range between 0 and 100, with higher values indicating a higher capability of the original time series to retain significant information content to be used in analyses such as phenological studies, crop dynamics analysis, and time series analysis, among others.

$$\forall x \in \{[1,16] \cup Gt_{16}\}$$

$$F_{IEI}(x) = \begin{cases} 100 - \left(1/N \sum_{x=1}^{m} f(x) * 100\right) \; if \; F_{IEI} \geq 0 \\ 0 \; if \; F_{IEI} < 0 \end{cases} \quad (1)$$

$$f(x) = \begin{cases} 100\, if\ x = N \\ W_t(x) * n(x)\ if\ 1 \le x\ \le 16 \\ W_t(16) + \frac{x*100}{N}\ if\ x > 16 \end{cases} \quad (2)$$

where the function $F_{IEI}(x)$ ranges from 0 to 100 and is based on $f(x)$ and N (time series length); $f(x)$ is a piecewise function that depends on x (gap) and W_t (the weight of x), and n (the number of gaps that range from 1 to m) for the two intervals defined in Table 2: for x in Equation (2) and N. G_{t16} represents the upper limit of the x values in the calculation of F_{IEI}. It is a list of values calculated from the expression of $f(x)$ for $x > 16$ and contains each gap longer than 16 in the original time series. The weights associated with the gaps in the IEI were defined based on the understanding that larger gaps have a detrimental impact on the overall series. This implies that larger gaps require greater weights to be applied.

Table 2. Weights (W_t) for the different numbers of consecutive invalid observations (gaps).

Gaps (x)	Weight (Wt)	Gaps (x)	Weight (Wt)
1	0.1	9	8
2	0.3	10	9
3	0.5	11	10
4	0.7	12	12
5	0.9	13	14
6	2	14	16
7	4	15	18
8	6	16	20

2.5. Time Series Interpolation and Filtering

Linear interpolation was performed to complete the time series by replacing non-valid observation values as defined in the Sentinel-2 SCL quality band.

In the second step, four classical filters were applied: Savitzky–Golay, Whittaker, Fast Fourier Transform, and Maximum Value. These filters modify the original data values to achieve their respective objectives, such as noise reduction or smoothing. The changes are a necessary consequence of the filtering processes used to extract or emphasize features of the data.

The Savitzky–Golay (SG) filter [38] is based on the least squares method and is one of the most used methods for reducing noise and smoothing remote sensing time series [62,63]. In our research, the width of the smoothing window and the degree of the smoothing polynomial were set to 7 and 2, respectively.

The Whittaker (WHI) filter [39,40] is based on the balance between fidelity to the original data (S) and roughness (R). The smoothing parameter chosen in our research was 10% of the length of the time series.

The Fast Fourier Transform (FFT) filter [49] is an algorithm that computes the Discrete Fourier transform (DFT) and its inverse efficiently [64]. The time domain data are decomposed into the frequency domain using Fourier analysis.

The Maximum Value Filter (MVF) is based on the MVC [29]. We set as a threshold value the mean calculated using a window width of seven observations. NDVI values lower or equal to the mean are substituted by the maximum value within the window width of seven observations. Finally, each observation is averaged with the previous and subsequent values.

2.6. Assessment of Similarity

The similarity between valid observations in the raw and filtered time series was estimated by two indictors: first, the Pearson Correlation Coefficient (CC) to measure the strength of the adjustment, where high CC values represent a better adjustment to the original valid data [43]; second, the Root Mean Square Error (RMSE) based on the standard deviation of the residuals in the relationship between the original and filtered valid observations; lower RMSE values represent a better adjustment [65].

2.7. Assessment of Time Series Information Content

In order to evaluate the information content of the time series, both the intra-annual and interannual dynamics in the original and processed time series were considered.

Assessment of NDVI Dynamics

Time series analysis was used to study the dynamics of the original (including noise) and the interpolated and filtered time series, including short-term, seasonal phenological patterns, and interannual variability. Three tools of time series analysis were used:

(1) The average year was calculated by applying Buys-Ballot tables [66] to study the intra-annual dynamics of the interpolated and the filtered time series.

(2) The Q-Ljung-Box test (L-B Q) (Equation (3)) [67] was originally defined to test model adequacy in time series. In this work, this test was used to assess the amount of significant information content in a time series by the estimation of the autocorrelation function [17,68,69]. The test was carried out under the classical null hypothesis that the time series were white noise.

The statistic for the test is as follows:

$$L - B\;Q(M) = N(N+2)\sum_{k=1}^{M}(N-k)^{-1}\hat{r}_k^2 \;\sim\; x^2\;(M\;d.f.) \tag{3}$$

where

N is the number of time series observations,

$\hat{r}_k$ is the autocorrelation coefficient at lag k,

M is the number of lags up to which the sum of autocorrelation coefficients is calculated in the Q test and defines the number of degrees of freedom (d.f.) for the X^2 probability distribution. The Q-test was calculated for several time horizons (M), short-term analysis (lags 1 to 7), seasonal analysis (lags 36 and 73), and biannual analysis (lag 146).

(3) The periodogram analysis [70] was performed to identify the presence of periodic components of the NDVI time series [71]. The presence of a seasonal component and its significance were evaluated using Fisher's Kappa test (Fk) [72] calculated as a ratio of the maximum periodogram ordinate to the mean of all ordinates. The null hypothesis is the lack of a periodic component.

The statistical analyses were performed using Python 3 and SAS software (SAS 9.4 Software, SAS Institute Inc., Cary, NC, USA).

For the calculation of the IEI and the assessment of similarity, the complete time series from January 2017 was used. However, for the assessment of time series information content, data from the years when Sentinel-2b was operational (starting in March 2017) were more suitable. Therefore, we used the series from 2018 onwards.

3. Results

3.1. Assessment of Pixels Quality

The invalid observations and gap frequency and length due to clouds for each pixel along the study period (SCL band) are shown in Figures 4 and 5. Figure 4a shows the percentage of valid observations for each pixel in the study area. Figure 4b shows the number of pixels for each number of valid observations (the far extremes are not shown).

The number of missing images was 21, 1, 2, 1, and 2 in the years 2017, 2018, 2020, 2022, and 2023, respectively. Low frequencies of valid observations in buildings and water bodies are not shown in the histogram.

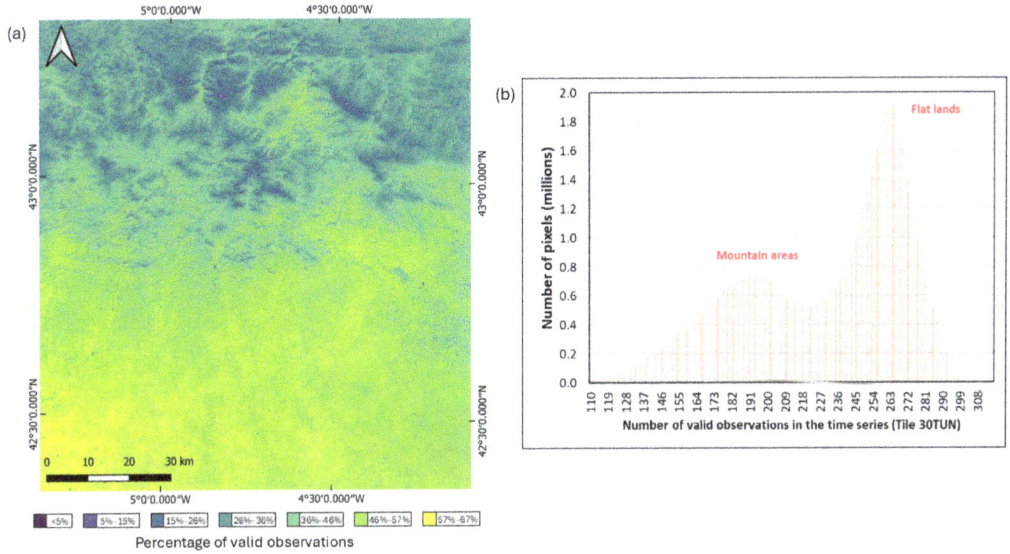

Figure 4. Valid observations according to the SCL band of Sentinel-2. (**a**) Spatial distribution in percentage and (**b**) number of pixels for each level of valid observations.

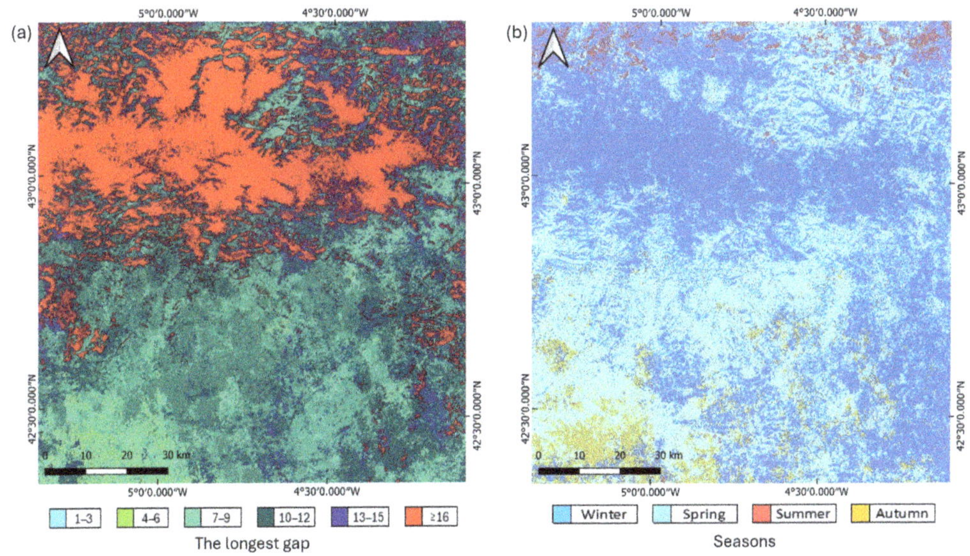

Figure 5. (**a**) Length of the longest gap within time series; (**b**) season in which the longest gap occurs.

The spatial distribution of the percentage of valid observations shows a significant spatial coherence with two clear regions and a significant gradient related to latitude and elevation (Figure 4a). Time series with a greater number of valid observations were mainly in the southern flat areas (Koppen Cs, temperate with dry summer) while in the northern mountainous areas (Koppen, temperate, fully humid Cfb), the percentage of valid observation was lower.

This pattern is shown clearly in the number of pixels for each number of valid observations (Figure 4b), with two distinct populations corresponding to the mountainous areas in the north and the flat areas in the south, with approximately 110 to 236 and 245 to 299 valid observations, respectively.

Figure 5a,b show the length of the longest gap in the time series and season when it occurs, both on a pixel basis. Table 3 shows the total area in which the longest gap in the time series occurs at a given season, including the statistics of the length of the gaps for each season. The maps show a consistent spatial distribution; the longest gaps with more than 16 consecutive observations make up around 25% of the tile, and most of these gaps are in the mountainous region situated in the northern part of the tile, occurring more frequently in winter and secondarily in spring. Most of the surface shows the longest gap between 7 and 9 observations (36% of the tile), and the range of 10 to 12 observations represents 26%; these are in the southern part of the tile.

Table 3. Statistics of the longest gap length values and percentage of the total surface where the longest gap is in each season.

Season	Length of the Longest Gap		
	Mean	Standard Deviation	Percentage (%)
Winter	9.50	4.34	46.78
Spring	7.87	2.33	34.31
Summer	4.26	1.78	2.14
Autumn	7.16	3.06	16.78

Dates of each season: Winter (21 Dec.–19 Mar.), Spring (20 Mar.–19 Jun.), Summer (20 Jun.–21 Sep.), and Autumn (22 Sep.–20 Dec.).

3.2. Interpolating Efficiency Indicator (IEI)

Figure 6a shows the IEI for Tile 30TUN with high values in light yellow and low values in dark purple and black; Figure 6b shows the frequency distribution.

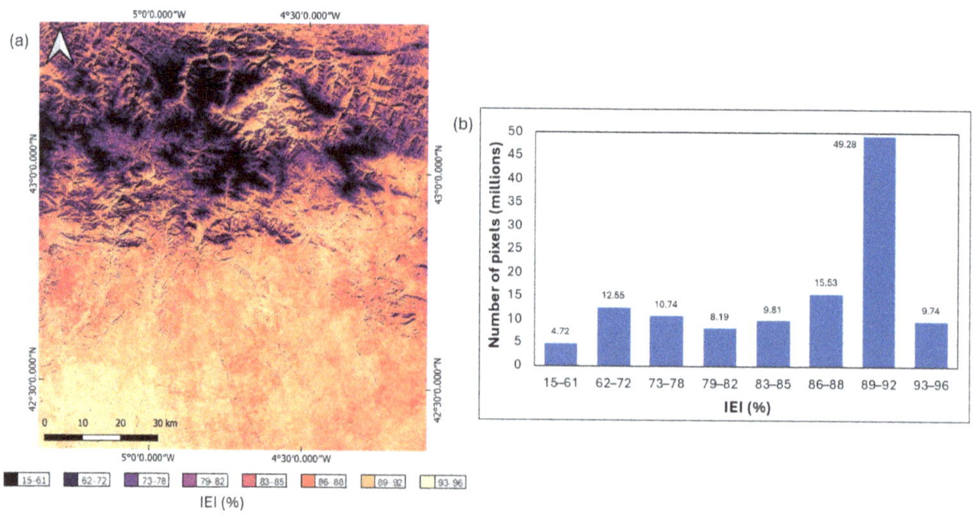

Figure 6. Interpolating Efficiency Indicator for Tile 30TUN: (**a**) spatial distribution and (**b**) frequency distribution.

In the flat areas, values of up to 96% were observed, with the gap length ranging from 1 to 8 and having a smaller impact in the interpolation and filtering process than in the mountainous areas, where maximum values were around 78%. In 40.9% of the surface (4928.13 km^2), IEI values ranged between 89% and 92%, indicating that the interpolation and filtering process will be able to retain more information content for subsequent analysis.

Table 4 shows the IEI values by season and land use, both annually and seasonally. IEI for beech forests, located in Cfb climate areas, presents the lowest annual values, whereas maize, in Cs areas, shows the highest values; the pine forests in the transitional zones show intermediate values. IEI is lowest in winter and highest in summer across all land covers.

Table 4. IEI values by vegetation species and season.

Vegetation Species	Annual	Winter	Spring	Summer	Autumn
Wheat	91.23	86.36	91.73	97.49	91.51
Barley	91.78	87.91	92.30	97.58	91.53
Maize	92.85	91.40	91.55	97.78	92.23
Alfalfa	92.25	88.93	91.98	97.58	92.05
Beech forest	74.90	56.10	81.89	92.74	72.57
Pine forest	88.15	84.34	87.16	96.48	88.05

Crops (Cs), Beech Forest (Cf), and Pine Forest (Transition).

3.3. Examples of Interpolation and Filtering Data

Examples of the interpolated and filtered NDVI time series are shown in Figures 7 and 8 for crops and forests, respectively; time series and the average year corresponding to each land cover are shown. For all types of vegetation, the SG is the one that best fits the interpolated series, while the MVF seems to overestimate the NDVI values.

Figure 9 shows an example of the evolution of one image (10/06/20) after the interpolation and Whittaker filtering processes. Figure 9a (10/06/20), Figure 9b (15/06/20), and Figure 9c (20/06/20) show three consecutive dates of the raw NDVI time series, where invalid values according to the SCL band are indicated in purple; these low-quality values were replaced after linear interpolation (Figure 9d) and filtering (Figure 9e), showing an improvement in the spatial consistency of the scene.

3.4. Similarity between Original and Filtered Time Series in Pure Plots

Pure plots of different vegetation species were selected to evaluate the performance of the filters in terms of the Correlation Coefficient (CC) and the Root Mean Square Error (RMSE) (Table 5). In the crops, alfalfa showed the worst values for both indicators (CC = 0.83 and RMSE = 0.15) and the highest variability among filters. The SG filter showed the minimum RMSE and the maximum CC value, whereas the MVF showed the worst performance with the opposite pattern. For the forest species, beech forests have higher CC values than pines but similar values of RMSE.

Table 5. Average values of Correlation Coefficient and the Root Mean Square Error for the plots of each filter and vegetation species.

	Crops								Forest Species			
	Wheat		Barley		Maize		Alfalfa		Beech		Pine	
Filters	CC	RMSE	CC	RMSE	CC	RMSE	CC	RMSE	CC	RMSE	CC	RMSE
SG	0.97	0.04	0.98	0.04	0.98	0.05	0.95	0.07	0.94	0.08	0.89	0.08
WHI	0.94	0.06	0.95	0.06	0.96	0.07	0.87	0.10	0.85	0.13	0.76	0.11
FFT	0.93	0.07	0.94	0.06	0.95	0.08	0.87	0.10	0.84	0.13	0.73	0.12
MVF	0.91	0.10	0.92	0.10	0.93	0.12	0.83	0.15	0.80	0.17	0.66	0.15

SG: Savitzky–Golay (SG), Whittaker (WHI), Fast Fourier Transform (FFT), and Maximum Value Filter (MVF).

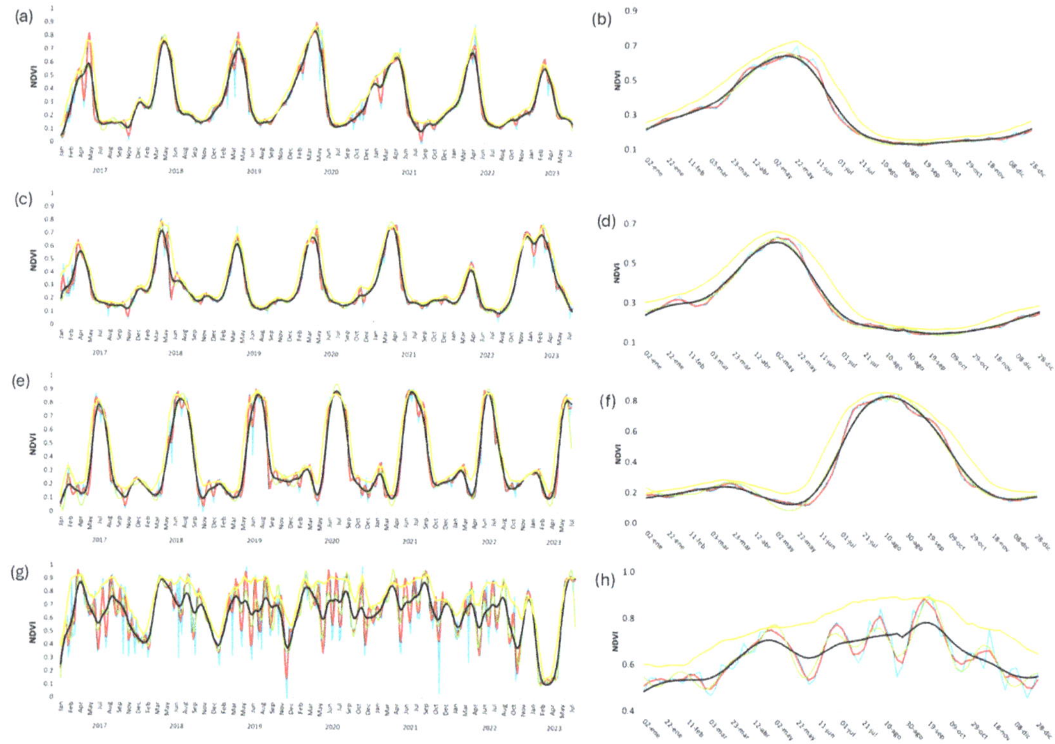

Figure 7. Time series of pure pixels for a wheat (**a**), barley (**c**), maize (**e**), and alfalfa (**g**) crop. Time series of the interpolated NDVI (—) and its four filters: SG (—), FFT (—), WHI (—), and MVF (—), and (**b**,**d**,**f**,**h**) the average year, i.e., Buys-Ballot. The colors of each filter and interpolation are the same as those of the entire time series.

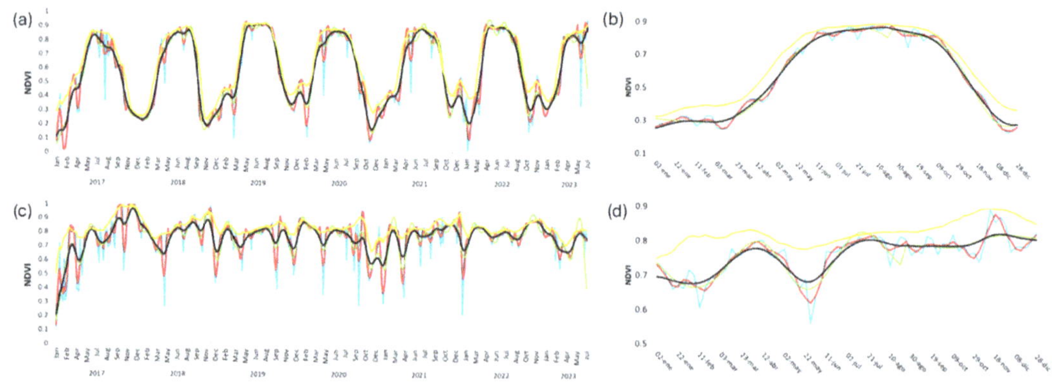

Figure 8. Time series of pure pixels for a beech forest (**a**) and a pine forest (**c**). Time series of the interpolated NDVI (—) and its four filters: SG (—), FFT (—), WHI (—), and MVF (—), and (**b**,**d**) the average year, i.e., Buys-Ballot. The colors of each filter and interpolation are the same as those of the entire time series.

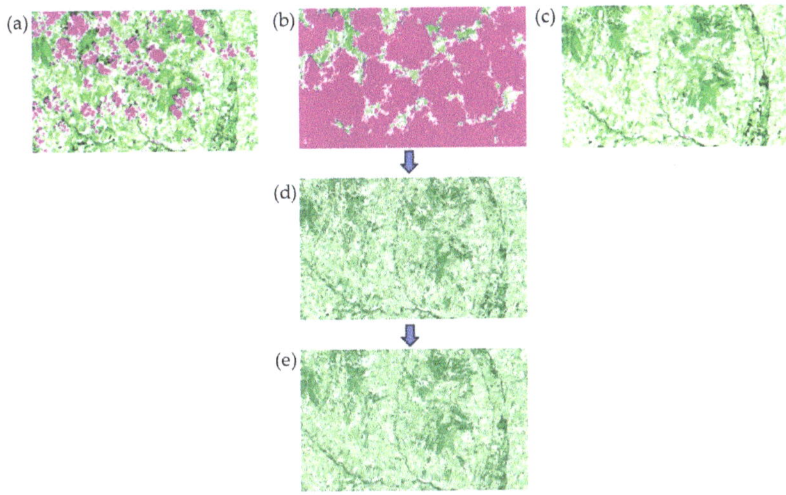

Figure 9. Representation of the evolution at the image level of the interpolation and filtering processes of the Sentinel-2 NDVI time series using the Whittaker filter. (**a**) NDVI image (10/06/20) without interpolation, (**b**) NDVI image (15/06/20) without interpolation, (**c**) NDVI image (20/06/20) without interpolation, (**d**) NDVI image (15/06/20) interpolated, and (**e**) NDVI image (15/06/20) filtered using the Whittaker filter.

3.5. Assessment of NDVI Time Series Information Content and Dynamics

The information content in the NDVI time series from the raw to the filtered data was estimated by the Q-Ljung-Box test, calculated for the short term (lags 1, 2, 3, 4, 5, 6, and 7) and at lags 36 (seasonal), 73 (annual), and 146 (interannual).

To clearly illustrate the results, Q-test values of raw NDVI time series (including low-quality observations) are shown in Table 6. Figure 10 (short term) and Figure 11 (seasonal and annual terms) show the Q-Ljung-Box test values at different lags (1 to 146) for the raw data and the interpolated and filtered time series. Q-values in all land covers were higher than the respective critical values at the 5% significance level for lags 1 to 146; consequently, the null hypothesis of white noise could be rejected in all cases. The largest Q-value increase was after the interpolation, while the filtering procedure resulted in less improvement; the maximum increase was observed when applying the MVF (i.e., wheat Q73 = 5043) and the lowest when using the SG (i.e., wheat Q73 = 4058). Beech forest showed the highest increase while pine forest the lowest at all lags.

Table 6. Mean Q-test for time series of vegetation species plots (raw).

Vegetation Species	Q1	Q2	Q3	Q4	Q5	Q6	Q7	Q36	Q73	Q146
Wheat	65.71	132.21	171.03	212.49	243.69	269.16	291.74	504.05	718.86	954.10
Barley	70.50	140.43	178.93	221.23	251.81	275.36	295.50	500.39	787.05	1158.08
Maize	155.56	302.68	428.80	539.56	629.75	705.33	763.33	1377.24	2650.96	4611.12
Alfalfa	26.22	48.81	58.03	68.61	77.56	84.20	92.33	174.69	266.44	424.26
Beech forest	24.06	47.16	65.80	80.92	96.38	109.45	122.64	342.62	677.62	1227.30
Pine forest	12.38	22.21	27.26	30.51	34.22	38.56	42.11	100.06	180.82	336.58

The critical values of a χ^2 (Chi-Square distribution) at 5% for the Q-Ljung-Box test with 1, 2, 3, 4, 5, 6, 7, 36, 73 and 146 degrees of freedom are 3.84, 5.99, 7.81, 9.49, 11.07, 12.59, 14.07, 51.00, 93.95, and 175.20, respectively.

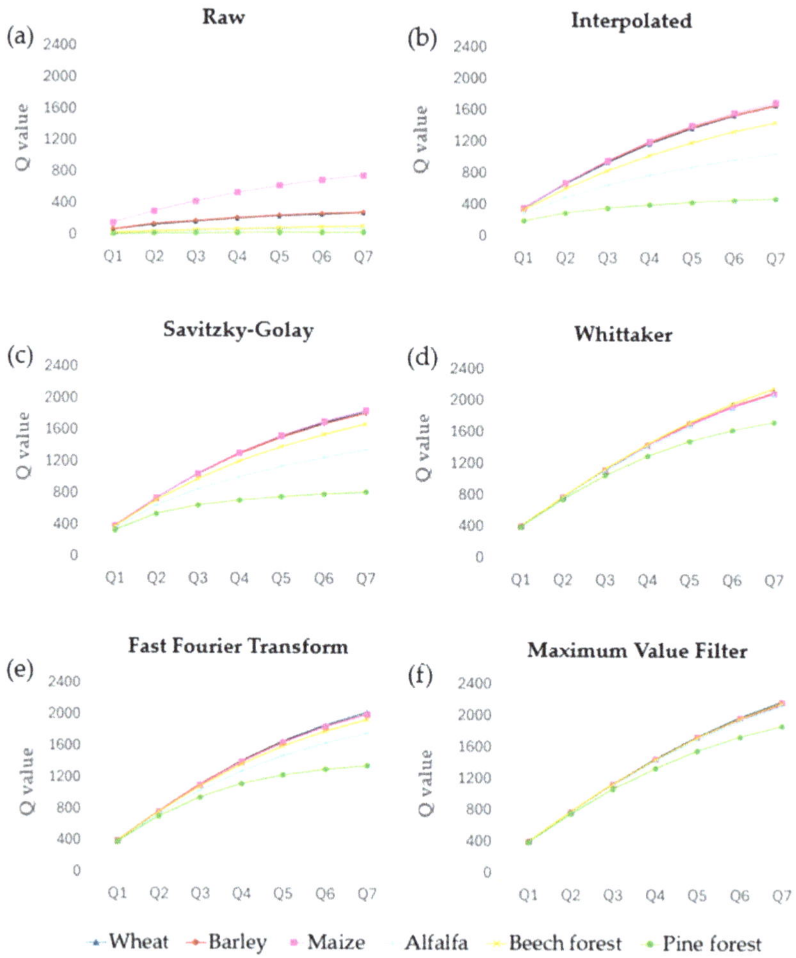

Figure 10. Values of the Q-test for the short term (lags 1, 2, 3, 4, 5, 6, and 7).

Temporal dependency until lag 7 (35 days) could be divided in four levels: (1) maize, (2) wheat and barley, (3) alfalfa and beech forest and, (4) pine forest. The patterns of the seasonal and annual terms (Q36, Q73, and Q146) were similar to those of the short term, except for the beech forest, where NDVI time series presented a high information content compared to the short term (i.e., Q73/Q1 = 28).

Maize presented the highest values for the raw data in the short and long terms, while beech forests reached the highest values when filtered. The lowest values were for pine forests. It was observed that after interpolation, at seasonal and annual horizons, there were three groups with similar patterns: beech and maize with high values, wheat and barley with medium values, and alfalfa and pines with low values.

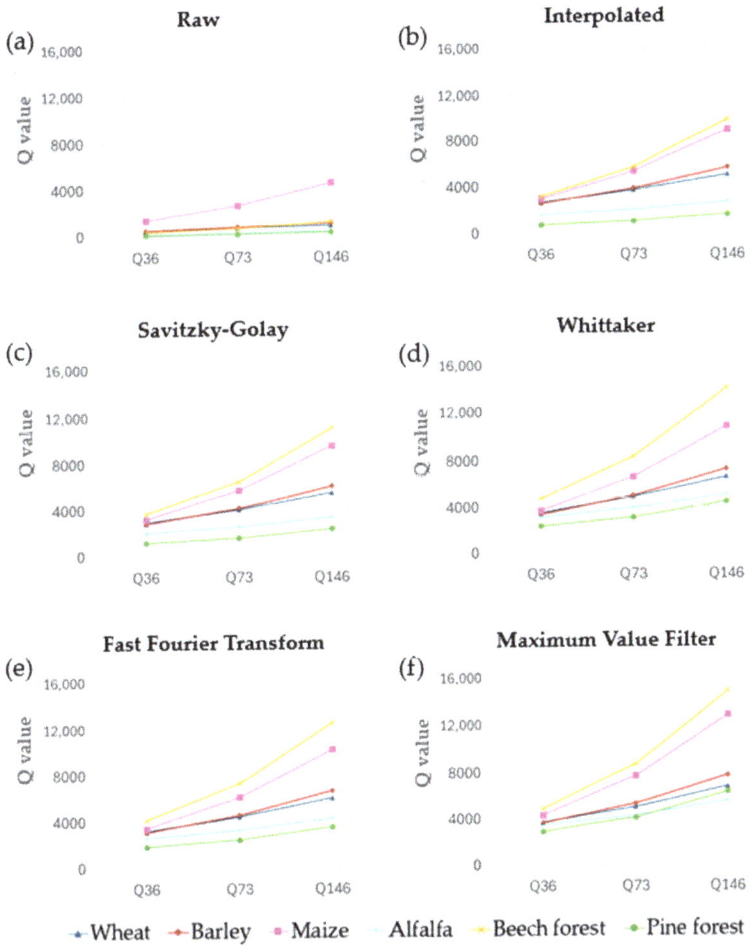

Figure 11. Values of the Q-test at lag 36 (6 months), lag 73 (one year), and lag 146 (two years).

Table 7 shows the percentage of the surface for each land cover with the NDVI time series main periodicities at the periods 73 (annual component or phenological cycle), 36 (6 months), and 203 (representing the presence of a trend). The rest of the possible periodicities are not shown because none of them represented more than 1% of the surface. The periodicity at the annual period (period 73) was the most widely extended across all vegetation types but with different surface percentages. With the raw series, more than 90% of the surface had the main periodicity at one year in maize, barley, and beech forest, while in alfalfa and pine forest, the annual cycle was dominant only in 74% of the surface. In alfalfa, 13% of the time series with the maximum periodicity at 203 may indicate the presence of a trend while in the pine forest's 5% of the surface shows the presence of a second intra-annual cycle.

In the interpolated and filtered time series, the percentage of the surface with an annual cycle decreased in all cases except in the beech forest case, where it reached 99%. In the rest of cases, the surface with the semi-annual cycle increased; in addition, in alfalfa and pine, there was a significant increase in the surface with a dominant cycle at a period of 203.

Table 7. Percentage of surface in which the maximum ordinate of the NDVI periodogram corresponds to periods 36, 73, 203, and others for different filters and land covers.

Land Covers	Period	Fraction of Surface (%)					
		Raw	Interpolated	SG	WHI	FFT	MVF
Wheat	36	0.25	1.33	1.33	1.04	1.33	1.13
	73	87.31	84.65	84.60	84.96	84.60	83.70
	203	0.28	0.78	0.78	0.82	0.78	1.24
	Others	12.16	13.24	13.29	13.18	13.29	13.93
Barley	36	1.12	4.21	4.20	3.38	4.18	3.19
	73	94.97	88.76	88.72	89.50	88.74	88.54
	203	0.08	0.31	0.32	0.33	0.32	0.46
	Others	3.83	6.72	6.76	6.79	6.76	7.81
Maize	36	2.19	4.91	4.90	4.13	4.92	1.19
	73	90.98	86.28	86.24	87.07	86.23	90.08
	203	0.11	0.78	0.78	0.80	0.77	0.96
	Others	6.72	8.03	8.08	8.00	8.08	7.77
Alfalfa	36	1.17	8.37	8.38	7.29	8.42	5.92
	73	74.00	49.41	49.43	50.34	49.45	50.34
	203	13.12	33.56	33.56	34.20	33.52	37.11
	Others	11.71	8.66	8.63	8.17	8.61	6.63
Beech forest	36	0.00	0.17	0.18	0.16	0.18	0.18
	73	91.19	98.76	98.71	98.83	98.71	98.78
	203	0.01	0.24	0.24	0.26	0.24	0.36
	Others	8.80	0.83	0.87	0.75	0.87	0.68
Pine forest	36	5.45	8.23	8.31	8.15	8.51	3.92
	73	73.20	48.02	48.30	52.85	48.31	69.90
	203	0.86	15.82	15.85	17.32	15.57	18.04
	Others	20.49	27.93	27.54	21.68	27.61	8.14

Figure 12 shows the average Fk values for the periodicity of 73 for each vegetation type. The Fk values of the raw time series are represented on the left axis, and the percentage of increase of the Fk after NDVI time series interpolation and filtering is shown on the right axis.

In the raw time series, maize showed the highest Fk values; however, when interpolating and filtering, the percentage of increase was lower than in the other land uses (wheat, barley, and beech forest). Beech forest also showed high Fk values and a high increase. In wheat and barley, a better identification of the cycle was obtained when applying the filters after interpolation. Alfalfa also showed a high increase but not as much as other crops; this was probably due to cuts during the vegetative cycle. Pine forest showed the lowest improvement. Generally, the MVF and WHI filters showed higher improvements, whereas the SG filter presented the lowest.

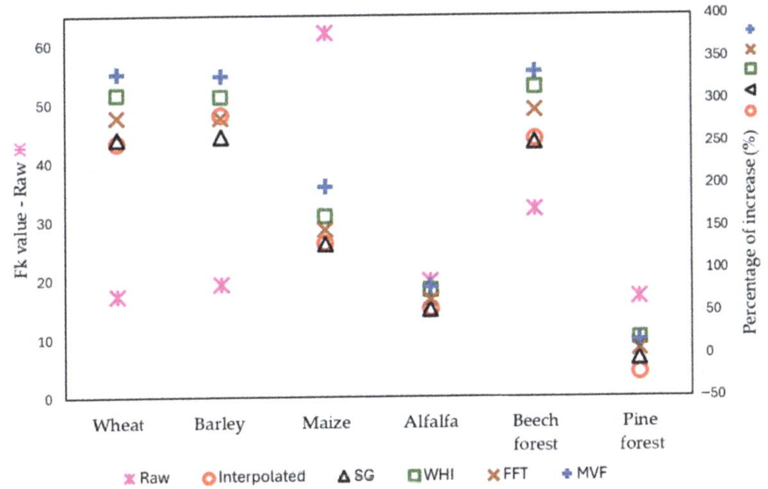

Figure 12. Average Fk value for period 73 (one year) for each vegetation type (left axis) and percentage of increase after interpolation and filtering (right axis).

4. Discussion

The information content in terms of the temporal dependency of Sentinel-2 NDVI time series from 2018 to 2023 for a tile (1,200,000 ha) in central Spain was evaluated. Five scenarios with different climate and distinct phenological patterns were assessed: (1) wheat and barley, (2) maize, (3) alfalfa, (4) beech forest, and (5) pine forest. In these scenarios, the time series information content was quantified at three processing levels: raw, interpolated, and filtered. The IEI summarized information on atmospheric conditions based on the frequency of invalid observations and number and length of gaps due to clouds. For clarity purposes, general patterns are discussed first, and then specific issues for each land cover are addressed.

In the raw time series, Q-test values were higher than the test critical values in all scenarios, indicating that NDVI Sentinel-2 time series were not white noise and contained significant information related to vegetation dynamics. In more than 87% of the surface of wheat, barley, maize, and beech, and around 74% of alfalfa and pine forest, one seasonal cycle per year was found (Table 7, column Raw); this was expected, given that only fields with the same land cover in five years were evaluated.

After interpolation, the information content (i.e., Q and Fk values) increased in all land covers and all terms (Figure 10b, Figure 11b and Figure 12), indicating an increase in time series coherence; particularly, the increase in the Q1 to Q7 slope (i.e., 35 days, Figure 10b) demonstrated a higher consistency in the short term, which could be useful for phenological studies. While Chen et al. [73] used linear interpolation to reduce noise, maintaining the time series dynamics, in this study, the surface with one annual cycle decreased (Table 7, column Interpolated), except in beech forest. These results suggest that the filtering procedure emphasized certain dynamics not identified previously, such as a subtle second cycle due to weeds.

After filtering, crops showed better adjustment than forests (Table 5) located in cloudier areas with lower IEI values. The SG filter showed the highest similarity between the filtered values and the original high-quality ones, especially in maize (0.98), due to the high IEI value and information content in the interpolated time series. These results agree with those of Li et al. [43] who compared different filters in a forestry district in China with MODIS data in five years and obtained the highest CC and lowest RMSE values with the SG filter. In addition, Fontana et al. [74] showed excellent results with the SG filter in AVHRR images evaluating grassland phenology. The MVF and WHI filters showed a high smoothing capability (visual inspection, Figures 7 and 8); however, the MVF overestimated NDVI values, which is in accordance with other studies [75–77]. The lowest CC value (0.68) was found when applying the MVF filter to the pine forest time series in agreement with the intermediate IEI values and lowest information content. The WHI filter showed slightly better results in adjustment than the MVF, which did not agree with Liu et al. [78] who found that the WHI filter did not have a good overall performance in filling the gaps of 8-day MODIS time series on a global scale. On the other hand, the WHI filter was faster than the others in computation time, as other authors have reported [79,80]. The FFT showed better performance in capturing small decreases in alfalfa NDVI time series due to cuts during summer (Figure 7).

The temporal dependency (i.e., Q values) of NDVI time series was slightly higher in all terms than in the interpolated data, especially when using the MVF and WHI, which showed similar Q values in the short term in all land covers, except in pine. This is consistent with their higher smoothing capability and the lower values in similarity measures. In the annual term (Figure 11c–f), they also showed the highest Q values, especially in beech forest.

When applying the MVF filter in pine forest and maize, the surface where the main periodicity was the annual cycle increased (Table 7), as well as the significance of the Fk test value, again indicating the highest smoothing capability of this filter. Figure S1 shows the surface where the maximum ordinate after the interpolating and filtering procedures changed. Patel and Oza [81] were able to identify double cycles in their time series through effective filtering techniques.

In the spatial context (Figure 9), NDVI images generated during the interpolation and filtering process represented the field structure with spatial consistency and detail due to the high spatial resolution of Sentinel-2 time series. In this regard, Reza et al. [82] also achieved good results when filtering contaminated images using the HANTS filter. Similar results were obtained by other researchers in areas with regular NDVI patterns, such as crops or urban areas [3,83,84].

Case 1. Wheat and Barley. Rainfed crops grown in winter–spring in a Mediterranean climate.

These rainfed crops presented significant random short-term variability in NDVI time series (i.e., lower Q1–Q7 values and slope) due to highly variable atmospheric conditions and frequent cloud presence (IEI values: 86% in winter and 91% in spring, see Table 4).

The temporal dependency under seasonal, annual, and interannual dynamics (i.e., Q36, Q73, and Q146 values) was lower than in maize and beech, with a low ratio between memory in one year and in the short term (Q73/Q1 = 11) suggesting the low stability of the annual cycle and the presence of high interannual variability. This was expected in Mediterranean regions where temperature and precipitation show a high interannual variability [85], which could be increased by farmers who decide on the date of sowing every year depending on precipitation and soil status [86].

Barley presented a predominant annual cycle in a larger surface than wheat, with 95% vs. 87%, also showing higher accumulated information in the annual terms (i.e., Q73, 787.05 vs. 718.86) and slightly higher Fk values (19.23 vs. 17.30). In the wheat area, 12.16% of the time series did not show any defined annual cycle (i.e., class "others") versus 3.83% in barley fields, suggesting a higher stability of the growing cycle in barley than wheat [87].

Interpolated time series significantly increased Q values in the short term, and this result may be important in cases involving identifying phenological parameters, such as in Zhu et al. [88] who observed that time series dynamics and phenological stages remained constant after linear interpolation. In our case, after interpolation, there was a small percentage of time series with significant annual dynamics that were re-allocated to the classes "period 36" (two intra-annual cycles) and "others". It is probable that there was an enhancement of subtle information content related to the presence of weeds or short-cycle crops used as forage that were previously masked. Figures S2 and S3 show examples of time series with different main periodicities.

Case 2. Maize. An irrigated crop grown in summer in a Mediterranean climate.

The raw time series of maize showed the highest accumulated information in all terms, with 91% of the surface displaying an annual cycle; the average maize Fk value (Fk = 62.10) was the highest among land covers, which agrees with the high Q value in one year (Q73 = 2050.96) and in two years (Q146) (Figure 11). These results are consistent with the stable and repetitive annual growing period in the irrigated areas during the Mediterranean summer, with optimal conditions of temperature and stable moisture availability. The high IEI values in the maize fields during summer (97.78%) explain the fast increase in Q values from 1 to 7 lags, indicating a low high-frequency variability due to clouds, which results in a regular NDVI evolution within the growing cycle.

After interpolation, the temporal dependency increase was the lowest among land covers (i.e., $Q_{int}/Q_{raw} \cong 2$), likely because it was already high from the outset. Also, the surface with two intra-annual cycles increased slightly, suggesting that the noise elimination made it possible to uncover subtle variability sources such as the one due to the growth of weeds before preparing the soil between the growing seasons; this is shown in the exploratory analyses (Figure S4a). In addition, a significant surface percentage with other cycles suggested the presence of two crops in some years, such as vetch until the end of July and then, maize, resulting in a double greening cycle; this was not stated by the farmer, so it probably was not a common practice every year, and, therefore, the periodicity could not be defined.

After filtering, there was a slight increase in temporal dependency for all lags and in Fk values, with the lowest values for the SG filter. The MVF filter showed the highest increase at the interannual scale, together with an increase in the area with an annual cycle (90.08%), suggesting that the strong smoothing capability of this filter emphasized the annual cycle over seasonal patterns.

Case 3. Alfalfa. A permanent irrigated crop (5-6 years) in a Mediterranean climate.

Alfalfa showed low Q values, both in the short, seasonal, and annual terms, despite growing in summer with no water limitations and optimal atmospheric conditions (IEI_{summer} = 97.58%). Crop management including several cuts through the growing period may explain the short-term high variability (Figure 7g) that could hide the annual cycle information (Q73/Q1 = 10.16); this was corroborated by the fact that only 74.00% of the surface of alfalfa was shown in the annual cycle, with low significance (Fk value = 19.73). The other 26% of the surface was mainly distributed between the classes "others" (11.71%) and "period 203" (13.12%), suggesting the presence of a trend.

After interpolation, there was still a low amount of cumulated information in all terms, suggesting that noise due to atmospheric effects was not the main cause of high frequency variability, but rather the decrease in NDVI due to the cuts during the growing period. In addition, there was a significant increase in the surface percentage with a possible trend (33.56%); since the interpolation process tends to emphasize subtle patterns and decrease high-frequency variability, and the alfalfa crop usually lasts five to six years, this result suggested that these time series corresponded to fields where alfalfa was at the first growing stages in 2018 (Figure 13) or at the last in 2023. This hypothesis is consistent with Figure S5 in the Supplementary Material, which shows these time series located in homogenous fields, indicating that it was due to management.

The rest of the alfalfa surface (17%) is divided between the classes "period 36" and "others", probably due to the high variability introduced by the cuts. The MVF and WHI significantly increased temporal dependency in the short term; however, the similarity indicators (Table 5) showed low values, which was probably due to the low adjustment of filtered NDVI time series to the original NDVI values when alfalfa is cut (Figure 7h).

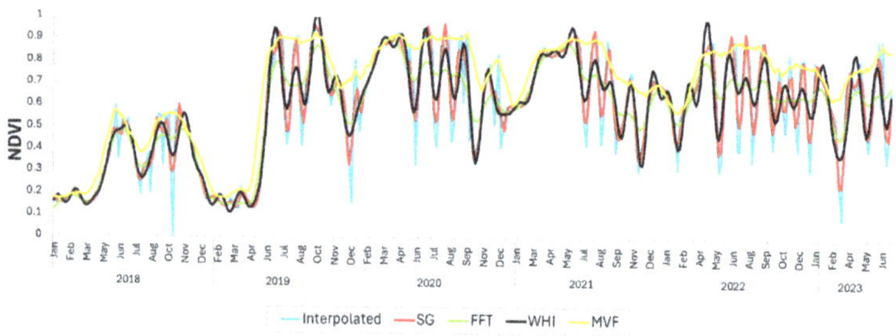

Figure 13. Time series of a pixel declared as alfalfa implemented during the first year.

Case 4. Beech forest. A deciduous forest in a temperate climate.

Beech forest showed significant but remarkably low values of temporal dependency in the short term (Q1–Q7), indicating an important source of high-frequency random variability. Annual information was highly relevant in relation to the short term (i.e., Q73/Q1 = 28); in addition, Q146 was higher than Q73 (i.e., Q146/Q73 = 1.81), suggesting the stability of inter-annual dynamics. Growing in the northern area with high cloudiness (i.e., IEI = 74.90%) resulted in high short-term variability. However, the indicators on the annual and inter-annual scales suggested that the distinct phenological pattern of the Fagus species growing in summer (IEI_{summer} = 92.74%) provided enough consistent information to override short-term variability. Similar results were found by Recuero et al. [89] who were able to map maize crop dynamics in Ecuador despite the high presence of clouds due to the significant annual cycle.

Ninety one percent of the beech surface presented one seasonal cycle with a high Fk value (Figure 12). On the other hand, 8.80% of the surface, specifically located in the highest latitudes, did not show a dominant periodicity; in this area, the average Q value of raw time series was significantly lower than in the whole beech forest areas (Q73 = 98 vs. Q73 = 677.62) probably due to the extremely high presence of clouds that resulted in an IEI_{winter} = 56.10%. Still, even in this area, the information content in the annual and bi-annual terms showed Q values greater than the critical.

After interpolation, temporal dependency increased, especially at seasonal, annual, and interannual lags (i.e., Q36, Q73, and Q146), reaching the highest values among land covers, and suggesting that these periodic components were highly significant and regular in intensity. In addition, the percentage of the area with one annual cycle increased to almost 99% (Table 7), and the significance of the annual cycle increased significantly (i.e., Fk increase percentage, Figure 12). This suggests that the interpolation process resulted in a relevant enhancement of significant annual information by diminishing the short-term random variability derived from the frequent cloud coverage.

Case 5. Pine forests. An evergreen forest in a transition area from mediterranean to temperate climate.

The lowest Q values (in the limit of the critical values) in the short term indicate the presence of high-frequency variability; this is consistent with the relatively high NDVI values in the year of evergreen species and the fast response to changes in environmental conditions; fast NDVI especially increases when optimal conditions of temperature are reached [90]. In addition, the pine forest was distributed in a wide area with a significant climate gradient and an average annual IEI value of 88.15% due to cloudy conditions. This, together with their irregular NDVI dynamics, entailed the lowest Fk values, suggesting a lack of a stable annual cycle.

In the interpolated time series, the Q values were still in the limit of significance, despite the increase in information content. In the beech forest, the short-term variability due to clouds was effectively ameliorated; in the pine forest, the variability introduced by physiological responses resulted in low-temporal-dependency information. In the annual term, the Q values remained low due to the unstable phenological dynamics, which has also been reported by other authors; specifically, Aragones et al. [91] found high interannual variability when working with a 17-year MODIS NDVI data set. Furthermore, in our case, the Fk values even decreased after interpolation and SG filtering, indicating an even lower significance in the annual cycle. In addition, in 34% of the surface associated with the annual cycle's "period 73", other seasonal components were uncovered (Table 7), such as the presence of a trend of "period 203" and "other" cycles. The NDVI time series showing these changes were distributed throughout all the pine forest areas and presented high local variability (see Figure S7). This is consistent with the significant differences found by Aragones et al. [91] among pine species in terms of phenological patterns. The MVF filter showed an increase in the surface with an annual cycle and the presence of trends, which is probably related to the high smoothing capability of this filter. Mantas et al. [92] demonstrated the capability of Sentinel-2 imagery and the importance of identifying trends in pine forests to analyze forest decline due to disease.

5. Conclusions

This work quantitatively demonstrates that Sentinel-2 NDVI time series contains relevant information on land cover dynamics based on temporal dependency (Q-test) in five land covers, even in areas with significant atmospheric disturbance. It was also possible to identify the main periodic components and trends by the classic Fk test.

Temporal dependency increased significantly in all terms after interpolation and filtering; in addition, the surface with an annual cycle decreased, making it possible to find fields with other cycles as main periodicity, which was not possible with the raw time series.

The Q and Fk tests, together with the similarity indicators CC and RMSE, showed that a better fit (i.e., high similarity values) does not always imply an improvement in the information content in terms of dynamics assessment.

The IEI adequately represented atmospheric disturbance in the region on the annual and seasonal scales, and it was highly related to short-term variability. Furthermore, it was demonstrated that the enhancement derived from interpolating and filtering was more notable in time series located in areas with lower IEI values. It is expected that the IEI could provide the first insight into the final information content based on a quality dataset.

Significant differences in information content patterns among scenarios were found due to interactions between vegetation dynamics, cloud disturbances, and specific management practices.

Specific insights into the information content and the impact of interpolation and filtering for the scenarios assessed can be summarized as follows:

In rainfed wheat and barley, highly variable cloud cover (low IEI) in winter and spring, together with high dependence on precipitation and temperature, resulted in significant high-frequency and interannual variabilities. Interpolation and filtering reduced short-term variability; however, the impact on annual and inter-annual dynamics was lower.

Irrigated crops, such as maize, growing in summer with a significant annual cycle and low cloud coverage expressed the largest amount of information content, so that interpolating and filtering did not result in a significant increase in information (Q-values); however, this slight increase and the use of the Fk test made it possible to find fields where the main periodicity corresponded to subtle cycles such as those caused by weeds. In addition, reducing high short-term variability could facilitate phenological analyses.

In irrigated alfalfa, a multiyear crop, all filters, especially the MVF, made it possible to identify fields with trends, caused by low NDVI values at the beginning or at the end of the study period, due to crop implantation or removal, respectively. On the other hand, the MVF and WHI filters masked the small NDVI decreases due to cuts during summer.

In beech forest, which is temperate deciduous, diminishing the cloud disturbance through interpolating and filtering combined with the strong and stable annual cycle could facilitate an accurate dynamics assessment. Thus, in these areas, the Sentinel-2's interpolated time series could provide a significant amount of ecological information.

In the pine forest, which is evergreen coniferous in Mediterranean areas, the combination of high physiological short-term variability and variable cloud coverage resulted in complex time series. Despite these difficulties, after interpolation and filtering, the Fk test was able to identify areas with trends that would not appear in the raw time series, and this would be useful for decaying studies.

Supplementary Materials: The following supporting information can be downloaded at: https://www.mdpi.com/article/10.3390/rs16162980/s1.

Author Contributions: Conceptualization, C.S. and A.P.-O.; methodology, C.S., J.L., G.G., D.M., L.R., and A.P.-O.; software, C.S., G.G., D.M., L.R., and A.B.-S.; formal analysis, C.S., A.P.-O., V.C., and J.L.; investigation, C.S., V.C., G.G., D.M., L.R., A.B.-S., J.L., I.d.l.C., and A.P.-O.; resources, C.S., J.L., and A.P.-O.; writing—original draft preparation, C.S., V.C., L.R., A.B.-S., J.L., and A.P.-O.; writing—review and editing, C.S., V.C., A.P.-O., and J.L.; supervision, A.P.-O.; funding acquisition, A.P.-O. and C.S. All authors have read and agreed to the published version of the manuscript.

Funding: This research was supported by a predoctoral scholarship awarded by the Community of Madrid and Quasar Science Resources, S.L. (No. IND2020/AMB-17747). It was also conducted in the framework of the I+D+i Spanish National Project PID2020-115509RB-I00 project (INFOLANDYN) funded by the Ministerio de Ciencia e Innovación of Spain MCIN/AEI/ 10.13039/501100011033. VC was supported by a postdoctoral Juan de la Cierva fellowship (FJC2021-046735-I) funded by the Spanish Ministerio de Ciencia e Innovación MCIN/AEI/ 10.13039/501100011033 and by the European Union ≪NextGenerationEU≫/PRTR. LR was supported by a UPM grant as part of the national research program Recualificación del Sistema Universitario Español, financed by the Recovery and Resilience Package—NextGenerationEU (European Commission).

Data Availability Statement: The original contributions presented in the study are included in the article, further inquiries can be directed to the corresponding author.

Acknowledgments: We would like to thank the European Space Agency (ESA) for the images provided, which were necessary for this study. The authors would also like to acknowledge the participation of Quasar Science Resources, S.L., in this study.

Conflicts of Interest: César Sáenz and Ignacio de la Calle were employed by the company Quasar Science Resources, S.L. The remaining authors declare that the research was conducted in the absence of any commercial or financial relationships that could be construed as a potential conflict of interest. The funders had no role in the design of the study; in the collection, analyses, or interpretation of data; in the writing of the manuscript; or in the decision to publish the results.

References

1. Harmanny, K.S.; Malek, Ž. Adaptations in Irrigated Agriculture in the Mediterranean Region: An Overview and Spatial Analysis of Implemented Strategies. *Reg. Environ. Chang.* **2019**, *19*, 1401–1416. [CrossRef] [PubMed]
2. Gumma, M.K.; Tummala, K.; Dixit, S.; Holecz, F.; Kolli, R.N.; Whitbread, A.M.; Krishna, M.; Tummala, K.; Dixit, S. Crop Type Identification and Spatial Mapping Using Sentinel-2 Satellite Data with Focus on Field-Level Information. *Geocarto Int.* **2020**, *37*, 1833–1849. [CrossRef]
3. Tran, K.H.; Zhang, H.K.; Mcmaine, J.T.; Zhang, X.; Luo, D. 10 m Crop Type Mapping Using Sentinel-2 Reflectance and 30 m Cropland Data Layer Product. *Int. J. Appl. Earth Obs. Geoinf.* **2022**, *107*, 102692. [CrossRef]
4. Rujoiu-Mare, M.-R.; Olariu, B.; Mihai, B.-A.; Nistor, C.; Săvulescu, I. Land Cover Classification in Romanian Carpathians and Subcarpathians Using Multi-Date Sentinel-2 Remote Sensing Imagery. *Eur. J. Remote Sens.* **2017**, *50*, 496–508. [CrossRef]
5. Wakulinska, M.; Marcinkowska-Ochtyra, A. Multi-Temporal Sentinel-2 Data in Classification of Mountain Vegetation. *Remote Sens.* **2020**, *12*, 2696. [CrossRef]
6. Wang, M.; Li, M.; Wang, F.; Ji, X. Exploring the Optimal Feature Combination of Tree Species Classification by Fusing Multi-Feature and Multi-Temporal Sentinel-2 Data in Changbai Mountain. *Forests* **2022**, *13*, 1058. [CrossRef]
7. Solano-Correa, Y.T.; Bovolo, F.; Member, S.; Bruzzone, L.; Fernández-prieto, D. A Method for the Analysis of Small Crop Fields in Sentinel-2 Dense Time Series. *IEEE Trans. Geosci. Remote Sens.* **2020**, *58*, 2150–2164. [CrossRef]
8. Pan, L.; Xia, H.; Yang, J.; Niu, W.; Wang, R.; Song, H.; Guo, Y.; Qin, Y. Mapping Cropping Intensity in Huaihe Basin Using Phenology Algorithm, All Sentinel-2 and Landsat Images in Google Earth Engine. *Int. J. Appl. Earth Obs. Geoinf.* **2021**, *102*, 102376. [CrossRef]

9. Amin, E.; Belda, S.; Pipia, L.; Szantoi, Z.; El Baroudy, A.; Moreno, J.; Verrelst, J. Multi-Season Phenology Mapping of Nile Delta Croplands Using Time Series of Sentinel-2 and Landsat 8 Green LAI. *Remote Sens.* **2022**, *14*, 1812. [CrossRef]
10. Dimov, D.; Uhl, J.H.; Löw, F.; Negash, G. Sugarcane Yield Estimation through Remote Sensing Time Series and Phenology Metrics. *Smart Agric. Technol.* **2022**, *2*, 100046. [CrossRef]
11. Kosczor, E.; Forkel, M.; Hernández, J.; Kinalczyk, D.; Pirotti, F.; Kutchartt, E. Assessing Land Surface Phenology in Araucaria-Nothofagus Forests in Chile with Landsat 8/Sentinel-2 Time Series. *Int. J. Appl. Earth Obs. Geoinf.* **2022**, *112*, 102862. [CrossRef]
12. Ye, J.; Bao, W.; Liao, C.; Chen, D.; Hu, H. Corn Phenology Detection Using the Derivative Dynamic Time Warping Method and Sentinel-2 Time Series. *Remote Sens.* **2023**, *15*, 3456. [CrossRef]
13. Del Pozo, A.; Brunel-saldias, N.; Engler, A.; Ortega-farias, S.; Acevedo-opazo, C.; Lobos, G.A.; Jara-rojas, R.; Molina-montenegro, M.A. Climate Change Impacts and Adaptation Strategies of Agriculture in Mediterranean-Climate Regions (MCRs). *Sustainability* **2019**, *11*, 2769. [CrossRef]
14. Aguilera, E.; Díaz-gaona, C.; García-laureano, R.; Reyes-palomo, C.; Guzmán, G.I.; Ortolani, L.; Sánchez-rodríguez, M.; Rodríguez-estévez, V. Agroecology for Adaptation to Climate Change and Resource Depletion in the Mediterranean Region. A Review. *Agric. Syst.* **2020**, *181*, 102809. [CrossRef]
15. Ryan, J.; Singh, M.; Pala, M. Long-Term Cereal-Based Rotation Trials in the Mediterranean Region: Implications for Cropping Sustainability. *Adv. Agron.* **2008**, *97*, 273–319.
16. Jacobsen, S.; Jensen, C.R.; Liu, F. Field Crops Research Improving Crop Production in the Arid Mediterranean Climate. *Field Crops Res.* **2012**, *128*, 34–47. [CrossRef]
17. Recuero, L.; Wiese, K.; Huesca, M.; Cicuéndez, V.; Litago, J.; Tarquis, A.M.; Palacios-orueta, A. Fallowing Temporal Patterns Assessment in Rainfed Agricultural Areas Based on NDVI Time Series Autocorrelation Values. *Int. J. Appl. Earth Obs. Geoinf.* **2019**, *82*, 101890. [CrossRef]
18. Yang, C.; Fraga, H.; Van Ieperen, W.; Trindade, H.; Santos, J.A. Effects of Climate Change and Adaptation Options on Winter Wheat Yield under Rainfed Mediterranean Conditions in Southern Portugal. *Clim. Chang.* **2019**, *154*, 159–178. [CrossRef]
19. Oweis, T.; Zhang, H.; Pala, M. Water Use Efficiency of Rainfed and Irrigated Bread Wheat. *Agron. J.* **2000**, *92*, 231–238. [CrossRef]
20. Katerji, N.; Mastrorilli, M.; Rana, G. Water Use Efficiency of Crops Cultivated in the Mediterranean Region: Review and Analysis. *Eur. J. Agron.* **2008**, *28*, 493–507. [CrossRef]
21. Pradipta, A.; Soupios, P.; Kourgialas, N.; Doula, M.; Dokou, Z.; Makkawi, M.; Alfarhan, M.; Tawabini, B.; Kirmizakis, P. Remote Sensing, Geophysics, and Modeling to Support Precision Agriculture—Part 2: Irrigation Management. *Water* **2022**, *14*, 1157. [CrossRef]
22. Liu, X.; Ji, L.; Zhang, C.; Liu, Y. A Method for Reconstructing NDVI Time-Series Based on Envelope Detection and the Savitzky-Golay Filter. *Int. J. Digit. Earth* **2022**, *15*, 553–584. [CrossRef]
23. Pelletier, C.; Valero, S.; Inglada, J.; Champion, N.; Marais Sicre, C.; Dedieu, G. Effect of Training Class Label Noise on Classification Performances for Land Cover Mapping with Satellite Image Time Series. *Remote Sens.* **2017**, *9*, 173. [CrossRef]
24. Kong, D.; Zhang, Y.; Gu, X.; Wang, D. A Robust Method for Reconstructing Global MODIS EVI Time Series on the Google Earth Engine. *ISPRS J. Photogramm. Remote Sens.* **2019**, *155*, 13–24. [CrossRef]
25. Hemmerling, J.; Pflugmacher, D.; Hostert, P. Mapping Temperate Forest Tree Species Using Dense Sentinel-2 Time Series. *Remote Sens. Environ.* **2021**, *267*, 112743. [CrossRef]
26. Abdollahi, A.; Liu, Y.; Pradhan, B.; Huete, A.; Dikshit, A.; Tran, N.N. Short-Time-Series Grassland Mapping Using Sentinel-2 Imagery and Deep Learning-Based Architecture. *Egypt. J. Remote Sens. Space Sci.* **2022**, *25*, 673–685. [CrossRef]
27. Wang, Y.; Fang, S.; Zhao, L.; Huang, X.; Jiang, X. Parcel-Based Summer Maize Mapping and Phenology Estimation Combined Using Sentinel-2 and Time Series Sentinel-1 Data. *Int. J. Appl. Earth Obs. Geoinf.* **2022**, *108*, 102720. [CrossRef]
28. Marino, S. Assessing the Agronomic Subfield Variability by Sentinel-2 NDVI Time-Series and Landscape Position. *Agronomy* **2023**, *13*, 44. [CrossRef]
29. Holben, B.N. Characteristics of Maximum-Value Composite Images from Temporal AVHRR Data. *Int. J. Remote Sens.* **1986**, *7*, 1417–1434. [CrossRef]
30. Narasimhan, R.; Stow, D. Daily MODIS Products for Analyzing Early Season Vegetation Dynamics across the North Slope of Alaska. *Remote Sens. Environ.* **2010**, *114*, 1251–1262. [CrossRef]
31. Nagai, S.; Saitoh, T.M.; Suzuki, R.; Nasahara, K.N.; Lee, K.; Son, Y.; Muraoka, H. The Necessity and Availability of Noise-Free Daily Satellite-Observed NDVI during Rapid Phenological Changes in Terrestrial Ecosystems in East Asia. *For. Sci. Technol.* **2011**, *7*, 174–183. [CrossRef]
32. Zeng, L.; Wardlow, B.; Hu, S.; Zhang, X.; Zhou, G.; Peng, G.; Xiang, D.; Wang, R.; Meng, R.; Wu, W. A Novel Strategy to Reconstruct NDVI Time-Series with High Temporal Resolution from MODIS Multi-Temporal Composite Products. *Remote Sens.* **2021**, *13*, 1397. [CrossRef]
33. Yang, Z.; Li, J.; Shen, Y.; Miao, H.; Yan, X. A Denoising Method for Inter-Annual NDVI Time Series Derived from Landsat Images. *Int. J. Remote Sens.* **2018**, *39*, 3816–3827. [CrossRef]
34. Yu, W.; Li, J.; Liu, Q.; Zhao, J.; Dong, Y.; Zhu, X.; Lin, S.; Zhang, H.; Zhang, Z. Gap Filling for Historical Landsat NDVI Time Series by Integrating Climate Data. *Remote Sens.* **2021**, *13*, 484. [CrossRef]
35. Gao, F.; Masek, J.; Schwaller, M.; Hall, F. On the Blending of the Landsat and MODIS Surface Reflectance: Predicting Daily Landsat Surface Reflectance. *IEEE Trans. Geosci. Remote Sens.* **2006**, *44*, 2207–2218. [CrossRef]

36. Zhu, X.; Chen, J.; Gao, F.; Chen, X.; Masek, J.G. An Enhanced Spatial and Temporal Adaptive Reflectance Fusion Model for Complex Heterogeneous Regions. *Remote Sens. Environ.* **2010**, *114*, 2610–2623. [CrossRef]
37. Zhou, J.; Jia, L.; Menenti, M.; Gorte, B. On the Performance of Remote Sensing Time Series Reconstruction Methods—A Spatial Comparison. *Remote Sens. Environ.* **2016**, *187*, 367–384. [CrossRef]
38. Savitzky, A.; Golay, M.J.E. Smoothing and Differentiation of Data by Simplified Least Squares Procedures. *Anal. Chem.* **1964**, *36*, 1627–1639. [CrossRef]
39. Whittaker, E. On a New Method of Graduation. *Proc. Edinb. Math. Soc.* **1923**, *41*, 63–75. [CrossRef]
40. Eilers, P.H.C. A Perfect Smoother. *Am. Chem. Soc.* **2003**, *75*, 3631–3636. [CrossRef]
41. Tucker, C.J. Red and Photographic Infrared Linear Combinations for Monitoring Vegetation. *Remote Sens. Environ.* **1979**, *8*, 127–150. [CrossRef]
42. Hird, J.N.; Mcdermid, G.J. Noise Reduction of NDVI Time Series: An Empirical Comparison of Selected Techniques. *Remote Sens. Environ.* **2009**, *113*, 248–258. [CrossRef]
43. Li, S.; Xu, L.; Jing, Y.; Yin, H.; Li, X.; Guan, X. High-Quality Vegetation Index Product Generation: A Review of NDVI Time Series Reconstruction Techniques. *Int. J. Appl. Earth Obs. Geoinf.* **2021**, *105*, 102640. [CrossRef]
44. Htitiou, A.; Boudhar, A.; Lebrini, Y.; Hadria, R.; Lionboui, H.; Elmansouri, L.; Tychon, B.; Benabdelouahab, T. The Performance of Random Forest Classification Based on Phenological Metrics Derived from Sentinel-2 and Landsat 8 to Map Crop Cover in an Irrigated Semi-Arid Region. *Remote Sens. Earth Syst. Sci.* **2019**, *2*, 208–224. [CrossRef]
45. Stendardi, L.; Karlsen, S.R.; Niedrist, G.; Gerdol, R.; Zebisch, M.; Rossi, M.; Notarnicola, C. Exploiting Time Series of Sentinel-1 and Sentinel-2 Imagery to Detect Meadow Phenology in Mountain Regions. *Remote Sens.* **2019**, *11*, 542. [CrossRef]
46. Goldberg, K.; Herrmann, I.; Hochberg, U.; Rozenstein, O. Generating Up-to-Date Crop Maps Optimized for Sentinel-2 Imagery in Israel. *Remote Sens.* **2021**, *13*, 3488. [CrossRef]
47. Huang, X.; Huang, J.; Li, X.; Shen, Q.; Chen, Z. Early Mapping of Winter Wheat in Henan Province of China Using Time Series of Sentinel-2 Data. *GIsci. Remote Sens.* **2022**, *59*, 1534–1549. [CrossRef]
48. Khanal, N.; Matin, M.A.; Uddin, K.; Poortinga, A. A Comparison of Three Temporal Smoothing Algorithms to Improve Land Cover Classification: A Case Study from NEPAL. *Remote Sens.* **2020**, *12*, 2888. [CrossRef]
49. Cooley, B.J.W.; Tukey, J.W. An Algorithm for the Machine Calculation Complex Fourier Series. *Math. Comput.* **1965**, *19*, 297–301. [CrossRef]
50. Peng, N.; Yang, S.; Tao, Y.; Zhai, D.; Fan, W.; Liu, Q. Lai Time Series Reconstruction from Sentinel-2 Imagery Using Vegetation Growing Phenology Feature. In Proceedings of the IGARSS 2023–2023 IEEE International Geoscience and Remote Sensing Symposium, Pasadena, CA, USA, 16–21 July 2023; pp. 3046–3049. [CrossRef]
51. Padhee, S.K.; Dutta, S. Spatio-Temporal Reconstruction of MODIS NDVI by Regional Land Surface Phenology and Harmonic Analysis of Time-Series. *GIsci. Remote Sens.* **2019**, *56*, 1261–1288. [CrossRef]
52. Son, N.; Chen, C.; Chen, C.; Guo, H. Classification of Multitemporal Sentinel-2 Data for Field-Level Monitoring of Rice Cropping Practices in Taiwan. *Adv. Space Res.* **2020**, *65*, 1910–1921. [CrossRef]
53. Beck, P.S.A.; Jönsson, P.; Høgda, K.A.; Karlsen, S.R. A Ground-Validated NDVI Dataset for Monitoring Vegetation Dynamics and Mapping Phenology in Fennoscandia and the Kola Peninsula. *Int. J. Remote Sens.* **2007**, *28*, 4311–4330. [CrossRef]
54. Maignan, F.; Bréon, F.; Bacour, C.; Demarty, J.; Poirson, A. Interannual Vegetation Phenology Estimates from Global AVHRR Measurements Comparison with In Situ Data and Applications. *Remote Sens. Environ.* **2008**, *112*, 496–505. [CrossRef]
55. AEMET/IMP. *Atlas Climático Ibérico Iberian Climate Atlas*; AEMET—Agencia Estatal de Meteorología, IMP—Instituto de Meteorología de Portugal, Ministerio de Medio Ambiente Rural y Marino, Eds.; Closas Orcoyen S.L.: Madrid, Spain, 2011; ISBN 978-84-7837-079-5.
56. Beck, H.E.; Zimmermann, N.E.; McVicar, T.R.; Vergopolan, N.; Berg, A.; Wood, E.F. Present and Future Köppen-Geiger Climate Classification Maps at 1-Km Resolution. *Sci. Data* **2018**, *5*, 180214.
57. FAO. *FAO/Unesco Soil Map of the World, Revised Legend, with Corrections and Updates*; World Soil Resources Report 60; FAO: Rome, Italy, 1988.
58. MITECO Ministerio Para La Transición Ecológica y El Reto Demográfico. *Mapa Forestal de España 50000*; 2013. Available online: https://www.miteco.gob.es/es/biodiversidad/servicios/banco-datos-naturaleza/informacion-disponible/mfe50.html (accessed on 20 October 2023).
59. Ministerio de Agricultura, Pesca y Alimentación. *Calendario de Siembra, Recolección y Comercialización 2014–2016*; Madrid, 2021. Available online: https://www.mapa.gob.es/es/estadistica/temas/estadisticas-agrarias/agricultura/calendarios-siembras-recoleccion/ (accessed on 20 October 2023).
60. Drusch, M.; Del Bello, U.; Carlier, S.; Colin, O.; Fernandez, V.; Gascon, F.; Hoersch, B.; Isola, C.; Laberinti, P.; Martimort, P.; et al. Sentinel-2: ESA's Optical High-Resolution Mission for GMES Operational Services. *Remote Sens. Environ.* **2012**, *120*, 25–36. [CrossRef]
61. Zekoll, V.; Main-Knorn, M.; Louis, J.; Frantz, D.; Richter, R.; Pflug, B. Comparison of Masking Algorithms for Sentinel-2 Imagery. *Remote Sens.* **2021**, *13*, 137. [CrossRef]
62. Cao, R.; Chen, Y.; Shen, M.; Chen, J.; Zhou, J.; Wang, C.; Yang, W. A Simple Method to Improve the Quality of NDVI Time-Series Data by Integrating Spatiotemporal Information with the Savitzky-Golay Filter. *Remote Sens. Environ.* **2018**, *217*, 244–257. [CrossRef]

63. Cai, Y.; Lin, H.; Zhang, M. Mapping Paddy Rice by the Object-Based Random Forest Method Using Time Series Sentinel-1/Sentinel-2 Data. *Adv. Space Res.* **2019**, *64*, 2233–2244. [CrossRef]
64. Menenti, M.; Azzali, S.; Verhoef, W.; Swol, R. Van Mapping Agroecological Zones and Time Lag in Vegetation Growth by Means of Fourier Analysis of Time Series of NDVI Images. *Adv. Space Res.* **1993**, *13*, 233–237. [CrossRef]
65. Geng, L.; Ma, M.; Wang, X.; Yu, W.; Jia, S.; Wang, H. Comparison of Eight Techniques for Reconstructing Multi-Satellite Sensor Time-Series NDVI Data Sets in the Heihe River Basin, China. *Remote Sens.* **2014**, *6*, 2024–2049. [CrossRef]
66. Buys-Ballot, C.H.D. *Les Changements Périodiques de Température, Dépendants de La Nature Du Soleil et de La Lune, Mis En Rapport Avec Le Pronostic Du Temps, Déduits d'observations Néerlandaises de 1729 à 1846*; Kemink: Utrecht, The Netherlands, 1847.
67. Ljung, G.M.; Box, G.E.P. On a Measure of Lack of Fit in Time Series Models. *Biometrika* **1978**, *65*, 297–303. [CrossRef]
68. Huesca, M.; Litago, J.; Merino-de-Miguel, S.; Cicuendez-López-Ocaña, V.; Palacios-Orueta, A. Modeling and Forecasting MODIS-Based Fire Potential Index on a Pixelbasis Using Time Series Models. *Int. J. Appl. Earth Obs. Geoinf.* **2014**, *26*, 363–376. [CrossRef]
69. Huesca, M.; Merino-de-Miguel, S.; Eklundh, L.; Litago, J.; Cicuéndez, V.; Rodríguez-Rastrero, M.; Ustin, S.L.; Palacios-Orueta, A. Ecosystem Functional Assessment Based on the "Optical Type" Concept and Self-Similarity Patterns: An Application Using MODIS-NDVI Time Series Autocorrelation. *Int. J. Appl. Earth Obs. Geoinf.* **2015**, *43*, 132–148. [CrossRef]
70. Hamilton, J.D. *Time Series Analysis*; Princeton University Press: Princeton, MA, USA, 1994.
71. Recuero, L.; Litago, J.; Pinzón, J.E.; Huesca, M.; Moyano, M.C.; Palacios-Orueta, A. Mapping Periodic Patterns of Global Vegetation Based on Spectral Analysis of NDVI Time Series. *Remote Sens.* **2019**, *11*, 2497. [CrossRef]
72. Fuller, W.A. *Introduction to Statistical Time Series*; Wiley: New York, NY, USA, 1976.
73. Chen, Y.; Cao, R.; Chen, J.; Liu, L.; Matsushita, B. A Practical Approach to Reconstruct High-Quality Landsat NDVI Time-Series Data by Gap Filling and the Savitzky–Golay Filter. *ISPRS J. Photogramm. Remote Sens.* **2021**, *180*, 174–190. [CrossRef]
74. Fontana, F.; Rixen, C.; Jonas, T.; Aberegg, G.; Wunderle, S. Alpine Grassland Phenology as Seen in AVHRR, VEGETATION, and MODIS NDVI Time Series—A Comparison with In Situ Measurements. *Sensors* **2008**, *8*, 2833–2853. [CrossRef] [PubMed]
75. Van Leeuwen, W.J.D.; Huete, A.R.; Laing, T.W. MODIS Vegetation Index Compositing Approach: A Prototype with AVHRR Data. *Remote Sens. Environ.* **1999**, *69*, 264–280. [CrossRef]
76. Maselli, F.; Battista, P.; Chiesi, M.; Rapi, B.; Angeli, L.; Fibbi, L.; Magno, R.; Gozzini, B. Int J Appl Earth Obs Geoinformation Use of Sentinel-2 MSI Data to Monitor Crop Irrigation in Mediterranean Areas. *Int. J. Appl. Earth Obs. Geoinf.* **2020**, *93*, 102216. [CrossRef]
77. Tong, X.; Tian, F.; Brandt, M.; Liu, Y.; Zhang, W.; Fensholt, R. Trends of Land Surface Phenology Derived from Passive Microwave and Optical Remote Sensing Systems and Associated Drivers across the Dry Tropics. *Remote Sens. Environ.* **2019**, *232*, 111307. [CrossRef]
78. Liu, R.; Shang, R.; Liu, Y.; Lu, X. Global Evaluation of Gap- Fi Lling Approaches for Seasonal NDVI with Considering Vegetation Growth Trajectory, Protection of Key Point, Noise Resistance and Curve Stability. *Remote Sens. Environ.* **2017**, *189*, 164–179. [CrossRef]
79. Eilers, P.H.C.; Pesendorfer, V.; Bonifacio, R. Automatic Smoothing of Remote Sensing Data. In Proceedings of the 9th International Workshop on the Analysis of Multitemporal Remote Sensing Images (MultiTemp), Brugge, Belgium, 27–29 June 2017; pp. 1–3.
80. Vuolo, F.; Atzberger, C. Exploiting the Classification Performance of Support Vector Machines with Multi-Temporal Moderate-Resolution Imaging Spectroradiometer (MODIS) Data in Areas of Agreement and Disagreement of Existing Land Cover Products. *Remote Sens.* **2012**, *4*, 3143–3167. [CrossRef]
81. Patel, J.H.; Oza, M.P. Deriving Crop Calendar Using NDVI Time-Series. *Int. Arch. Photogramm. Remote Sens. Spat. Inf. Sci.* **2014**, *40*, 869–873. [CrossRef]
82. Reza, H.; Malamiri, G.; Zare, H.; Rousta, I.; Olafsson, H.; Verdiguier, E.I.; Zhang, H.; Mushore, T.D. Comparison of Harmonic Analysis of Time Series (HANTS) and Multi-Singular Spectrum Analysis (M-SSA) in Reconstruction of Long-Gap Missing Data in NDVI Time Series. *Remote Sens.* **2020**, *12*, 2747. [CrossRef]
83. Cetin, M.; Aksoy, T.; Nihan, S.; Anil, M.; Kurkcuoglu, S.; Cabuk, A. Employing Remote Sensing Technique to Monitor the Influence of Newly Established Universities in Creating an Urban Development Process on the Respective Cities. *Land Use Policy* **2021**, *109*, 105705. [CrossRef]
84. Mehta, D.H.; Shukla, S.H.; Kalubarme, M.H.; Mehta, D.H. Winter Crop Growth Monitoring Using Multi-Temporal NDVI Profiles in Kapadvanj Taluka, Gujarat State. *Int. J. Environ. Geoinform.* **2021**, *8*, 33–38. [CrossRef]
85. Cicuéndez, V.; Litago, J.; Sánchez-Girón, V.; Recuero, L.; Sáenz, C.; Palacios-Orueta, A. Identification and Modeling Carbon and Energy Fluxes from Eddy Covariance Time Series Measurements in Rice and Rainfed Crops. *Eng. Proc.* **2021**, *9*, 9. [CrossRef]
86. Debaeke, P.; Aboudrare, A. Adaptation of Crop Management to Water-Limited Environments. *Eur. J. Agron.* **2004**, *21*, 433–446. [CrossRef]
87. Cossani, C.M.; Slafer, G.A.; Savin, R. Do Barley and Wheat (Bread and Durum) Differ in Grain Weight Stability through Seasons and Water-Nitrogen Treatments in a Mediterranean Location? *Field Crops Res.* **2011**, *121*, 240–247. [CrossRef]
88. Zhu, W.; Pan, Y.; He, H.; Wang, L.; Mou, M.; Liu, J. A Changing-Weight Filter Method for Reconstructing a High-Quality NDVI Time Series to Preserve the Integrity of Vegetation Phenology. *IEEE Trans. Geosci. Remote Sens.* **2012**, *50*, 1085–1094. [CrossRef]
89. Recuero, L.; Maila, L.; Cicuéndez, V.; Sáenz, C.; Litago, J.; Tornos, L.; Merino-de-Miguel, S.; Palacios-Orueta, A. Mapping Cropland Intensification in Ecuador through Spectral Analysis of MODIS NDVI Time Series. *Agronomy* **2023**, *13*, 2329. [CrossRef]

90. Cicuéndez, V.; Litago, J.; Sánchez-Girón, V.; Román-Cascón, C.; Recuero, L.; Saénz, C.; Yagüe, C.; Palacios-Orueta, A. Dynamic Relationships between Gross Primary Production and Energy Partitioning in Three Different Ecosystems Based on Eddy Covariance Time Series Analysis. *Front. For. Glob. Chang.* **2023**, *6*, 1017365. [CrossRef]
91. Aragones, D.; Rodriguez-Galiano, V.F.; Caparros-Santiago, J.A.; Navarro-Cerrillo, R.M. Could Land Surface Phenology Be Used to Discriminate Mediterranean Pine Species? *Int. J. Appl. Earth Obs. Geoinf.* **2019**, *78*, 281–294. [CrossRef]
92. Mantas, V.; Fonseca, L.; Baltazar, E.; Canhoto, J.; Abrantes, I. Detection of Tree Decline (Pinus Pinaster Aiton) in European Forests Using Sentinel-2 Data. *Remote Sens.* **2022**, *14*, 2028. [CrossRef]

remote sensing

MDPI

Article

Classifying Stand Compositions in Clover Grass Based on High-Resolution Multispectral UAV Images

Konstantin Nahrstedt [1], Tobias Reuter [2], Dieter Trautz [2], Björn Waske [1] and Thomas Jarmer [1,*]

[1] Remote Sensing Group, Institute for Computer Science, Osnabrück University, 49074 Osnabrück, Germany; konstantin.nahrstedt@uni-osnabrueck.de (K.N.); bjoern.waske@uni-osnabrueck.de (B.W.)

[2] Faculty of Agricultural Sciences and Landscape Architecture, Osnabrück University of Applied Sciences, 49090 Osnabrück, Germany; tobias.reuter@hs-osnabrueck.de (T.R.); d.trautz@hs-osnabrueck.de (D.T.)

* Correspondence: thomas.jarmer@uni-osnabrueck.de

Abstract: In organic farming, clover is an important basis for green manure in crop rotation systems due to its nitrogen-fixing effect. However, clover is often sown in mixtures with grass to achieve a yield-increasing effect. In order to determine the quantity and distribution of clover and its influence on the subsequent crops, clover plants must be identified at the individual plant level and spatially differentiated from grass plants. In practice, this is usually done by visual estimation or extensive field sampling. High-resolution unmanned aerial vehicles (UAVs) offer a more efficient alternative. In the present study, clover and grass plants were classified based on spectral information from high-resolution UAV multispectral images and texture features using a random forest classifier. Three different timestamps were observed in order to depict the phenological development of clover and grass distributions. To reduce data redundancy and processing time, relevant texture features were selected based on a wrapper analysis and combined with the original bands. Including these texture features, a significant improvement in classification accuracy of up to 8% was achieved compared to a classification based on the original bands only. Depending on the phenological stage observed, this resulted in overall accuracies between 86% and 91%. Subsequently, high-resolution UAV imagery data allow for precise management recommendations for precision agriculture with site-specific fertilization measures.

Keywords: remote sensing; grassland; clover; random forest; unmanned aerial vehicle; texture; feature selection

1. Introduction

Increasing nitrate pollution in the soil prompted the European Commission (EC) to adopt a nitrate directive to reduce the use of mineral fertilizers [1]. Taking into account the growing demand for food as a result of global population growth, this poses major challenges for agriculture. In the recent past, rising energy costs due to global political conflicts have also led to an increase in fertilizer prices [2]. Alternative, sustainable methods for stabilizing and increasing yields are needed to ensure long-term food security. Legumes such as red or white clover can provide an ecological solution. Primarily used as a forage crop, clover is often cultivated in combination with grass species as temporary grassland in a crop rotation. There, clover serves as a natural fertilizer for subsequent crops, as it can fix atmospheric nitrogen (N) [3,4]. Furthermore, it helps to increase biomass production by suppressing weeds and reducing soil degradation [5]. Clover is usually sown in mixtures with grass in order to achieve higher yield stability and improved feed quality compared to pure grass or clover stands [6,7]. Within a crop rotation, grasses can contribute to better rooting of the soil, while clover makes atmospheric nitrogen available to the grass plants [8,9]. Here, the amount of N fixation correlates strongly with legume biomass [10] and can be further increased by competition with non-legumes such as grasses [11]. Due to variable soil conditions, the spatial distribution of clover and grass plants can vary, which may result in differences in N availability in the soil [12].

Information on the spatial distribution of clover in clover grass mixtures can be used to derive the expected forage quality and the potential nitrogen input [4]. Conventional fertilization measures in temporary grasslands are usually based on a visual assessment of the crop structure. Although more reliable information is provided by in situ data acquisition, this can be time-consuming and costly [13]. In contrast, spatial data offer a more efficient and cost-effective way to determine the clover grass distribution for site-specific fertilization. The fertilizer requirement can be adapted to the determined quantity and distribution of the clover in order to achieve improved biomass production and at the same time reduce N leaching from the soil [14].

Image-based estimations of plant compositions in clover grass mixtures have been the subject of several studies. In particular, different methodologies for the classification of clover and grass from ground-based imagery have been developed. This includes the pixel-based analysis of high-resolution image data to identify clover and grass leaves [15], as well as the estimation of clover and grass proportions in comparison with field-measured dry biomass [16,17]. These studies focused on the analysis of stand compositions from ground-based imagery with a particularly high spatial resolution.

Site-specific management recommendations for fertilization measures require information on the field scale. This is especially important in temporary grassland with small-scale heterogeneity [18,19]. Unmanned aerial vehicle (UAV)-based camera systems can be used to generate high-resolution image data over a wide area [20]. They are, therefore, suitable for monitoring agricultural fields [21,22]. UAV-based imagery provides the possibility of a spatially differentiated analysis of temperate grassland. Monitoring of clover grass stands using UAV imagery has already been discussed in various studies. The main focus here was the determination of stand biomass [23–25]. However, UAV imagery has also been used for the differentiation of crop cover types. Abduleil et al. [26] showed the suitability of UAV-based RGB data for estimating the amount of clover. They captured images using a fixed-wing UAV and quantified the composition of the clover grass mixture using different image classification algorithms.

Random forest classifier (RF) has proven itself in several studies for identifying clover stands. To monitor compositions of arable landscapes on a larger scale, it was used to classify various crop types (including clover and grass) [27]. Hahn et al. [28] discussed RF for differentiating between clover, grass, and weeds based on high-resolution imagery, acquired in a simulated UAV setup. Here, a pixel-based approach was compared with object-based image analysis (OBIA). When classifying crop cover types, accuracies between 77% and 98% were achieved. They also demonstrated the added value of texture information, as these take into account not only individual pixels but also the surrounding pixels and can, thus, emphasize spatial features [29]. Grüner et al. [30,31] demonstrated the added value of texture features in quantifying biomass in clover grass mixtures. They used texture features based on the Haralick texture extraction from the Orfeo toolbox (OTB) [32].

Texture features must be calculated band by band. The resulting high data dimensionality can lead to overfitting and a high computational and memory cost when scaling the method to larger areas. Feature selection approaches can be used to reduce datasets to the most important features. Currently, no studies are known that use UAV-based spectral and textural information filtered by feature selection for clover grass classification. Therefore, this approach will be the subject of the present work. Furthermore, comparable studies are only focused on a single observation date. In order to evaluate the phenological development of stand composition, this methodology is applied at three different observation dates. In this work, clover and grass plants were distinguished at individual plant levels based on high-resolution UAV multispectral (MS) image data using an RF classifier. The classifier was trained in a practice area. Texture parameters were preselected via feature selection based on a wrapper analysis approach and integrated into the classification. Finally, classification results from three recording dates were used to evaluate the stand development.

This results in three main objectives: (I) The suitability of high-resolution UAV image data for differentiating clover and grass plants at the single-plant level has to be examined.

(II) Analyzing whether texture parameters contribute to an increase in classification accuracy. Both a stage-specific and a cross-stage selection will be evaluated in order to be able to assess the applicability of selected texture features independently of the phenological stage. (III) Investigating whether trends for the spatial development of stand composition can be derived from the classification of UAV image data.

2. Materials and Methods

2.1. Study Area

This study was carried out on the organically farmed field "Kiesschacht" (52°19′N, 8°09′E) near Osnabrück (Germany). In order to reduce the processing effort for field data sampling, the test field boundaries were limited to a smaller part (0.74 ha) of the original field. In 2021, the area was cultivated with a clover grass mixture as part of a crop rotation. Figure 1 shows an orthophoto and the experimental setup. The clover grass mixture at "Kiesschacht" was sown on 20 August 2020 with a seed mixture of 22% red clover (*Trifolium pratense*), 11% alsike clover (*Trifolium hybridum*), 11% white clover (*Trifolium repens*), and 56% perennial ryegrass (*Lolium perenne*). The observed growing period was between 16 July 2021 (second mowing) and 22 September 2021 (third mowing).

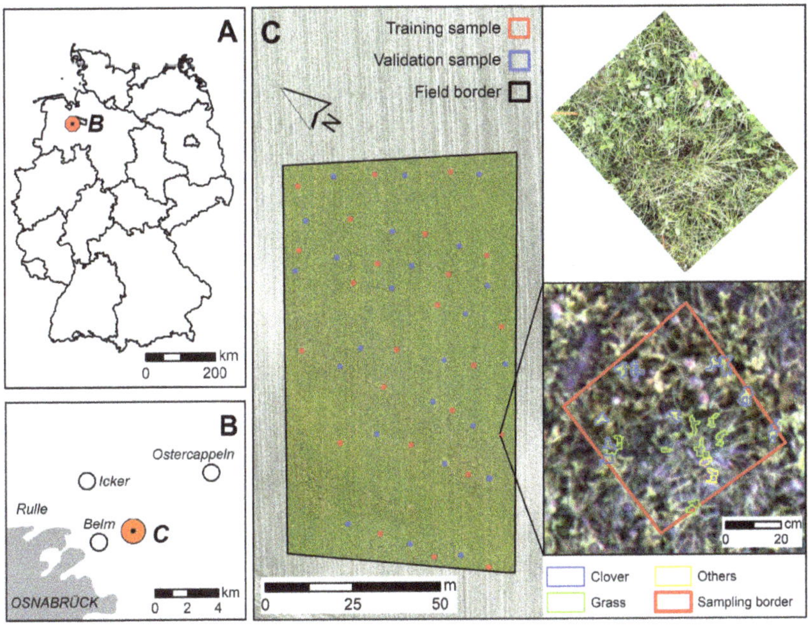

Figure 1. Study site; (**A**): location in Germany, (**B**): field location in suburban Osnabrück (Lower Saxony), (**C**): orthophoto with sampling area distribution (**left**) and example samples (**right**).

Climatic changes during the observation period have to be taken into account when analyzing phenological stages. Weather changes can influence crop development, as clover and grass plants react differently to drought. During the field campaign, climatic conditions were characterized by a long dry phase (Figure 2). The total precipitation of approx. 46 mm in August was lower than the long-term monthly average between 1990 and 2020 (83 mm). The average daily temperature was 17.1 °C, which was comparable to the long-term average for August (18.0 °C).

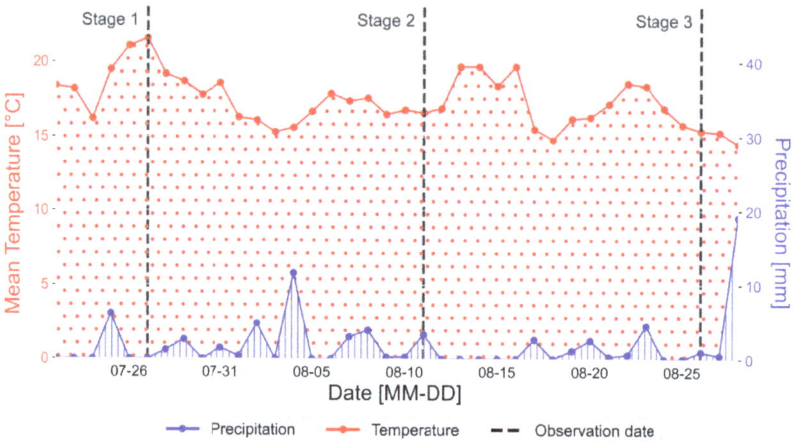

Figure 2. Climate constitutions during the field campaign and observation dates. Data were taken from the local weather station at Belm [33].

2.2. Data Collection

UAV-based MS images were captured using a DJI Phantom Multispectral. The attached camera records spectral properties of the surface in the wavelengths blue (450 nm $\pm$ 16 nm), green (560 nm $\pm$ 16 nm), red (650 nm $\pm$ 16 nm), RedEdge (720 nm $\pm$ 16 nm), and near-infrared (840 nm $\pm$ 26 nm). A flight altitude of 12 m provided a spatial resolution of 6 mm per pixel and was maintained across all observation dates. To ensure low-loss alignment of the individual images at this resolution, a front and side overlap of 85% was defined for all flights. This resulted in a number of 1040 to 1070 captured images per flight. The images were recorded under changeable conditions with sunny and cloudy days alternating irregularly. In order to analyze the development of the clover grass mixtures and possible changes in its spatial composition over time, image data were recorded at three timestamps between the second and third mowing (Table 1). An interval of about two weeks was chosen between the data recordings to be able to observe different stages of plant development and their influence on the UAV image data analysis.

Table 1. Dates from field campaign for data acquisition and corresponding phenological information.

Stage	Date	Days after Mowing	Phenological Stage Grass	Phenological Stage Clover
Stage 1	27 July	11	Begin of tillering	Begin of tillering
Stage 2	11 August	25	End of tillering	Begin of flowering
Stage 3	26 August	40	Ear emerging	Flowering

Field data were collected on each observation date at 48 sampling areas (0.5 m × 0.5 m), evenly distributed across the study area. High-resolution ground-based images were taken for each sampling area to reference the UAV-based training and validation data sampling. Dividing the sampling areas into training areas (see Figure 1, red squares) and validation areas (see Figure 1, blue squares) should ensure spatially independent sampling. To map spectral properties and variations of the crop in both datasets sampling areas for training and validation were spatially alternated. The location of every corner of the sampling area was recorded using a GNSS receiver (Stonex GNSS S9III).

2.3. Data Processing

To analyze the spatial distribution of clover and grass plants from MS UAV data from different phenological stages, a workflow was developed to (partially) automate raw image processing and analysis (Figure 3). UAV images were radiometrically and geometrically corrected using Agisoft Metashape (Version 1.7.2). Illumination differences between the observation dates could be recorded via the sunlight sensor of the Phantom Multispectral and normalized during image processing in raw data. The single images were radiometrically corrected and aligned to a point cloud at a high accuracy level. Based on the resulting sparse cloud, a dense cloud was calculated at a high-quality level using aggressive depth filtering. Finally, a terrain model and orthophoto were generated from the dense cloud. Geographical positioning was carried out using eight reference panels distributed evenly over the area. The center point of every panel was measured with the GPS and the resulting point coordinates were used as ground control points (GCPs) for georeferencing. The average deviation between the image and reference coordinates was between 3 cm and 6 cm.

2.4. Training and Validation Data

Training and validation were selected spatially, independently from the associated sampling areas for each stage individually. High-resolution ground-based images served as a reference for sampling. When selecting training and validation data, a distinction was made between three types of coverage. Perennial ryegrass was determined with the class "Grass". As the different clover species from the seed mixture hardly differ from each other in visual appearance and functionality, they are grouped in the class "Clover". Changes in phenological appearance due to the beginning of flowering (stage 2) and actual flowering (stage 3) were also included in the "Clover" class at the corresponding observation dates. Field components that could not be assigned to these surface types became part of the class "Others", which contains open soil and crop residues.

In order to obtain comparable classification results, the number of training and validation data samples was standardized across all stages. For each stage, 1000 pixels per class were randomly drawn from the training data. Validation was conducted by counting correctly identified pixels for each class in every classification image. For validation 2500 pixels were randomly selected per class and split with a five-time repeat without replacement to ensure comparability of the classification results. Overall Accuracy (OA) was calculated as a measure of quality in each replicate. To evaluate class-wise classification results additionally, the F1 score was determined from validation samples. The aforementioned accuracy measures served as the benchmark for evaluating stage-independent applicability.

2.5. Texture Feature Extraction

Textural characteristics were considered in addition to the pure spectral information of the original bands to optimize the identification of clover and grass plants. Textures can be interpreted as spatial gray-level variabilities of images [34]. They are therefore particularly suitable for reproducing structures and heterogeneities of vegetation [35]. Haralick et al. [36] developed a gray-level co-occurrence matrix (GLCM) consisting of 14 features to derive textures from image data. Referring to Grüner et al. [30,31], the eight features—energy, entropy, correlation, inverse difference moment (IDM), inertia, cluster shade, cluster prominence, and Haralick correlation—were used in the present work as well. Texture features were calculated in QuantumGIS (Version 3.24.1) with a window size of 2×2 and 32 histogram bins.

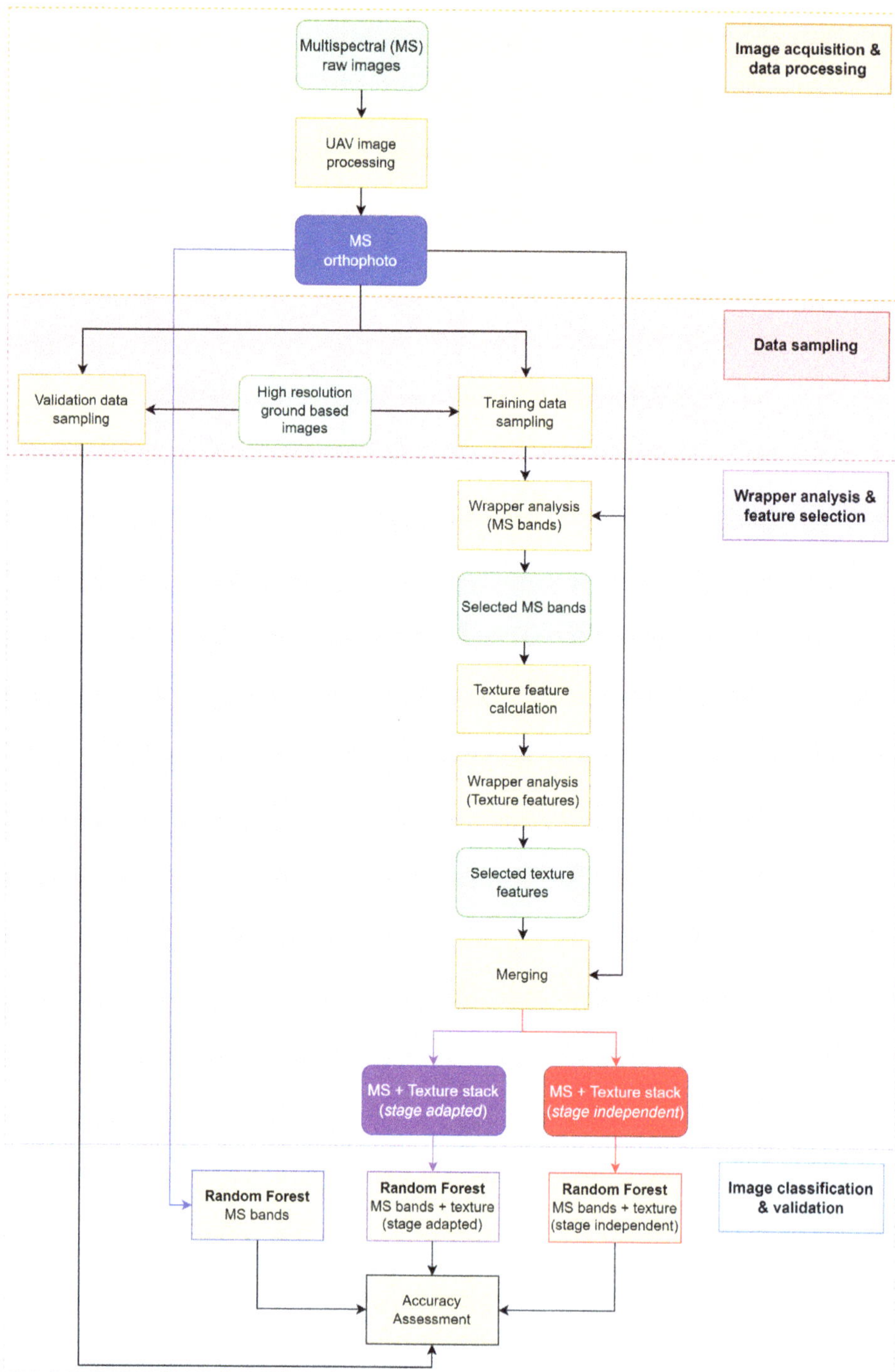

Figure 3. Workflow summary. Beige boxes describe processing steps. Key data sets resulting from individual processing steps are shown in green boxes. Boxes in blue, purple and red describe the final data sets for classification.

2.6. Random Forest Classifier

An RF classifier was trained to differentiate between "Clover", "Grass", and "Others". The RF is based on a defined number of decision trees that are arranged in an ensemble. To train the classifier for each decision tree, samples are randomly drawn from a training dataset [37]. Based on this, individual classification results are obtained for each tree. Finally, a classification result is defined from all trees using a majority vote [38]. While single decision trees tend to overfit the training data with increasing depth [39], this can be prevented by the classification decision from an ensemble of several uncorrelated trees within the forest. Due to its design, the RF is, therefore, considered to be robust against overfitting [40,41]. In addition, it is particularly suitable for high-dimensional data (e.g., multisource or hyperspectral) and needs less computation time than comparable machine learning approaches [42,43]. In this work, Python (Version 3.7) was used to implement an RF classifier with the software library scikit-learn [44]. The classifier parameters were kept in the default settings since the RF is basically evaluated as an algorithm that achieves good results in these settings [45].

2.7. Wrapper Analysis

To determine the essential texture features for classification, a feature selection based on wrapper analysis was carried out before classification. This approach avoids the need to individually calculate all texture bands for each spectral band, as well as data redundancy from highly correlated bands. In remote sensing, wrapper analyses are mainly used for the iterative selection of features from high-dimensional data [46]. They are, therefore, primarily suitable for hyperspectral imagery [47] and MS data when the dataset includes higher-dimensioned by-products derived from the original bands [48]. This approach makes it possible to estimate the importance of different bands for classification accuracy.

For this purpose, an RF classifier was parameterized as described in Section 2.6 and used to compare input bands by their classification impact. The training dataset was split into 50% training and 50% testing. For validation, the out-of-bag (OOB) score was calculated from test data. The OOB score describes the mean estimation error of all decision trees of an RF and is calculated using a subset of the data that is not part of the training ("out-of-bag") [49]. The wrapper designed in the present work is shown in Figure 4. In the first iteration, all bands were used individually to train the RF and then compared based on their OOB score. The band with the highest score was selected and combined with each of the remaining bands individually in the following iteration to determine the second most important band. This procedure is followed until the last iteration step. The number of steps depends on the number of single bands examined in the wrapper analysis. Finally, the development of the OOB scores indicates whether the accuracy level stagnates after a certain iteration.

Wrappers were used in two steps for feature selection at each phenological stage individually: (1) Since the original bands served as the basis for the texture calculation, feature selection was first carried out using the spectral bands. The number of bands selected was based on the observed saturation point in the OOB score development. All eight texture features were then calculated for these selected bands only. (2) With the resulting features, a second wrapper analysis was performed to finally determine the texture information essential for classification. For each phenological stage, all calculated features of all bands selected in step (1) were equally included in a single wrapper analysis. To determine their added value compared to the pure spectral information, all five original bands were used as the basis in each iteration and a single texture feature was added during each step according to the principle described in Figure 4. The first iteration starts with six bands (five original bands plus a single texture feature derived from one of the selected original bands).

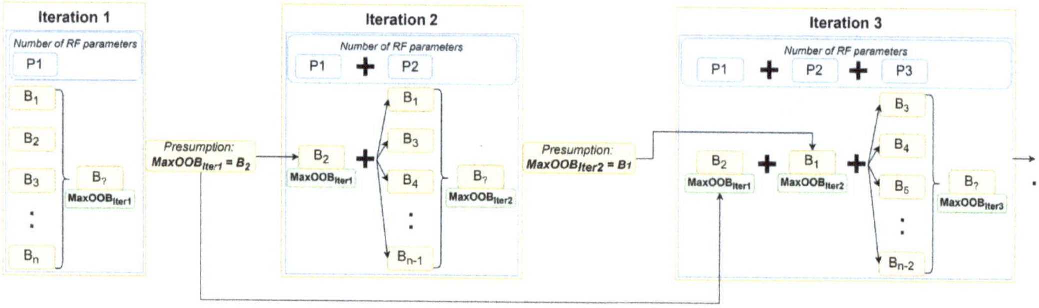

Figure 4. Scheme of the wrapper analysis. In the first iteration, RF classifiers are trained with one band each and validated using the test data from the training samples to identify the band with the highest OOB score ($MaxOOB_{Iter1}$). In the following step, ($MaxOOB_{Iter1}$) is combined with each of the remaining bands individually. The RF is then tested again on all band combinations. B1 to Bn (beige boxes) represents the individual bands tested in one iteration step. P (blue boxes) describes the number of bands considered in the respective iteration step.

2.8. Classification Setup

Three different datasets were generated for each phenological stage. These are referred to below as "Original bands (OB)", "Texture bands: stage-adapted (TBSA)", and "Texture bands: stage-independent (TBSI)". First, RF classification was carried out based on all original bands only (OB). Secondly, the most relevant texture features were selected based on the wrapper analysis depending on the OOB scores in each phenological stage individually and were combined stagewise with all corresponding original bands (TBSA). Thirdly, it should be tested whether a set of texture features could be identified that would improve the classification accuracy, regardless of the observed phenological stage. For this purpose, the most relevant texture features that occurred most frequently across all stages were selected and combined with the original bands (TBSI). Training and validation samples were used identically for all datasets within a stage.

3. Results

3.1. Wrapper Analysis

The wrapper analysis of the original bands showed an increase in the OOB score after the first two iterations, independent of the stage (Figure 5). From the third iteration onward, a saturation effect was observed in the development of the maximum OOB score in stages 1 and 2. This led to the conclusion that only the bands with the highest OOB score determined in the first two iterations should form the basis for the texture analysis.

In stage 3, this effect occurred from the fourth iteration onward. In order to ensure comparability of the classification results, the selection of bands is nevertheless limited to the first two iterations in this stage. It should also be noted that the first wrapper analysis is only used to determine suitable bands as the basis for texture feature calculation and that all of the original bands are still taken into account in the classification.

In all three phenological stages, the NIR band produced the highest OOB score after the first iteration (Table 2). The same was observed for the red band in the second iteration. A combination of NIR and red contained the essential information for clover grass differentiation. These bands were, therefore, used for the texture feature computation in all three phenological stages.

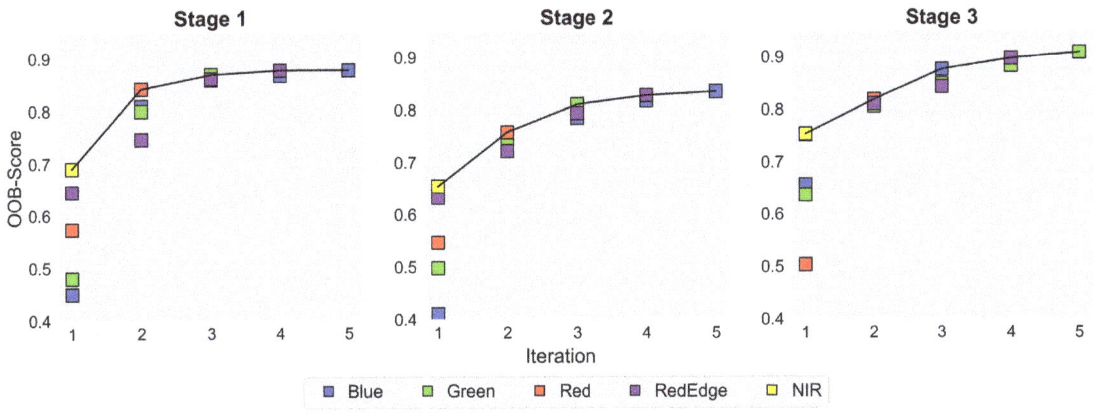

Figure 5. Results of wrapper analysis of the original bands. OOB scores of the individual bands or band combinations are shown. The iteration step is equal to the number of combined bands. Lines represent the development of the maximum OOB score. In stage 3, the RedEdge band is covered by the NIR band in the first iteration.

Table 2. Features with the max OOB score of wrapper analysis of the original bands for each iteration. The feature with the highest (additive) OOB score in the corresponding iteration is shown.

Iteration	Stage 1	Stage 2	Stage 3
1	NIR	NIR	NIR
2	Red	Red	Red
3	Green	Green	Blue
4	RedEdge	RedEdge	RedEdge
5	Blue	Blue	Green

A feature selection from the resulting texture features was then carried out individually for each phenological stage, which led to three further wrapper analyses. In two of the three wrapper analyses, the calculated maximum OOB scores again indicated a saturation effect from the sixth iteration at the latest (Figure 6). The only exception was noticeable during the wrapper analysis of the features in stage 3. A further slight increase was observed here up to the seventh iteration. Nevertheless, only the six most important texture features were included in the classification to maintain comparability between the datasets.

In contrast to the wrapper analysis of the original bands, the feature selection of the six most important texture features was not unique. Therefore, datasets were first generated individually and adapted to the phenological stages under consideration (TBSA). For each TBSA dataset, the texture features with the highest OOB score from the first six iterations were selected (Table 3) and combined with the original bands. Here, a dominance of NIR-based features was noticeable in the first iteration. Beyond this, no significant differences in the OOB score were observed between red- and NIR-based features. Each TBSA dataset, therefore, consists of 11 features (five spectral bands plus six texture features).

Although the results of the wrapper analysis from the texture features differed fundamentally between the stages, individual features appeared particularly frequently in the first three iterations. This provides the opportunity to test uniform selected texture features across all phenological stages (TBSI). NIR-based cluster prominence achieved the highest OOB score in all three stages in the first iteration. In the two following iterations, the highest scores were also achieved across all stages with the same features (red- and NIR-based Haralick correlation). This trend was no longer confirmed in the following iterations, as different features achieved the highest OOB score between the observed stages. These can therefore not be generalized for a cross-stage application. Furthermore, from iteration four onward, the gap between the highest (colored) and lowest (gray) OOB score decreased compared to iteration three. This suggests that essential information for the classification may already be represented by the first three features. Therefore, the TBSI datasets for each phenological stage were created by combining the original bands with three features: red- and NIR-based Haralick correlation, and NIR-based cluster prominence.

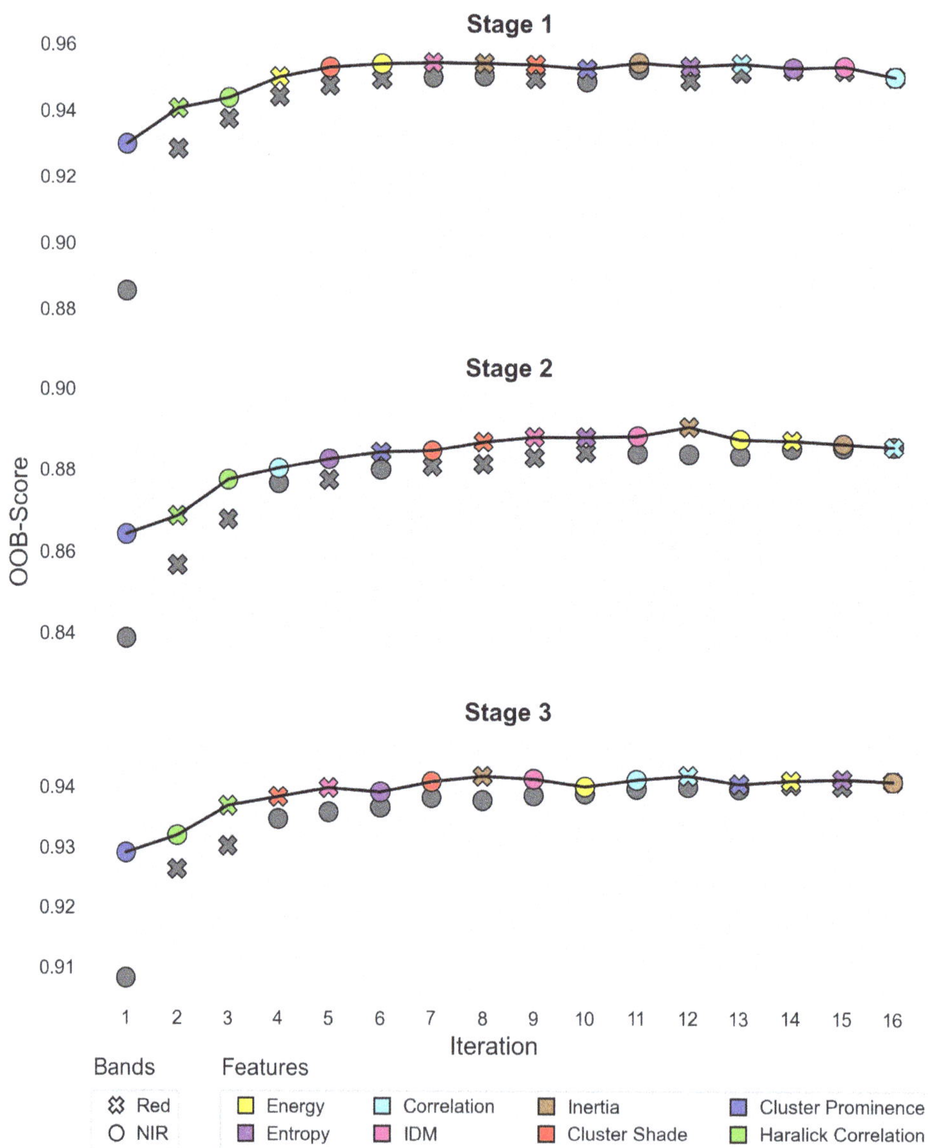

Figure 6. Results of the wrapper analysis of the Haralick texture features. The features were combined individually with the original bands. Added features with the highest OOB score in each iteration are shown in color. Gray symbols represent features with the lowest OOB score. The iteration step is equal to the number of combined features. Lines represent the development of the maximum OOB score.

Table 3. Features with maximum OOB score of wrapper analysis of the Haralick texture features. The feature with the highest (additive) OOB score in the corresponding iteration is shown. Only the first six iterations are shown.

Iteration	Stage 1		Stage 2		Stage 3	
	Feature	Band	Feature	Band	Feature	Band
1	Cluster Prominence	NIR	Cluster Prominence	NIR	Cluster Prominence	NIR
2	Haralick Correlation	Red	Haralick Correlation	Red	Haralick Correlation	NIR
3	Haralick Correlation	NIR	Haralick Correlation	NIR	Haralick Correlation	Red
4	Energy	Red	Correlation	NIR	Cluster Shade	Red
5	Cluster Shade	NIR	Entropy	NIR	IDM	Red
6	Energy	NIR	Cluster Prominence	Red	Entropy	NIR

3.2. Classification Results

Validation of the classification results revealed differences in accuracy between the datasets based on the original bands and the texture-optimized datasets (Figure 7). Mean OAs of classifications based on the original bands varied between 82.5% (stage 1), 82.8% (stage 2), and 85.7% (stage 3).

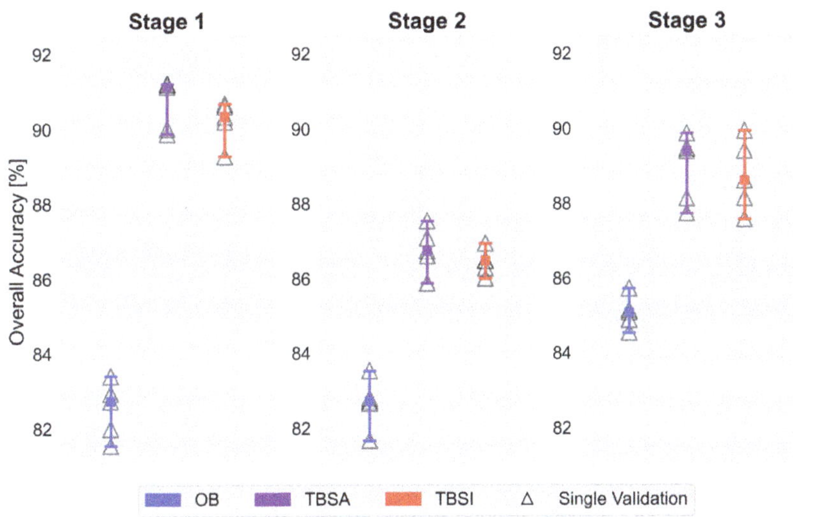

Figure 7. Comparison of overall accuracies for classifications out of the five-time repeat based on the three datasets. Overall accuracies of the five single validations are marked with white triangles.

By adding the stage-adapted texture features, the OAs could be significantly increased in all observed stages. The added value of the texture features was most clearly recognizable in the first stage. Here, the classification accuracy was increased to 90.7% (+8.2% compared to original bands) on average, while 86.8% (+4.0%) and 88.9% (+3.4%) were achieved in stages 2 and 3, respectively. The differences between TBSA and TBSI dataset-based classifications were minor. In stage 1, the overall accuracy decreased to 90.2% (−0.4%), in stage 2 to 86.4% (−0.4%) in stage 3 to 88.7% (−0.2%).

Analyzing the F1 scores, differences in classification accuracies between the individual classes could be recognized independently of the underlying dataset (Table 4). With the texture-optimized datasets, a consistently higher F1 score was achieved for each class. Samples of the "Others" class were generally identified correctly with very high accuracy. At least 90% of the pixels in this class were correctly assigned across all phenological stages.

Table 4. Mean F1 scores and interval limits of classes "Clover", "Grass", and "Others" for the classifications based on the original bands (OBs), the original bands combined with the stage-adapted texture bands (TBSA), and the original bands combined with the stage-independent texture bands (TBSI).

Stage	Class	F1 Score [%]					
		OB		TBSA		TBSI	
		Mean	±	Mean	±	Mean	±
S1	Clover	77.8	1.5	89.4	1.3	88.7	1.1
	Grass	77.4	1.2	87.9	0.5	87.5	1.4
	Others	92.8	0.7	94.8	0.7	94.6	0.6
S2	Clover	79.6	1.2	86.3	1.1	86.8	0.7
	Grass	75.6	1.3	80.6	1.3	79.1	1.4
	Others	92.8	0.6	92.9	1.2	92.6	0.7
S3	Clover	86.2	1.3	91.8	1.2	91.7	1.3
	Grass	79.0	1.3	83.8	1.9	83.8	1.8
	Others	90.0	1.1	91.0	1.9	90.6	1.0

The added value of the texture features was particularly evident for the "Clover" and "Grass" classes. In stage 1, the F1 score for the identification of clover increased from 77.8% (OB) to 89.4% (TBSA). Grass pixels were correctly assigned based on the original bands with an accuracy of 77.4%. Adding the texture features, the F1 score was increased to 87.9%. Similar trends were observed in stages 2 and 3. Significant improvements between OB and TBSA were measurable for clover (stage 2: +6.7%; stage 3: +5.6%) and grass (stage 2: +5.0%; stage 3: +4.8%). F1 scores between the classifications based on the TBSA and TBSI datasets did not differ noticeably from each other in any of the classes in any of the stages. Deviations in the F1 score varied between 0.0% and 0.7%. The interval limits of the F1 scores revealed significant differences between the classification results based on the original bands and the datasets extended by the texture features. In contrast, no significant differences were found when comparing the datasets extended with stage-adapted (TBSA) and stage-independent (TBSI) texture features.

The results of the classification maps (Figure 8) showed a heterogeneous distribution of clover and grass plants typical of clover grass meadows. In the first phenological stage, the dominance of grasses (44%) and bare soil (39%) was noticeable, especially in the south-western and northeastern parts of the study area. An increase in the proportion of clover could be observed between the first and second stages (+27%), while the proportion of grass decreased slightly (−1%). The increase in clover proportion was mainly associated with a decrease in bare soil and crop residues (−26%) and was distributed very heterogeneously across the entire study area. Increases in the proportion of clover were most clearly recognizable in the eastern, northeastern, and southwestern parts of the area. Between the second and third stages, a lower growth in clover proportion was observed (+9%). This was particularly evident in the northeastern part of the study area. The degree of soil cover continued to decrease (−5%) as well as the grass proportion (−4%).

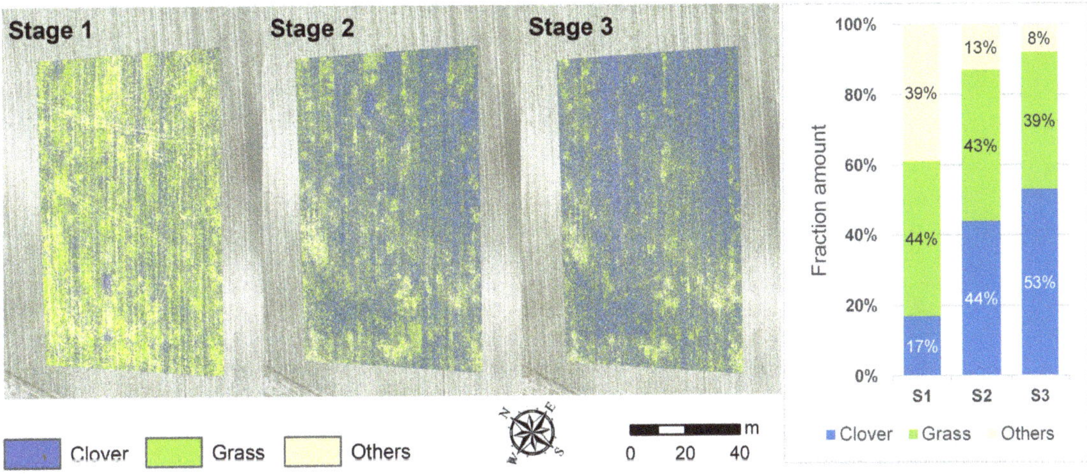

Figure 8. Classification results for the three phenological stages based on the combined dataset of original bands and the selected texture features (TBSI).

4. Discussion

4.1. Objective I: Single Plant Clover Grass Classification

In the present study, an RF classifier was used to distinguish between clover and grass plants in an organically managed practice area. A low flight altitude was chosen to enable high-resolution images in order to distinguish plant species with a high level of detail. Despite the still lower spatial resolution compared to ground-based image acquisition methods, spectral information from the UAV-based MS camera system proved to be an efficient basis for clover and grass classification. In the present study, OAs between 82% and 91% were achieved, depending on the underlying dataset. Although a direct comparison with the above-mentioned studies is challenging (e.g., due to different numbers of test samples, sampling design, size of field plots, and image resolution), these results are in line with the pixel-based approaches developed for detecting clover in clover grass [26,28]. However, these studies were limited to single observation dates and evaluated the features used in the classification post-classificatory.

Therefore, in the present study, an approach using spectral and textural information was developed and tested in different stages of phenology. In a pre-classification step, texture features that significantly improve the classification accuracy were determined by a wrapper analysis. With this step, data redundancy and processing time can be reduced, which seems particularly interesting for operational applications. Moreover, RF is considered to be more computationally efficient in complex classification approaches compared to other machine learning algorithms such as support vector machines [50] or deep learning methods [43]. Since RF classification usually requires only a few training samples, the sampling effort is also significantly lower than with conventional deep learning approaches [51,52].

4.2. Objective II: Texture Features and Feature Selection

Texture features have proven to be very useful for monitoring clover grass stands, but usually have to be calculated band by band. With increasing pixel resolution, field size, and number of bands, this can lead to increasing demands on storage and computing capacity. Furthermore, as high correlations between the original bands can cause data redundancies, feature selection is required for data reduction. This can additionally contribute to the robustness of the approach against overfitting and thus improve its transferability [53,54].

A two-stage feature selection based on a wrapper analyses approach was implemented in this study. During the first feature selection process, it was shown that not all spectral bands are necessary for texture feature calculation to achieve high accuracies. The wrapper analyses of the original bands revealed the NIR band and the red band as the most important features in the first and second iterations, respectively. This could be observed regardless of the phenological stage. NIR bands are sensitive to chlorophyll content and are, therefore, particularly suitable in dense vegetation stands [55]. Since the red band also has a comparatively low spectral correlation to the NIR band [56], a combination of both bands provides high information content for differentiating soil and plant as well as plant species. They, therefore, serve as a suitable basis for calculating texture features to identify clover and grass.

The second feature selection revealed that the most important texture features for classification can be fundamentally variable depending on the phenological stage. However, as some features were selected particularly frequently in the first iterations, trends could be identified. In particular, the Haralick correlation and cluster prominence are well suited to distinguish between clover, grass, and soil, regardless of the phenological stage observed. The Haralick correlation is a measure of the correlation between neighboring pixels with similar gray values. High values mean a strong correlation and, therefore, high homogeneity [36]. Cluster prominence describes the asymmetry between pixel values. Here, high values are associated with lower gray value symmetry and therefore higher variation [57]. Clover plants are more susceptible to variable lighting than grass plants due to their leaf mass and size. The resulting differences in homogeneity and gray value variation between samples with pixels of clover and grass plants could also be observed in the present study.

Furthermore, the wrapper analysis of the texture parameters confirmed the necessity of using both spectral information from the red and NIR bands for calculating texture information. In the initial iterations, the highest OOB scores were alternately achieved with texture features based on the Red and NIR bands.

Texture features were included in the classification to highlight spatial differences between clover and grass plants. Based on the high spatial resolution of the original bands, the spectral reflectance of the plants could be derived in high detail. Information on the shape and structure of individual and contiguous plants was determined using texture features. By integrating these spatial parameters a significant improvement of the classification results could be achieved. As a result, OAs could be improved by up to 8%. Furthermore, no significant differences were found between a stage-adapted (TBSA) and a stage-independent (TBSI) selection of texture features. This is due to the fact that all features selected for the TBSI datasets were consistently the most important features at all stages examined. Thus, a cross-stage selection of texture features could be realized in the present work.

4.3. Objective III: Development of Spatial Composition of Clover Grass Mixture

By monitoring the crop at three different timestamps during phenology, it was possible to evaluate the development of the stand composition. The classification maps revealed spatial and quantitative changes in the occurrence and quantity of clover. With respect to the spatial distribution of clover and grass based on RF classification, an increase in clover proportion was observed between the stages (see Figure 8). Here, the increase from the first to the second observation date was more noticeable than at the follow-up date in stage 3. At the same time, a significant decrease in the amount of soil pixels was observed. An increasing proportion of clover over time can have a direct influence on the N-fixing capacity of the soil and must also be considered in a spatially differentiated manner with regard to its impact on subsequent crops.

In particular, climatic influences on the crop can make it necessary to monitor the spatial development of clover grass over several points in time. N-fixing plants like *Trifolium repens* or *Trifolium pratense* have been shown to be more resistant to drought stress than non-N-fixing plants like *Lolium perenne* [58]. Additionally, the root system of the plants must be taken into account. Deep-rooted plant species (e.g., *Trifolium pratense*) are considered more resistant to drought than shallow-rooted species (e.g., *Lolium perenne*) [59,60]. This was also observed in the present study. Between the first and second observation dates, little precipitation, as well as temperature peaks in the middle of August, were measured (see Figure 2). Analysis of the plant species distribution based on the classification maps showed a sharp increase in the proportion of clover during this period while the amount of grass was sinking. A similar dry phase was also observed between the second and third observation dates. This led to a further increase in the proportion of clover, although it was lower than between stages 1 and 2. Due to the more favorable weather conditions for clover, its phenological development also progressed faster. As a result, an increase in clover leaf mass and the beginning of flowering were already evident at the second observation time, which must be taken into account when considering the stand composition development.

4.4. Limitations

Based on the F1 scores of the classification results, it was found that clover plants were better recognized than grass plants at all phenological stages. This is likely due to the plant structure and leaf area in conjunction with the recording geometry of the UAV camera. Despite the high spatial resolution (6 mm), grass plants were more difficult to identify, as they have a much smaller leaf mass and width than clover plants. Consequently, the risk of mixed signals in the training data is increased. Furthermore, clover leaves are easier to identify due to their size based on spatial structures that should be represented by the texture features. This has also been taken into account when analyzing the resulting classification maps and the shown amounts of crop fractions.

The methodology presented requires a high spatial resolution. Depending on the camera system, this is associated with a fixed maximum flight altitude. The camera used in this study provided a lower resolution than UAV-based camera systems currently available. Consequently, applications of the methodology with higher-resolution camera systems are required in order to be able to guarantee higher flight altitudes and higher area performance. However, an increasing volume of data must be taken into account here. This can be compressed using the above-described pre-classificatory wrapper analysis to be able to identify the most important parameters for classification.

Orthophotos showed strong differences in illumination in parts of the study area. Despite the illumination sensor attached to the UAV, these could not be fully corrected. Negative effects on the classification accuracy were always evident when the plants exhibited very atypical spectral signatures due to these effects. This might reason for the slight drop in classification accuracy at the second acquisition date, as differences in illumination were very high during data acquisition.

5. Conclusions

The present study has shown that high-resolution UAV imagery is suitable for identifying clover and grass plants on a single-plant level. Using RF classifier and texture feature analysis, accuracies between 86% and 91% could be achieved depending on the observed phenological stage. By determining stand compositions over three observation timestamps from the classification results, phenological features in crop development can be inferred. Due to longer dry phases in the observation period, an increasing dominance of clover crystallized at the observed site. However, classification inaccuracies must always be taken into account. It was found that including texture features is accompanied by significant improvement in OA. In particular, features for deriving structural homogeneity (Haralick correlation) and gray value variation (cluster prominence) enabled precise differentiation between clover and grass independent of the observed stage.

The results can be used for site-specific management recommendations (N-fertilization, reseeding, and crop rotation management). Nevertheless, future research should focus on scaling the methodology to larger study areas. Furthermore, measured yield proportions should be included in order to be able to allow statements on the (fractional) crop biomass. A combination of area-wide determination of the proportion of clover and grass and spatially differentiated biomass estimation could be used to improve yield predictions.

Author Contributions: Conceptualization, K.N., T.R. and T.J.; Methodology, K.N., B.W. and T.J.; Software, K.N. and T.R.; Validation, K.N. and T.R.; Formal Analysis, K.N., T.R. and T.J.; Investigation, K.N. and T.R.; Resources, D.T., B.W. and T.J.; Data Curation, K.N.; Writing—Original Draft Preparation, K.N. and T.R.; Writing—Review and Editing, K.N., T.J., B.W., T.R. and D.T.; Visualization, K.N. and T.J.; Supervision, T.R., K.N. and T.J.; Project Administration, T.J., K.N., T.R. and D.T.; Funding Acquisition, T.J. All authors have read and agreed to the published version of the manuscript.

Funding: We acknowledge the support of the Federal Ministry of Food and Agriculture (BMEL) and the Federal Office for Agriculture and Food (BLE) as part of the project "Experimentierfeld Agro-Nordwest" (funding code [28DE103C22]), as well as the Open Access Publishing Fund of Osnabrück University.

Data Availability Statement: The data presented in this study are available upon request from the corresponding author.

Acknowledgments: Special thanks are due to our colleagues at Osnabrück University and the Osnabrück University of Applied Science who actively supported us in the acquisition of data and their processing and evaluation during our field campaign.

Conflicts of Interest: The authors declare no conflicts of interest.

Abbreviations

The following abbreviations are used in this manuscript:

UAV	unmanned aerial vehicle
MS	multispectral
RF	random forest
OOB	out-of-bag
OA	overall accuracy
IDM	inverse difference moment
OBs	original bands
TBSA	texture bands: stage-adapted
TBSI	texture bands: stage-independent

References

1. EEC. Council Directive of 12 December 1991 Concerning the Protection of Waters against Pollution Caused by Nitrates from Agricultural Sources (91/676/EEC). *Off. J. Eur. Communities* **1991**, *375*. Available online: https://eur-lex.europa.eu/legal-content/EN/TXT/PDF/?uri=CELEX:31991L0676 (accessed on 19 February 2024).
2. Snapp, S.; Sapkota, T.B.; Chamberlin, J.; Cox, C.M.; Gameda, S.; Jat, M.L.; Marenya, P.; Mottaleb, K.A.; Negra, C.; Senthilkumar, K.; et al. Spatially differentiated nitrogen supply is key in a global food–fertilizer price crisis. *Nat. Sustain.* **2023**, *6*, 1268–1278. [CrossRef]
3. Eriksen, J.; Askegaard, M.; Søegaard, K. Residual effect and nitrate leaching in grass-arable rotations: Effect of grassland proportion, sward type and fertilizer history. *Soil Use Manag.* **2008**, *24*, 373–382. [CrossRef]
4. Ledgard, S.; Schils, R.; Eriksen, J.; Luo, J. Environmental impacts of grazed clover/grass pastures. *Ir. J. Agric. Food Res.* **2009**, *48*, 209–226. Available online: https://www.jstor.org/stable/20720369 (accessed on 19 February 2024).
5. Gaudin, A.C.M.; Westra, S.; Loucks, C.E.S.; Janovicek, K.; Martin, R.C.; Deen, W. Improving resilience of northern field crop systems using inter-seeded red clover: A review. *Agronomy* **2013**, *3*, 148–180. [CrossRef]
6. Arturi, M.J.; Aulicino, M.B.; Ansìn, O.; Gallinger, G.; Signorio, R. Combining Ability in Mixtures of Prairie Grass and Clovers. *Am. J. Plant Sci.* **2012**, *3*, 1355–1360. [CrossRef]
7. Zarza, R.; Rebuffo, M.; La Manna, A.; Balzarini, M. Red clover (Trifolium pratense L.) seedling density in mixed pastures as predictor of annual yield. *Field Crop. Res.* **2020**, *256*, 107925. [CrossRef]
8. Kirwan, L.; Lüscher, A.; Sebastià, M.T.; Finn, F.A.; Collins, R.P.; Porqueddu, C.; Helgadottir, A.; Baadshaug, O.H.; Brophy, C.; Coran, C.; et al. Evenness drives consistent diversity effects in intensive grassland systems across 28 European sites. *J. Ecol.* **2007**, *95*, 530–539. [CrossRef]
9. Nyfeler, D.; Huguenin-Elie, O.; Suter, M.; Frossard, E.; Lüscher, A. Grass–legume mixtures can yield more nitrogen than legume pure stands due to mutual stimulation of nitrogen uptake from symbiotic and non-symbiotic sources. *Agric. Ecosyst. Environ.* **2011**, *140*, 150–163. [CrossRef]
10. Suter, M.; Connolly, J.; Finn, J.A.; Loges, R.; Kirwan, L.; Sebastià, M.T.; Lüscher, A. Nitrogen yield advantage from grass-legume mixtures is robust over a wide range of legume proportions and environmental conditions. *Glob. Chang. Biol.* **2015**, *21*, 2424–2438. [CrossRef] [PubMed]

11. Rasmussen, J.; Søegaard, K.; Pirhofer-Walzl, K.; Eriksen, J. N2-fixation and residual N effect of four legume species and four companion grass species. *Eur. J. Agron.* **2012**, *36*, 66–74. [CrossRef]
12. Bloor, J.M.G.; Tardif, A.; Pottier, J. Spatial Heterogeneity of Vegetation Structure, Plant N Pools and Soil N Content in Relation to Grassland Management. *Agronomy* **2020**, *10*, 716. [CrossRef]
13. Wachendorf, M.; Fricke, T.; Möckel, T.; Remote sensing as a tool to assess botanical composition, structure, quantity and quality of temperate grasslands. *Grass Forage Sci.* **2018**, *73*, 1–14. [CrossRef]
14. Skovsen, S.; Dyrmann, M.; Mortensen, A.K.; Steen, K.A.; Green, O.; Eriksen, J.; Gislum, R.; Jørgensen, R.N.; Karstoft, H. Estimation of the Botanical Composition of clover grass Leys from RGB Images Using Data Simulation and Fully Convolutional Neural Networks. *Sensors* **2017**, *17*, 2930. [CrossRef] [PubMed]
15. Bonesmo, H.; Kaspersen, K.; Bakken, A.K. Evaluating an Image Analysis System for Mapping White Clover Pastures. *Acta Agric. Scand. Sect. Soil Plant Sci.* **2004**, *54*, 76–82. [CrossRef]
16. Himstedt, M.; Fricke, T.; Wachendorf, M. Determining the Contribution of Legumes in Legume–Grass Mixtures Using Digital Image Analysis. *Crop. Sci.* **2009**, *49*, 1910–1916. [CrossRef]
17. Mortensen, A.K.; Karstoft, H.; Søegaard, K.; Gislum, R.; Jørgensen, R.N. Preliminary Results of Clover and Grass Coverage and Total Dry Matter Estimation in Clover-Grass Crops Using Image Analysis. *J. Imaging* **2017**, *3*, 59. [CrossRef]
18. Fricke, T.; Richter, F.; Wachendorf, M. Assessment of forage mass from grassland swards by height measurement using an ultrasonic sensor. *Comput. Electron. Agric.* **2011**, *79*, 142–152. [CrossRef]
19. Lussem, U.; Bolten, A.; Menne, J.; Gnyp, M.; Martin, L.; Schellberg, J.; Bareth, G. Estimating biomass in temperate grassland with high resolution canopy surface models from UAV-based RGB images and vegetation indices. *J. Appl. Remote Sens.* **2019**, *13*, 034525. [CrossRef]
20. Torres-Sánchez, J.A.; López-Granados, F.; De Castro, A.I.; Peña-Barragán, J.M. Configuration and Specifications of an Unmanned Aerial Vehicle (UAV) for Early Site Specific Weed Management. *PloS ONE* **2013**, *8*, e58210. [CrossRef] [PubMed]
21. Delavarpour, N.; Koparan, C.; Nowatzki, J.; Bajwa, S.; Sun, X. A Technical Study on UAV Characteristics for Precision Agriculture Applications and Associated Practical Challenges. *Remote Sens.* **2021**, *13*, 1204. [CrossRef]
22. Tsouros, D.C.; Bibi, S.; Sarigiannidis, P.G. A Review on UAV-Based Applications for Precision Agriculture. *Information* **2019**, *10*, 349. [CrossRef]
23. Michez, A.; Lejeune, P.; Bauwens, S.; Herinaina, A.A.L.; Blaise, Y.; Castro Muñoz, E.; Lebeau, F.; Bindelle, J. Mapping and Monitoring of Biomass and Grazing in Pasture with an Unmanned Aerial System. *Remote Sens.* **2019**, *11*, 473. [CrossRef]
24. Li, K.; Burnside, N.G.; de Lima, R.S.; Peciña, M.V.; Sepp, K.; Yang, M.; Raet, J.; Vain, A.; Selge, A.; Sepp, K. The Application of an Unmanned Aerial System and Machine Learning Techniques for Red Clover-Grass Mixture Yield Estimation under Variety Performance Trials. *Remote Sens.* **2021**, *13*, 1994. [CrossRef]
25. Albert, P.; Saadeldin, M.; Narayanan, B.; Fernandez, J.; Mac Namee, B.; Hennessey, D.; O'Conner, N.E.; McGuinness, K. Detection and quantification of broadleaf weeds in turfgrass using close-range multispectral imagery with pixel- and object-based classification. *Int. J. Remote Sens.* **2022**, *42*, 8035–8055. [CrossRef]
26. Abduleil, A.M.; Taylor, G.W.; Moussa, M. An Integrated System for Mapping Red Clover Ground Cover Using Unmanned Aerial Vehicles, A Case Study in Precision Agriculture. In Proceedings of the 12th Conference on Computer and Robot Vision, Halifax, NS, Canada, 3–5 June 2015; IEEE: Piscataway, NJ, USA, 2015; pp. 277–284. [CrossRef]
27. Böhler, J.E.; Schaepman, M.E.; Kneubühler, M. Crop Classification in a Heterogeneous Arable Landscape Using Uncalibrated UAV Data. *Remote Sens.* **2018**, *10*, 1282. [CrossRef]
28. Hahn, D.S.; Roosjen, P.; Morales, A.; Njip, J.; Beck, L.; Cruz, C.V.; Leinauer, B. Detection and quantification of broadleaf weeds in turfgrass using close-range multispectral imagery with pixel- and object-based classification. *Int. J. Remote Sens.* **2021**, *42*, 8035–8055. [CrossRef]
29. Li, S.; Yuan, F.; Ata-Ul-Karim, S.T.; Zheng, H.; Cheng, T.; Liu, X.; Tian, Y.; Zhu, Y.; Cao, W.; Cao, Q. Combining Color Indices and Textures of UAV-Based Digital Imagery for Rice LAI Estimation. *Remote Sens.* **2019**, *11*, 1763. [CrossRef]
30. Grüner, E.; Wachendorf, M.; Astor, T. The potential of uav-borne spectral and textural information for predicting aboveground biomass and n fixation in legume-grass mixtures. *PloS ONE* **2021**, *15*, e0234703. [CrossRef] [PubMed]
31. Grüner, E.; Astor, T.; Wachendorf, M. Prediction of biomass and n fixation of legume–grass mixtures using sensor fusion. *Front. Plant Sci.* **2021**, *11*, 603921. [CrossRef] [PubMed]
32. Grizzonet, M.; Michel, J.; Poughon, V.; Inglada, J.; Savinaud, M.; Cresson, R. Orfeo ToolBox: Open source processing of remote sensing images. *Open Geospat. Data Softw. Stand.* **2017**, *2*, 15. [CrossRef]
33. Deutscher Wetterdienst (DWD). Open Data Bereich des Climate Data Center. Available online: https://www.dwd.de/DE/leistungen/cdc/climate-data-center.html?nn=17626 (accessed on 16 December 2023).
34. Petrou, M.; Sevilla, P.G. *Image Processing: Dealing with Texture*; John Wiley & Sons: Chilchester, UK, **2006**; pp. 1–10.
35. Gallardo-Cruz, J.A.; Meave, J.A.; González, E.J.; Lebrija-Trejos, E.E.; Romero-Romero, M.A.; Pérez-García, E.A.; Gallardo-Cruz, R.; Hernández-Stefanoni, J.L.; Martorell, C. Predicting Tropical Dry Forest Successional Attributes from Space: Is the Key Hidden in Image Texture? *PLoS ONE* **2012**, *7*, e30506. [CrossRef] [PubMed]
36. Haralick, R.M.; Shanmugam, K.; Dinstein, I. Textural features for image classification. *IEEE Trans. Syst. Man Cybern.* **1973**, *SMC-3*, 610–621. [CrossRef]
37. Cutler, A.; Cutler, D.R.; Stevens, J.R. Random Forests. *Mach. Learn.* **2012**, *45*, 157–175. [CrossRef]

38. Breiman, L. Random Forests. *Mach. Learn.* **2001**, *45*, 5–32. [CrossRef]
39. Bramer, M. Avoiding overfitting of decision trees. *Princ. Data Min.* **2007**, 119–134. [CrossRef]
40. Liaw, A.; Wiener, M. Classification and Regression by randomForest. *R News* **2022**, *2*, 18–22.
41. Robnik-Šikonja, M. Improving Random Forests. In *European Conference on Machine Learning*; Boulicaut, J.F., Esposito, F., Giannotti, F., Pedreschi, D., Eds.; Springer, Berlin/Heidelberg, Germany, **2004**; Volume 3201, pp. 359–370. [CrossRef]
42. Belgiu, M.; Drăgut, L. Random forest in remote sensing: A review of applications and future directions. *ISPRS J. Photogramm. Remote Sens.* **2016**, *114*, 24–31. [CrossRef]
43. Lan, T.; Hu, H.; Jiang, C.; Yang, C. A comparative study of decision tree, random forest, and convolutional neural network for spread-F identification. *Adv. Space Res.* **2020**, *65*, 2056–2061. [CrossRef]
44. Pedregosa, F.; Varoquaux, G.; Gramfort, A.; Michel, V.; Thirion, B. Scikit-learn: Machine Learning in Python. *J. Mach. Learn. Res.* **2011**, *12*, 2825–2830.
45. Fernandez-Delgado, M.; Cernadas, E.; Barro, S.; Amorim, D. Do we Need Hundreds of Classifiers to Solve Real World Classification Problems? *J. Mach. Learn. Res.* **2014**, *15*, 3133–3181.
46. Kohavi, R.; John, G.H. Wrappers for feature subset selection. *Artif. Intell.* **1997**, *97*, 273–324. [CrossRef]
47. Cao, X.; Wei, C.; Han, J.; Jiao, L. Hyperspectral Band Selection Using Improved Classification Map. *IEEE Geosci. Remote Sens. Lett.* **2017**, *14*, 2147–2151. [CrossRef]
48. Su, J.; Yi, D.; Coombes, M.; Liu, C.; Zhai, X.; MacDonald-Maier, K.; Chen, W. Spectral analysis and mapping of blackgrass weed by leveraging machine learning and UAV multispectral imagery. *Comput. Electron. Agric.* **2022**, *192*, 106621. [CrossRef]
49. Breiman, L. Out-of-Bag Estimation. Technical Report, Statistics Department, University. **1996**. Available online: https://www.stat.berkeley.edu/pub/users/breiman/OOBestimation.pdf (accessed on 10 April 2024).
50. Adugna, T.; Xu, W.; Fan, J. A Comparison of Random Forest and Support Vector Machine Classifiers for Regional Land Cover Mapping Using Coarse Resolution FY-3C Images. *Remote Sens.* **2022**, *14*, 574. [CrossRef]
51. Shetty, S.; Gupta, P.; Belgiu, M.; Srivastav, S.K. Assessing the Effect of Training Sampling Design on the Performance of Machine Learning Classifiers for Land Cover Mapping Using Multi-Temporal Remote Sensing Data and Google Earth Engine. *Remote Sens.* **2021**, *13*, 1433. [CrossRef]
52. Maxwell, A.E.; Warner, T.A.; Guillén, L.A. Accuracy Assessment in Convolutional Neural Network-Based Deep Learning Remote Sensing Studies—Part 1: Literature Review. *Remote Sens.* **2021**, *13*, 2450. [CrossRef]
53. Meyer, H.; Reudenbach, C.; Hengl, T.; Katurj, M.; Nauss, T. Improving performance of spatio-temporal machine learning models using forward feature selection and target-oriented validation. *Environ. Model. Softw.* **2018**, *101*, 1–9. [CrossRef]
54. Freiesleben, T.; Grote, T. Beyond generalization: A theory of robustness in machine learning. *Synthese* **2023**, *202*, 109. [CrossRef]
55. Ashapure, A.; Jung, J.; Chang, A.; Oh, S.; Maeda, M.; Landivar, J. A Comparative Study of RGB and Multispectral Sensor-Based Cotton Canopy Cover Modelling Using Multi-Temporal UAS Data. *Remote Sens.* **2019**, *11*, 2757. [CrossRef]
56. Peng, Y.; He, M.; Zheng, Z.; He, Y. Enhanced Neural Network for Rapid Identification of Crop Water and Nitrogen Content Using Multispectral Imaging. *Remote Sens.* **2023**, *13*, 2464. [CrossRef]
57. Yang, X.; Tridandapani, S.; Beitler, J.J.; Yu, D.S.; Yoshida, E.J.; Curran, W.J.; Liu, T. Ultrasound GLCM texture analysis of radiation-induced parotid-gland injuryin head-and-neck cancer radiotherapy: An in vivo study of late toxicity. *Med. Phys.* **2012**, *39*, 5732–5739. [CrossRef] [PubMed]
58. Hoekstra, N.J.; Suter, M.; Finn, J.A.; Husse, S.; Lüscher, A. Do belowground vertical niche differences between deepand shallow-rooted species enhance resource uptake and drought resistance in grassland mixtures? *Plant Soil* **2015**, *394*, 21–34. [CrossRef]
59. Hofer, D.; Suter, M.; Haughey, E.; Finn, J.A.; Nyncke, J.; Hoekstra, N.J.; Buchmann, N.; Lüscher, A. Yield of temperate forage grassland species is either largely resistant or resilient to experimental summer drought. *J. Appl. Ecol.* **2016**, *53*, 1023–1034. [CrossRef]
60. Tahir, M.; Li, C.; Zeng, T.; Xin, Y.; Chen, C.; Javed, H.H.; Yang, W.; Yan, Y. Mixture Composition Influenced the Biomass Yield and Nutritional Quality of Legume–Grass Pastures. *Agronomy* **2022**, *12*, 1449. [CrossRef]

remote sensing

Article

An Angle Effect Correction Method for High-Resolution Satellite Side-View Imaging Data to Improve Crop Monitoring Accuracy

Jialong Gong [1,2,3], Xing Zhong [1,2,3], Ruifei Zhu [3,*], Zhaoxin Xu [3], Dong Wang [3] and Jian Yin [4]

1 Changchun Institute of Optics, Fine Mechanics and Physics, Chinese Academy of Sciences, Changchun 130033, China; gongjialong@jl1.cn (J.G.); zhongxing@jl1.cn (X.Z.)
2 University of Chinese Academy of Sciences, Beijing 100049, China
3 Chang Guang Satellite Technology Co., Ltd., Changchun 130102, China; xuzhaoxin@jl1.cn (Z.X.); wangdong@jl1.cn (D.W.)
4 College of Civil Engineering, Jilin University of Architecture and Technology, Changchun 130114, China; yinjian@jluat.edu.cn
* Correspondence: zhuruifei@jl1.cn

Abstract: In recent years, the advancement of CubeSat technology has led to the emergence of high-resolution, flexible imaging satellites as a pivotal source of information for the efficient and precise monitoring of crops. However, the dynamic geometry inherent in flexible side-view imaging poses challenges in acquiring the high-precision reflectance data necessary to accurately retrieve crop parameters. This study aimed to develop an angular correction method designed to generate nadir reflectance from high-resolution satellite side-swing imaging data. The method utilized the Anisotropic Flat Index (AFX) in conjunction with a fixed set of Bidirectional Reflectance Distribution Function (BRDF) parameters to compute the nadir reflectance for the Jilin-1 GP01/02 multispectral imager (PMS). Crop parameter retrieval was executed using regression models based on vegetation indices, the leaf area index (LAI), fractional vegetation cover (FVC), and chlorophyll (T850 nm/T720 nm) values estimated based on angle corrected reflectance compared with field measurements taken in the Inner Mongolia Autonomous Region. The findings demonstrate that the proposed angular correction method significantly enhances the retrieval accuracy of the LAI, FVC, and chlorophyll from Jilin-1 GP01/02 PMS data. Notably, the retrieval accuracy for the LAI and FVC improved by over 25%. We expect that this approach will exhibit considerable potential to improve crop monitoring accuracy from high-resolution satellite side-view imaging data.

Keywords: angle effect correction; side-view imaging; bidirectional reflectance distribution function (BRDF); Jilin-1; crop parameters retrieval; leaf area index (LAI); fraction of vegetation cover (FVC); leaf chlorophyll concentration

1. Introduction

The rapid advancement of remote sensing technology has established it as a crucial tool in the realm of agricultural research. This is attributable to its capability to offer extensive and periodic observations of terrestrial surface data [1–3]. Nonetheless, the inherent high spatial heterogeneity observed in agricultural lands, coupled with the dynamic nature of crop development, imposes stringent demands on the timeliness, precision, and sequential integrity of satellite-derived remote sensing data [4]. Within the context of contemporary agricultural practices, the accurate derivation of crop parameters, including the vegetation index, leaf area index (LAI), and chlorophyll concentration, is paramount for the monitoring of crop growth, the estimation of yields, and the stewardship of agricultural resources [5,6].

High-resolution satellites are capable of delivering detailed imagery, which enables the capture of surface features at a finer scale. This enhanced resolution facilitates more precise identification and monitoring of crop growth patterns. However, these high-resolution satellites are inevitably constrained by factors such as optical limitations, bandwidth restrictions, and orbital configurations, which can impede the simultaneous attainment of high spatial and high temporal resolution data [7]. In recent years, the challenge of acquiring satellite data with both a high spatial resolution (less than 10 m) and a high temporal resolution (less than 5 days) has been addressed through the establishment of satellite constellations utilizing extensive satellite networks [8]. The 'Jilin-1' constellation, developed by Chang Guang Satellite Technology Co., Ltd., is designed to enable rapid updates of global surface information via a network comprising 138 high-resolution satellites. As a provisional measure towards this goal, the flexibility to adjust the satellite's angle of inclination can be exploited to reduce its revisit time.

The Jilin-1 GP01/02 satellites constitute pivotal elements within the 'Jilin-1' constellation. Equipped with a 5 m multispectral imager, these satellites possess the capability for side-swing imaging up to $\pm 45^{\circ}$, enabling daily revisits to local areas through side-view imaging techniques. When juxtaposed with traditional medium-resolution satellites such as the Landsat series and Sentinel-2, the enhanced imaging flexibility of the Jilin-1 satellites presents heightened challenges to satellite data processing and the inversion of vegetation parameters [8–10]. Among these challenges, angle correction emerges as a critical step for achieving more precise surface albedo measurements. The angle effect in satellite imagery is predominantly influenced by the sensor's observation angle and the solar angle at the time of image acquisition [11]. The Bidirectional Reflectance Distribution Function (BRDF) delineates the relationship between bidirectional reflectance and the spectral, as well as the structural, characteristics of the terrestrial target [12]. It serves as a tool to mitigate discrepancies in observed reflectance that arise from varying incidence and observation angles [13].

MODIS BRDF products have garnered widespread application across a spectrum of satellite sensors, including the Landsat series [14–16], Sentinel-2 MSI [17,18], GF-1 WFV [19,20], GF-4 PMS [21], and SPOT [22], and have even been extended to correct for angle effects in UAV observation data [13]. To streamline the computational process, researchers have formulated a suite of fixed BRDF model parameters predicated on MODIS BRDF, which has been effectively utilized to compute nadir reflectance for both Landsat and Sentinel-2 satellites [17,18,23]. The Vermote–Justice–Breon (VJB) method, as developed by Vermote et al., is instrumental in correcting the angle effect in time-series MODIS data. This approach adeptly adjusts BRDF parameters via the linear regression relationship with the normalized difference vegetation index (NDVI) [24,25]. Additionally, an Anisotropic Flat Index (AFX) was devised to extract BRDF prototypes from MODIS data, encapsulating the variability of the fundamental dome–bowl anisotropic reflectance patterns of the terrestrial surface. The AFX method augments the understanding of vegetation structure and other anisotropic reflectance properties of the land surface [26]. Roy et al. [23] have devised a comprehensive method for normalizing Landsat reflectance data to BRDF adjusted reflectance, leveraging MODIS BRDF products. Within the scope of their research, a consistent set of BRDF spectral model parameters was established. Subsequently, this fixed parameter set was effectively utilized to produce Nadir BRDF-adjusted Reflectance (NBAR) for Sentinel-2 satellites [17,18].

Retrieval methods for vegetation parameters such as the LAI, fractional vegetation cover (FVC), chlorophyll, and others, based on remote sensing, are generally categorized into two primary approaches: physical model methods grounded in radiative transfer models, and empirical model methods predicated on vegetation indices [27–29]. The physical model approach executes a cost function match by simulating the radiative transfer processes within the vegetation canopy and leaf structure to deduce the physical and chemical properties of the vegetation. This method is characterized by its robust universality, as it is not constrained by environmental conditions or crop types. However, it should be noted that the radiative transfer processes in natural settings are influenced by various factors, including the characteristics of the canopy architecture, and are contingent upon several assumed parameters for simulation [30]. Empirical model methods, on the other hand, are predominantly founded upon establishing regression relationships between observed vegetation parameters and corresponding vegetation indices. These indices include, but are not limited to, the NDVI, the ratio vegetation index (RVI), and the enhanced vegetation index 2 (EVI2), among others [31–33]. The efficacy of empirical models is more reliant on the accuracy of the regression relationships and the reliability of the satellite observation data [34]. Prior research has demonstrated that the Jilin-1 GP01/02 satellites possess an absolute radiometric accuracy that is comparable to that of the Sentinel-2A satellite [35], rendering them suitable for high-precision inversion of the crop LAI [36]. Nevertheless, given the unique flexible side-swing imaging capabilities of the Jilin-1 GP01/02 satellites, further investigation is warranted into the evaluation and correction of the angle effect on crop parameters retrieval by these satellites.

This study is primarily concentrated on the correction of the angle effect for high-resolution satellites equipped with side-swing imaging capabilities. We have developed an angle correction methodology that is specifically tailored for the Jilin-1 GP01/02 satellites, enabling accurate reflectance correction under diverse observation geometric conditions. The efficacy of the proposed angle correction method in enhancing the retrieval precision of vegetation indices and crop parameters has been rigorously assessed using an extensive collection of synchronous field measurements. This paper posits that it is imperative to address the angle effect for side-view imaging data obtained from high-resolution satellites. Furthermore, the application of a set of suitably calibrated Anisotropic Flat Index (AFX) parameters can significantly augment the accuracy of crop monitoring endeavors.

2. Materials

2.1. Study Area and Field Measurements

The study area encompassed the Hangjin Rear Banner (40°26′N–41°13′N, 106°34′E–107°24′E), Urat Front Banner (40°28′N–41°16′N, 108°11′E–109°54′E), Zhenglan Banner (42°21′N–42°56′N, 115°30′E–116°13′E), and Horqin Right Front Banner (45°50′N–46°21′N, 121°44′E–122°25′E) in the Inner Mongolia Autonomous Region (Figure 1). The predominant crops in the regions are corn, rice, and sunflower. Field measurements were conducted from 17 July to 18 September 2021, encompassing the flowering, grouting, and ripening phases of the crops (Table 1). Parameters including canopy spectrum, LAI, and FVC were systematically collected. Each measurement was executed in a homogenous field with dimensions no smaller than 100 m × 100 m. Within the central area of each field, five 10 m × 10 m quadrats were established, and the mean of these quadrats was used to represent the field's canopy spectrum, LAI, and FVC measurements. All measurements were performed under clear sky conditions. Crop canopy spectral data were acquired using an ASD FieldSpec4 spectrometer, which boasts a spectral range from 350 to 2500 nm. LAI measurements were obtained using an LAI-2200C plant canopy analyzer. The FVC was assessed through vertical photography. In vivo chlorophyll, flavanol, and anthocyanin content estimations were performed using an MPM-100 Multi-Pigment-Meter (OPTI-SCIENCES, USA) on fully expanded leaves [37]. For these measurements, adaxial leaf surfaces were selected, with care taken to avoid the leaf midribs when positioning the fluorescence detector for readings.

Table 1. Sites, time, crop type, and growing season information for field survey.

Sites	Crop Type	Time	Crop Growing Season	Number of Survey Points
Hangjin Rear Banner	Corn, Sunflower	17 July 2021–18 July 2021	Flowering Period	22
Urat Front Banner	Corn, Sunflower	19 July 2021–20 July 2021	Flowering Period	10
		20 August 2021–22 August 2021	Grouting Period	15
Zhenglan Banner	Corn, Sunflower	09 September 2021–11 September 2021	Grouting Period	22
Horqin Right Front Banner	Corn, Rice	15 September 2021–18 September 2021	Ripe Period	56

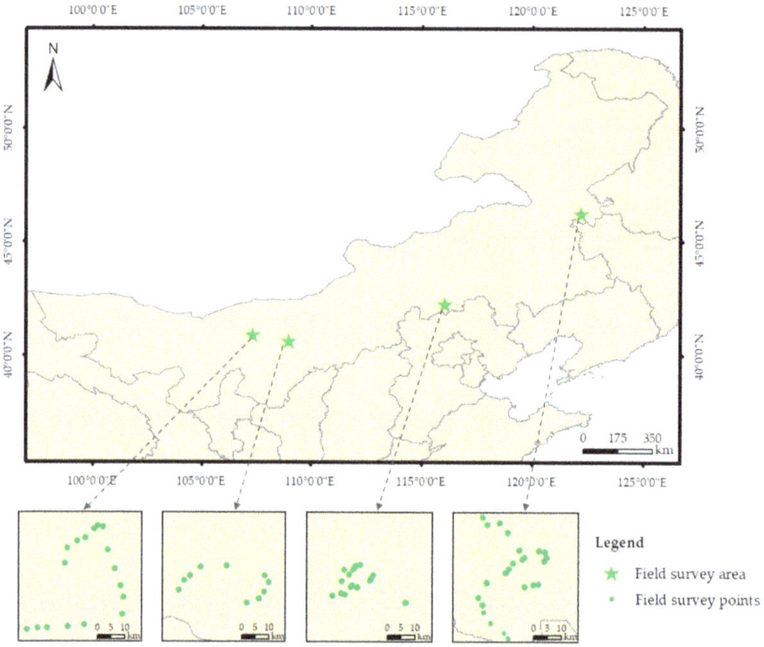

Figure 1. The distribution map of the study area and the field survey points.

2.2. *Jilin-1 GP01/02 Data*

The Jilin-1 GP01 and Jilin-1 GP02 satellites, launched on January 12, 2019, were equipped with identical sensor payloads, including a multispectral imager (PMS) and cameras sensitive to short-wave, medium-wave, and long-wave infrared spectra. Each satellite was furnished with a pair of multispectral imagers (PMS1 and PMS2) capable of acquiring remote sensing data across 20 spectral bands at spatial resolutions of 5 m, 10 m, and 20 m, respectively. The swath width for data capture was 60 km, with a lateral swing imaging capability of $\pm 45°$. The focus of this paper is on the data procured from the multispectral imager. Table 2 provides an overview of the pertinent technical specifications of the Jilin-1 GP01/02 PMS.

Due to the use of MODIS BRDF parameters, this paper selected B3 (450–515 nm), B4 (525–600 nm), B5 (630–680 nm), and B6 (784.5–899.5 nm) with a 5 m resolution similar to MODIS satellite settings for the angle effect correction study. Figure 2 illustrates the comparative spectral response functions of the Jilin-1 GP01/02 PMS and MODIS. Given that the blue band is susceptible to contamination from residual atmospheric influences, it was excluded from the accuracy assessment in the monitoring of crop parameters [38].

To ensure congruence between satellite observation data and corresponding field measurements, the Jilin-1 GP01/02 PMS data corresponding to 12 quasi-synchronized scenes were secured. Specifically, satellite and ground observations were synchronized on July 18, July 19, September 11, and September 17. For the remaining satellite data, the temporal discrepancy from the ground measurement acquisition was maintained within a 3-day window. The Jilin-1 GP01/02 PMS data underwent a series of processing steps, including radiometric calibration, atmospheric correction, and geometric correction. Additionally, the imaging geometry parameters for each pixel were calculated, encompassing the solar zenith angle (θ_s), view zenith angle (θ_v), and relative azimuth angle (ϕ). A summary of the imaging times and geometrical information for these data is presented in Table 3.

Table 2. Technical specifications of Jilin-1 GP01/02 PMS.

Band	Spectral Range (nm)	Spatial Resolution (m)
B0	450–800	5
B1	403–423	5
B2	433–453	5
B3	450–515	5
B4	525–600	5
B5	630–680	5
B6	784.5–899.5	5
B7	485–495	10
B8	615–625	10
B9	650–680	10
B10	698.75–718.75	10
B11	732.5–747.5	10
B12	773–793	10
B13	855–875	20
B14	660–670	20
B15	677.5–685	20
B16	750–757.5	20
B17	758.75–762.75	20
B18	935–955	20
B19	1000–1040	20

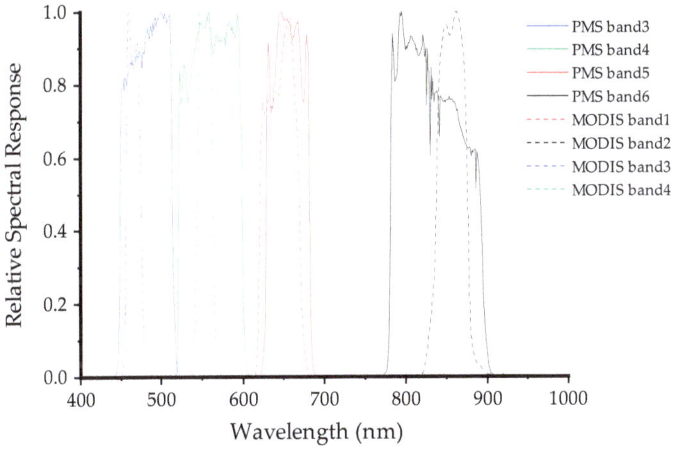

Figure 2. Spectral response functions of Jilin-1 GP01/02 PMS and MODIS in visible to near-infrared bands.

Table 3. A list of Jilin-1 GP01/02 PMS data quasi-synchronized with the field survey.

Date	Sensors	Solar Zenith Angle	View Zenith Angle	Relative Azimuth Angle
18/07/2021	Jilin-1 GP01 PMS2	20.33°	7.93°	19.73°
19/07/2021	Jilin-1 GP01 PMS1	20.06°	16.12°	56.17°
19/08/2021	Jilin-1 GP02 PMS2	28.31°	17.49°	57.41°
08/09/2021	Jilin-1 GP02 PMS1	36.53°	23.85°	66.24°
11/09/2021	Jilin-1 GP01 PMS1	37.73°	8.27°	21.20°
17/09/2021	Jilin-1 GP02 PMS2	44.58°	10.16°	43.51°

2.3. MODIS Data

The MCD43A1 V6.1 Bidirectional Reflectance Distribution Function and Albedo (BRDF/Albedo) Model Parameters dataset is a daily product with a 500 m resolution and a 16-day composite period. It offers parameter values that pertain to isotropic scattering, volumetric scattering, and geometric-optical scattering across MODIS bands 1 through 7, which correspond to the wavelengths of 645 nm, 859 nm, 469 nm, 555 nm, 1240 nm, 1640 nm, and 2130 nm, respectively. Additionally, the dataset encompasses three broad spectral bands, specified as 300–700 nm, 700–5000 nm, and 300–5000 nm. In the context of this paper, MCD43A1 data synchronized with the dates of the Jilin-1 GP01/02 PMS were secured. The concurrent MCD43A2 dataset was utilized for the quality control of BRDF parameters. Subsequently, a selection criterion was applied, wherein only BRDF parameters associated with pixels of the highest quality and good quality were deemed suitable for inclusion in the correction process. This study aimed to employ the MODIS BRDF parameters of band 1 (645 nm), band 2 (859 nm), and band 4 (555 nm) for the reflectance correction of Jilin-1 GP01/02 PMS green, red, and NIR bands.

3. Methods

3.1. RTLSR BRDF Model

The BRDF is a pivotal function that quantitatively delineates the relationship between ground cover reflectance and the geometry of incidence and observation [39]. Nearly all models for angle effect correction are predicated on the BRDF model framework [40]. The RossThick–LiSparse R (RTLSR) BRDF model [12] stands as one of the most esteemed and widely utilized semi-empirical models, and it is integral to the operational production algorithms for MODIS BRDF and albedo products [41].

Within the RTLSR BRDF model, a linear composite of three distinct scattering components is formulated, encompassing isotropic scattering, volumetric scattering, and geometric-optical scattering. The model's expression is articulated as follows [12,42]:

$$R_\lambda(\theta_s, \theta_v, \phi) = f_{iso}(\lambda) + f_{vol}(\lambda)K_{vol}(\theta_s, \theta_v, \phi) + f_{geo}(\lambda)K_{geo}(\theta_s, \theta_v, \phi) \quad (1)$$

where $R_\lambda(\theta_s, \theta_v, \phi)$ is the Bidirectional Reflectance Distribution Function of λ band; $f_{iso}(\lambda)$, $f_{vol}(\lambda)$, and $f_{geo}(\lambda)$ are the weights of isotropic scattering, volume scattering, and geometrically optical scattering, respectively; and $K_{vol}(\theta_s, \theta_v, \phi)$ and $K_{geo}(\theta_s, \theta_v, \phi)$ are the volume scattering kernel and the geometric optical kernel, respectively, which are both functions of the solar zenith (θ_s), view zenith (θ_v) and relative azimuth (ϕ).

3.2. MODIS-Based Parameters Method

The correction of angle effects in high-resolution satellite imagery, leveraging MODISBRDF parameters, is a prevalent and robust approach. This methodology has garnered widespread application in the rectification of BRDF effects for a spectrum of satellite platforms, including the Landsat series, Sentinel-2, and the GF series satellites [17–19,23].

$$R_\lambda(\theta_s, 0, \phi) = C_\lambda \times R_\lambda(\theta_s, \theta_v, \phi) \quad (2)$$

$$C_\lambda = \frac{f_{iso}(\lambda) + f_{vol}(\lambda)K_{vol}(\theta_s, 0, \phi) + f_{geo}(\lambda)K_{geo}(\theta_s, 0, \phi)}{f_{iso}(\lambda) + f_{vol}(\lambda)K_{vol}(\theta_s, \theta_v, \phi) + f_{geo}(\lambda)K_{geo}(\theta_s, \theta_v, \phi)} \quad (3)$$

where, $R_\lambda(\theta_s, 0, \phi)$ represents the nadir reflectance of the Jilin-1 GP01/02 PMS for band λ after BRDF correction, $R_\lambda(\theta_s, \theta_v, \phi)$ denotes the reflectance of the Jilin-1 GP01/02 PMS for band λ under the actual observation geometry, and C_λ signifies the correction coefficient applied to nadir reflectance to account for the differences in the actual observation direction. $f_{iso}(\lambda)$, $f_{vol}(\lambda)$, and $f_{geo}(\lambda)$ are derived from the MCD43A1 BRDF parameters, and θ_s, θ_v, and ϕ describe the observational geometry pertinent to the Jilin -1 GP01/02 PMS.

3.3. AFX Method

The AFX is defined as the ratio of the white sky albedo (WSA) to the isotropic parameter f_{iso}. The formulation of the model is as follows [26]:

$$AFX(\lambda) = \frac{WSA}{f_{iso}(\lambda)} = 1 + \frac{f_{vol}(\lambda)}{f_{iso}(\lambda)} H_{\text{ker\_}vol} + \frac{f_{geo}(\lambda)}{f_{iso}(\lambda)} H_{\text{ker\_}geo} \tag{4}$$

where, $H_{\text{ker\_}vol}$ and $H_{\text{ker\_}geo}$ are the bi-hemispherical integrals of the volumetric and geometric-optical kernels; here they are 0.189184 and −1.377622, respectively.

The AFX is utilized to discern the general characteristics of surface reflectance anisotropy, as outlined in [43]. Based on this attribute, six distinct classes of spectral BRDF prototypes were categorized. To mitigate the discrepancies in spectral reflectance amplitudes across various surfaces, the raw MODIS BRDF parameters were subjected to normalization. This process involved multiplying the parameters by a scaling factor K, where $K = 0.5/f_{iso}$. Table 4 shows the AFX ranges and the corresponding normalized BRDF parameters for the six spectral archetypes, specifically within the green, red, and near-infrared bands.

Table 4. AFX ranges of six BRDF archetypes in normalized (F_{iso}, F_{vol}, and F_{geo}) forms in the MODIS green, red, and NIR bands.

Band	Archetype Number	AFX Range	F_{iso}	F_{vol}	F_{geo}
Green	1	(0.381, 0.714)	0.5	0.0519	0.1384
	2	(0.714, 0.834)	0.5	0.1906	0.1055
	3	(0.834, 0.920)	0.5	0.2700	0.0806
	4	(0.920, 1.012)	0.5	0.3240	0.0587
	5	(1.012, 1.160)	0.5	0.4337	0.0329
	6	(1.160, 1.717)	0.5	0.7446	0.0072
Red	1	(0.382, 0.680)	0.5	0.0288	0.1426
	2	(0.680, 0.795)	0.5	0.1282	0.1134
	3	(0.795, 0.899)	0.5	0.2029	0.0845
	4	(0.899, 1.026)	0.5	0.3082	0.0585
	5	(1.026, 1.240)	0.5	0.4826	0.0274
	6	(1.240, 1.946)	0.5	1.0859	0.0088
NIR	1	(0.541, 0.804)	0.5	0.1218	0.1096
	2	(0.804, 0.896)	0.5	0.2377	0.0860
	3	(0.896, 0.966)	0.5	0.3135	0.0679
	4	(0.966, 1.042)	0.5	0.3521	0.0477
	5	(1.042, 1.142)	0.5	0.4321	0.0262
	6	(1.142, 1.361)	0.5	0.5657	0.0040

Utilizing the AFX method, the angular correction for the Jilin-1 GP01/02 PMS data was executed. Initially, the BRDF parameters derived from the MCD43A1 dataset were resampled to resolutions of 5 m. Subsequently, the AFX values were computed in accordance with Equation (4). The normalized BRDF parameters, F_{iso}, F_{vol}, and F_{geo} were ascertained based on the distribution of AFX values. Ultimately, the Jilin-1 GP01/02 PMS data were adjusted using these normalized BRDF parameters. The formula for the correction is presented as follows:

$$R_\lambda(\theta_s, 0, \varnothing) = R_\lambda(\theta_s, \theta_v, \varnothing) \times \frac{F_{iso}(\lambda) + F_{vol}(\lambda)K_{vol}(\theta_s, 0, \varnothing) + F_{geo}(\lambda)K_{geo}(\theta_s, 0, \varnothing)}{F_{iso}(\lambda) + F_{vol}(\lambda)K_{vol}(\theta_s, \theta_v, \varnothing) + F_{geo}(\lambda)K_{geo}(\theta_s, \theta_v, \varnothing)} \tag{5}$$

3.4. VJB BRDF Method

A novel BRDF decomposition algorithm, predicated on NDVI has been introduced for the estimation and rectification of the directional effect on satellite-derived surface reflectance time series [24]. This approach quantifies the BRDF profile through two distinct parameters: surface scattering (denoted as R) and volumetric scattering (denoted as V).

$$R^N(45,0,0) = R(\theta_s,\theta_v,\varnothing) \times \frac{1 + VK_{vol}(45,0,0) + RK_{geo}(45,0,0)}{1 + VK_{vol}(\theta_s,\theta_v,\varnothing) + RK_{geo}(\theta_s,\theta_v,\varnothing)} \quad (6)$$

The temporal variations in the parameters V and R demonstrated a significant correlation with the fluctuations in the NDVI across vegetated regions [24]. Both V and R were delineated as linear functions of the coefficients V_0, V_1, R_0, and R_1, respectively, each in conjunction with NDV. The aforementioned coefficients can be derived from the MCD43A BRDF parameters, which were resampled to a resolution of 5 m, in conjunction with the NDVI values observed from the Jilin-1 GP01/02 PMS.

3.5. Fixed Parameters Method

Table 5 delineates this set of fixed BRDF model parameters for the green, red, and near-infrared (NIR) bands of the Landsat satellites. Given that the Jilin-1 GP01/02 satellites share identical band settings with both Landsat and Sentinel-2 satellites, it was feasible to apply this same set of fixed parameters for related calculations.

Table 5. The fixed BRDF parameters for Landsat TM green, red, and NIR bands.

Landsat Band	f_{iso}	f_{vol}	f_{geo}
2 Green	0.1306	0.0580	0.0178
3 Red	0.1690	0.0574	0.0227
4 NIR	0.3093	0.1535	0.0330

3.6. Development of the Fixed-Jilin-1 Method

Although the Jilin-1 GP01/02 PMS and Landsat 5/7 TM share identical band configurations, discrepancies were observed in their spectral response functions, particularly within the red and NIR bands, Figure 3 illustrates the comparative spectral response curves of these two sensors. Furthermore, in contrast to the Landsat TM, which is limited to imaging at the subsatellite point, the Jilin-1 GP01/02 PMS is capable of a more expansive side-swing angle, extending up to ±45°. Consequently, the fixed parameters originally derived for the Landsat TM may not be directly applicable to the Jilin-1 GP01/02 PMS. It is thus imperative to develop a distinct set of fixed parameters that are tailored for the Jilin-1 GP01/02 PMS. In this study, we produced normalized BRDF parameters for the Jilin-1 GP01/02 PMS, employing both the AFX and fixed parameter methodologies. The procedure is outlined in the following steps: (1) AFXs of each band of the Jilin-1 GP01/02 PMS were calculated based on the Landsat TM's fixed parameters; (2) the normalized BRDF parameters (F_{iso}, F_{vol}, and F_{geo}) were determined according to the AFX range of the calculated AFX values of each band in Table 4; and (3) the NBAR of the Jilin-1 GP01/02 PMS was generated by bringing normalized BRDF parameters into the RTLSR BRDF model. Table 6 presents the fixed AFX values and the corresponding normalized BRDF parameters for the green, red, and NIR bands of the Jilin-1 GP01/02 PMS, as calculated by employing the proposed Fixed-Jilin-1 method.

Table 6. AFXs and normalized BRDF parameters for green, red, and NIR bands of Jilin-1 GP01/02 PMS.

Jilin-1 GP01/02 PMS Band	AFX	F_{iso}	F_{vol}	F_{geo}
4 Green	0.8963	0.5	0.2700	0.0806
5 Red	0.8792	0.5	0.2029	0.0845
6 NIR	0.9469	0.5	0.3135	0.0679

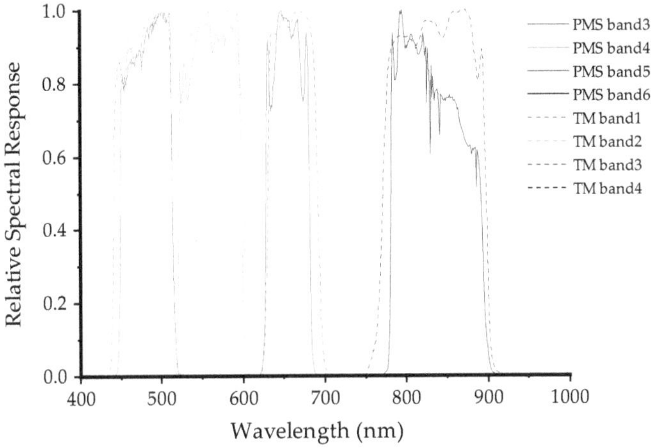

Figure 3. Spectral response functions of Jilin-1 GP01/02 PMS and Landsat TM in visible to near-infrared bands.

3.7. Vegetation Index Calculation Method

In the present study, empirical statistical regression models for the assessment of the crop LAI, FVC, and chlorophyll were developed utilizing a range of vegetation indices. Table 7 shows the vegetation indices selected for this analysis encompass the RVI [44], the NDVI [45], the Green NDVI (GNDVI) [46], the Green Chlorophyll Vegetation Index (GCVI) [31], EVI2 [47], the Modified Soil-Adjusted Vegetation Index (MSAVI) [48], and the Modified Triangular Vegetation Index 2 (MTVI2) [49].

Table 7. Vegetation indices used in this research (NIR, red, and green correspond to Jilin-1 GP01/02 PMS band 4, band 3, and band 2, respectively).

Vegetation Index	Equation
Ratio Vegetation Index (RVI)	$\frac{NIR}{Red}$
Normalized Difference Vegetation Index (NDVI)	$\frac{NIR-Red}{NIR+Red}$
Green NDVI (GNDVI)	$\frac{NIR-Green}{NIR+Green}$
Green Chlorophyll Vegetation Index (GCVI)	$\frac{NIR}{Green}-1$
Enhanced Vegetation Index 2 (EVI2)	$\frac{2.5(NIR-Red)}{NIR+2.4Red+1}$
Modified Soil-Adjusted Vegetation Index (MSAVI)	$\frac{2NIR+1-\sqrt{(2NIR+1)^2-8(NIR-Red)}}{2}$
Modified Triangular Vegetation Index 2 (MTVI2)	$\frac{1.5[1.2(NIR-Green)-2.5(Red-Green)]}{\sqrt{(2NIR+1)^2-(6NIR-5\sqrt{Red})-0.5}}$

A total of 315 in-situ measurements of the LAI, FVC, and chlorophyll (T850 nm/T720 nm) were conducted across corn, rice, and sunflower fields situated in the Inner Mongolia Autonomous Region. The dataset was apportioned, with 80% of the measurements allocated for model development and the remaining 20% reserved for validation purposes. The LAI estimation based on the RVI exhibited a root-mean-square error (RMSE) of 0.73 (determination (R^2) = 0.84, Table 8). In comparison, the RMSE for LAI estimations based on other vegetation indices (VIs) varied from 1.33 to 2.48, contingent upon the VI utilized (R^2 = 0.32–0.67, Figure 4). The FVC estimation derived from MTVI2 demonstrated an RMSE of 6.9% (R^2 = 0.81, Table 8), whereas the RMSE for other VI-based FVC estimations spanned from 12.1% to 32.4%, dependent on the VI in question (R^2 = 0.38–0.78, Figure 4). Furthermore, the chlorophyll (T850 nm/T720 nm) estimation predicated on the GCVI had an

RMSE of 0.15 (R^2 = 0.85, Table 8), and the RMSE for chlorophyll estimations based on other VIs ranged from 0.23 to 0.67, with variation attributed to the VI selected (R^2 = 0.45–0.73, Figure 4).

Table 8. Regression analysis between VIs and the LAI, FVC, chlorophyll (T850 nm/T720 nm).

Crop Parameters	Regression	R^2	RMSE
LAI	$LAI = e^{\frac{RVI-2.278}{5.289}}$	0.84	0.73
FVC	$FVC = 0.028 \times e^{4.52 \times MTVI2}$	0.81	6.9%
chlorophyll (T850 nm/T720 nm)	$CHL = e^{(0.26 \times GCVI-1)}$	0.85	0.15

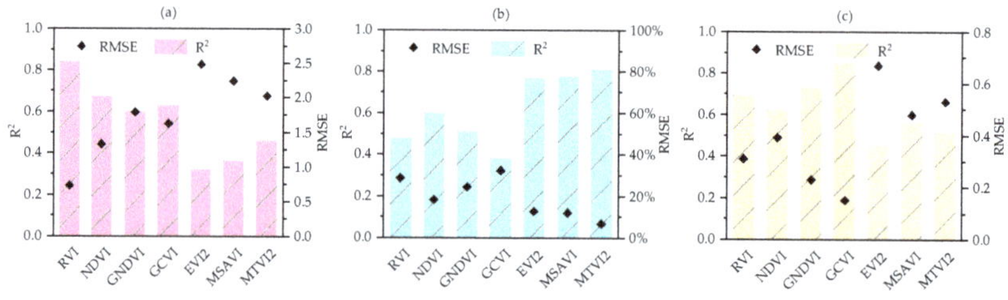

Figure 4. VIs-based LAI, FVC, and chlorophyll (T850 nm/T720 nm) estimations. The *X*-axis is the vegetation index, the left *Y*-axis is the R^2 value, and the right *Y*-axis is the RMSE value. (**a**) LAI, (**b**) FVC, and (**c**) chlorophyll (T850 nm/T720 nm).

4. Results

4.1. Jilin-1 GP01/02 PMS Angle Effect Correction Validation

The coefficient C_λ was employed to assess the efficacy of the angle effect correction across five distinct methodologies for the green, red, and near-infrared bands of the Jilin-1 GP01/02 PMS. Table 9 presents the mean and standard deviation values of C_λ for these five correction methods. Analysis of the data reveals that the average C_λ values obtained through the VJB method are consistently lower than those of the other four approaches, particularly when $\theta_s > 40°$. Under such conditions, the maximum discrepancy in C_λ values between the VJB method and the other algorithms can extend to -0.13, with the standard deviation being markedly higher—by several folds or even tens of folds—compared to the other four methods. This observation suggests a potential overcorrection issue associated with the VJB method.

Excluding the Fixed-Landsat method, wherein the mean C_λ values for the red band surpass 1, the remaining methods all yield values that are less than 1. This contrast implies that the fixed parameters optimized for the Landsat red band are not directly transferable for correcting the red band of the Jilin-1 GP01/02 PMS. Consequently, subsequent analyses will be restricted to the MODIS-based, AFX, and Fixed-Jilin-1 methods. As θ_s and θ_v increase for these three methods, there is a corresponding decrease in the mean C_λ value and an increase in the standard deviation. Notably, for the MODIS-based method, the mean C_λ value across the green, red, and NIR bands drops by more than 5.0%, 5.8%, and 1.8%, respectively, when $\theta_s > 40°$ as opposed to $\theta_s \approx 20°$. Furthermore, the C_λ values for both the MODIS-based and AFX methods in the NIR band are considerably higher than those in the green and red bands. In contrast, the Fixed-Jilin-1 method demonstrates greater stability, with C_λ standard deviation values that are markedly lower across all bands when compared to the other two methods.

To assess the precision of the angular correction effect across various methods, we conducted a verification using the nadir reflectance data from the Jilin-1 GP01/02 PMS, which were cross-referenced with the simulated green, red, and NIR reflectance values derived from measured spectral data. The reflectance correction accuracy for each method is depicted in Figures 5–7 for the green, red, and NIR bands, respectively. Post-correction reflectance values were found to be more closely aligned with actual values, as evidenced by higher coefficients of R^2, and a reduced root mean square error (RMSE) and mean relative error (MRE), compared to pre-correction values.

Table 9. Correction coefficient (C_λ) of the Jilin-1 GP01/02 PMS in the green, red, and NIR bands under five methods.

		MODIS-Based		AFX		VJB		Fixed-Landsat		Fixed-Jilin-1	
		Mean	Std	Mean	Std	Mean	Std	Mean	Std	Mean	Std
$C_\lambda(\theta_s = 20.33°$,	Green	0.960	0.008	0.958	0.006	0.855	0.042	0.964	0.003	0.957	0.004
$\theta_v = 7.93°$,	Red	0.961	0.010	0.957	0.007	0.848	0.056	1.031	0.002	0.959	0.003
$\phi = 19.73°)$	NIR	0.973	0.006	0.964	0.006	0.915	0.030	0.968	0.003	0.959	0.003
$C_\lambda(\theta_s = 20.06°$,	Green	0.977	0.003	0.972	0.001	0.848	0.034	0.977	0.000	0.972	0.000
$\theta_v = 16.12°$,	Red	0.980	0.005	0.974	0.001	0.836	0.046	1.044	0.000	0.975	0.000
$\phi = 56.17°)$	NIR	0.978	0.004	0.973	0.001	0.913	0.035	0.978	0.000	0.972	0.000
$C_\lambda(\theta_s = 28.31°$,	Green	0.955	0.012	0.949	0.010	0.909	0.033	0.964	0.000	0.956	0.000
$\theta_v = 17.49°$,	Red	0.954	0.015	0.943	0.022	0.901	0.045	1.061	0.001	0.961	0.000
$\phi = 57.41°)$	NIR	0.974	0.007	0.957	0.002	0.947	0.017	0.966	0.000	0.956	0.000
$C_\lambda(\theta_s = 36.53°$,	Green	0.956	0.008	0.959	0.007	0.901	0.017	0.965	0.000	0.957	0.000
$\theta_v = 23.85°$,	Red	0.962	0.013	0.964	0.009	0.905	0.022	1.102	0.000	0.964	0.000
$\phi = 66.24°)$	NIR	0.952	0.006	0.951	0.008	0.924	0.017	0.965	0.000	0.955	0.001
$C_\lambda(\theta_s = 37.73°$,	Green	0.940	0.010	0.942	0.007	0.915	0.015	0.950	0.002	0.939	0.003
$\theta_v = 8.27°$,	Red	0.938	0.013	0.935	0.018	0.913	0.019	1.091	0.003	0.944	0.003
$\phi = 21.20°)$	NIR	0.961	0.007	0.945	0.006	0.942	0.019	0.955	0.002	0.941	0.003
$C_\lambda(\theta_s = 44.58°$,	Green	0.928	0.015	0.940	0.009	0.925	0.015	0.951	0.006	0.939	0.007
$\theta_v = 10.16°$,	Red	0.923	0.016	0.942	0.009	0.921	0.016	1.124	0.006	0.945	0.006
$\phi = 43.51°)$	NIR	0.961	0.011	0.944	0.007	0.960	0.011	0.955	0.005	0.941	0.007

Similar to the findings regarding the C_λ values, the VJB method exhibited the poorest correlation with the true reflectance, indicated by the lowest R^2. The Fixed-Landsat method appears to have exacerbated the impact of the angular effect on the red band, resulting in a corrected reflectance that deviated more significantly from the actual value than before the correction. In contrast, the Fixed-Jilin-1 method demonstrated superior correction accuracy, with R^2 values exceeding 0.85 for all three bands and a reduction in the MRE of over 2%.

For smaller viewing and solar angles ($\theta_s < 30°$ and $\theta_v < 20°$), the impact of the angular effect on reflectance was minimal, and the correction outcomes from the four methods were largely comparable, with the exception of the Fixed-Landsat method's performance in the red band. However, as angles increased ($\theta_s > 30°$ and $\theta_v > 20°$), the reflectance values corrected by the four methods converged, with the angular effect becoming more pronounced and leading to varying correction outcomes among the methods. Overall, the comparative effectiveness of the five methods for angular correction on Jilin-1 GP01/02 PMS data can be ranked as follows: Fixed-Jilin-1, AFX, MODIS-based, Fixed-Landsat, and VJB.

4.2. Crop Parameters Retrieval Validation

4.2.1. LAI Retrieval Validation

The LAI was retrieved from both the original and nadir reflectance measurements using a regression formula based on the RVI, followed by the application of five distinct methods for correcting the angle effect. Subsequently, the accuracy of the retrieved LAI was validated against field survey data. As depicted in Figure 8, except for the Fixed-Landsat algorithm, which significantly underestimated the LAI extraction results, the retrieval accuracy from the remaining four methods saw a marked enhancement. Specifically, these methods demonstrated coefficients of R^2 greater than or equal to 0.81 and RMSE less than or equal to 0.62, which are substantial improvements over the original reflectance metrics (R^2 = 0.76 and RMSE = 0.735). Notably, the Fixed-Jilin-1 method achieved the highest accuracy in LAI retrieval (R^2 = 0.82 and RMSE = 0.539). Additionally, the MODIS-based and AFX methods exhibited very similar levels of accuracy in LAI retrieval.

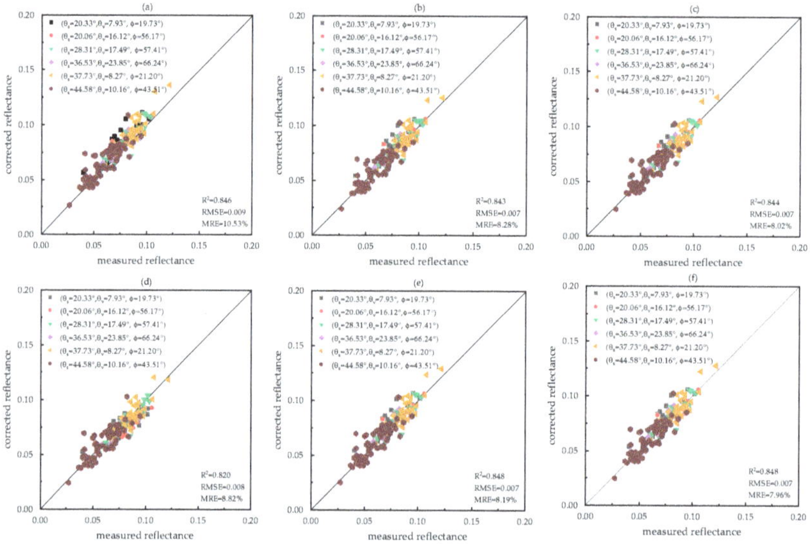

Figure 5. Green band reflectance validation based on field measured data. The *X*-axis is the measured reflectance, the *Y*-axis is the corrected reflectance, the black solid line is the 1:1 line, and the different color scatter points represent the different observation geometry. (**a**) Original, (**b**) MODIS-based, (**c**) AFX, (**d**) VJB, (**e**) Fixed-Landsat, And (**f**) Fixed-Jilin-1.

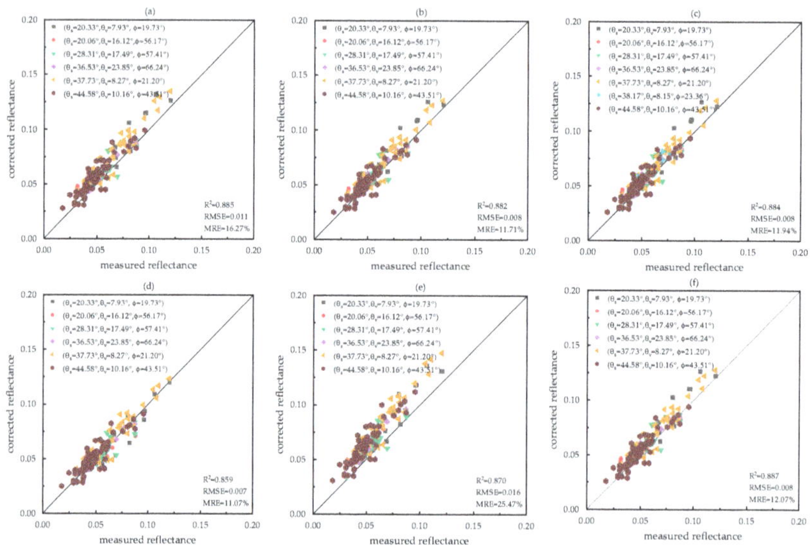

Figure 6. Red band reflectance validation based on field measured data. The *X*-axis is the measured reflectance, the *Y*-axis is the corrected reflectance, the black solid line is the 1:1 line, and the different color scatter points represent the different observation geometry. (**a**) Original, (**b**) MODIS-based, (**c**) AFX, (**d**) VJB, (**e**) Fixed-Landsat, And (**f**) Fixed-Jilin-1.

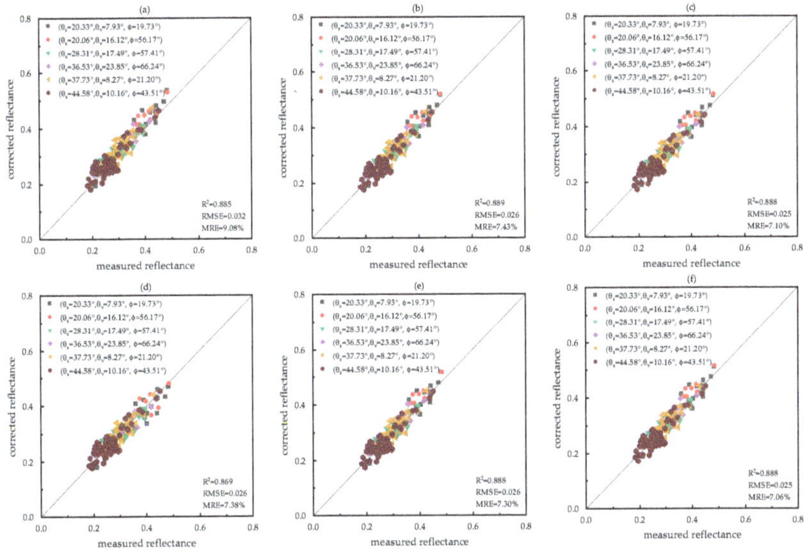

Figure 7. NIR band reflectance validation based on field measured data. The *X*-axis is the measured reflectance, the *Y*-axis is the corrected reflectance, the black solid line is the 1:1 line, and the different color scatter points represent the different observation geometry. (**a**) Original, (**b**) MODIS-based, (**c**) AFX, (**d**) VJB, (**e**) Fixed-Landsat, And (**f**) Fixed-Jilin-1.

4.2.2. FVC Retrieval Validation

Figure 9 presents the accuracy of FVC retrieval using various methods for angular correction. The performance of FVC retrieval surpassed that of LAI retrieval, with the coefficient of R^2 for FVC retrieval from the original reflectance already exceeding 0.8. Notably, the Fixed-Jilin-1 method significantly enhanced the accuracy of FVC retrieval when compared to the original reflectance. Specifically, the R^2 value rose from 0.8 to 0.87, the RMSE decreased from 9.54% to 7.58%, and the MRE reduced by over 35%. However, an underestimation of FVC was observed with the other four methods, despite an improvement in the R^2 value. Particularly, the AFX method exhibited the most pronounced underestimation, with an RMSE reaching 11.55%. When the crop coverage is low (FVC < 50%), the influence of the angular effect on FVC retrieval is small, and the application of the five angular correction methods did not lead to a significant improvement in the retrieval accuracy of FVC. Conversely, for higher crop coverage (FVC > 50%), the accuracy of FVC retrieval was markedly enhanced following angular correction.

4.2.3. Chlorophyll Retrieval Validation

Figure 10 illustrates the accuracy of chlorophyll retrieval, represented by the ratio T850 nm/T720 nm and based on the GCVI, utilizing various angular correction methods. When examining the overall chlorophyll retrieval accuracy across the five correction methods, it is observed that, apart from the VJB method, the distinctions among the remaining methods were not pronounced. Specifically, the variance in R^2 and RMSE among these four methods was minimal, with differences of only 0.003 and 0.01, respectively. The VJB method yielded the poorest accuracy for chlorophyll retrieval, which is even lower than that achieved using the original reflectance data. In contrast, the Fixed-Jilin-1 method outperformed the others, showing an increase in R^2 of 0.12 and a decrease in RMSE of 0.013 when compared to the chlorophyll retrieval results from the original reflectance.

In general, the angular effect exerted a negligible influence on the accuracy of chlorophyll extraction. However, when the chlorophyll ratio (T850 nm/T720 nm) exceeded 1.2, the chlorophyll estimates derived from the original reflectance tended to be higher than the measured values. This discrepancy was significantly mitigated following angular correction, particularly with the Fixed-Jilin-1 method.

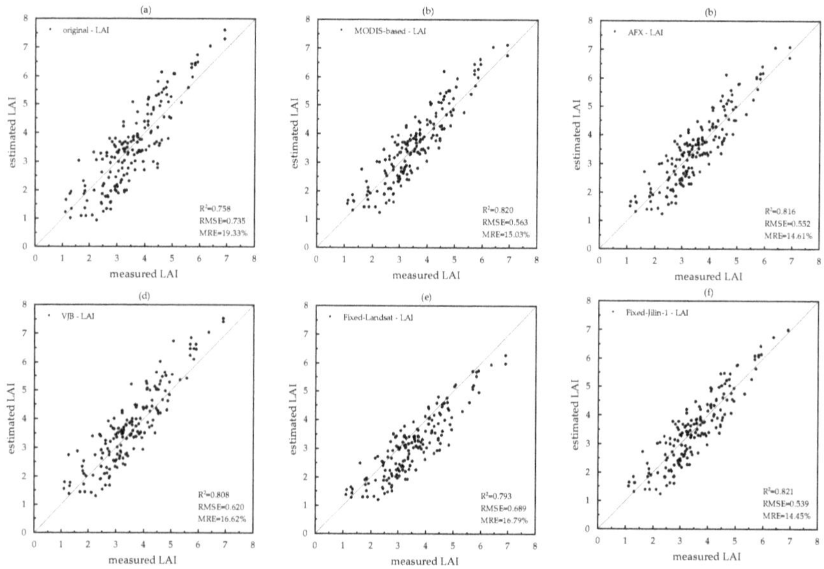

Figure 8. LAI validation based on field measured data. The *X*-axis is the measured reflectance, the *Y*-axis is the corrected reflectance, and the black solid line is the 1:1 line. (**a**) Original, (**b**) MODIS-based, (**c**) AFX, (**d**) VJB, (**e**) Fixed-Landsat, and (**f**) Fixed-Jilin-1.

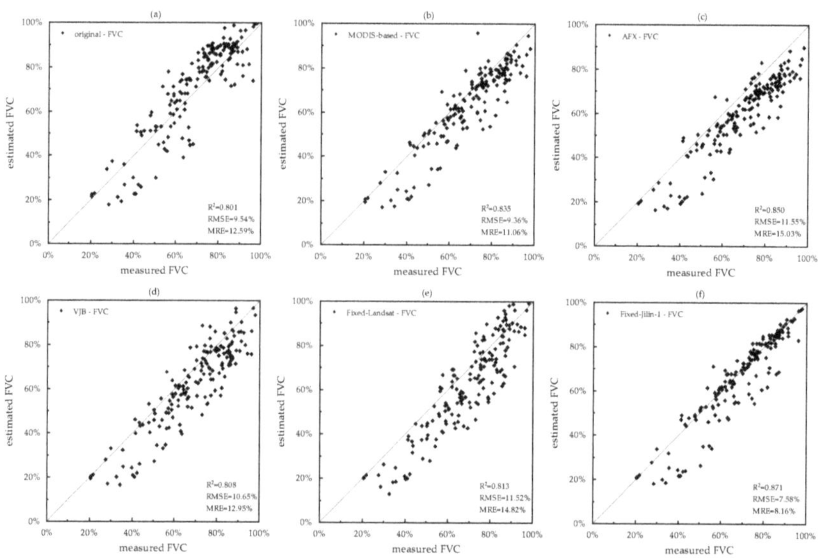

Figure 9. FVC validation based on field-measured data. The *X*-axis is the measured reflectance, the *Y*-axis is the corrected reflectance, and the black solid line is the 1:1 line. (**a**) Original, (**b**) MODIS-based, (**c**) AFX, (**d**) VJB, (**e**) Fixed-Landsat, and (**f**) Fixed-Jilin-1.

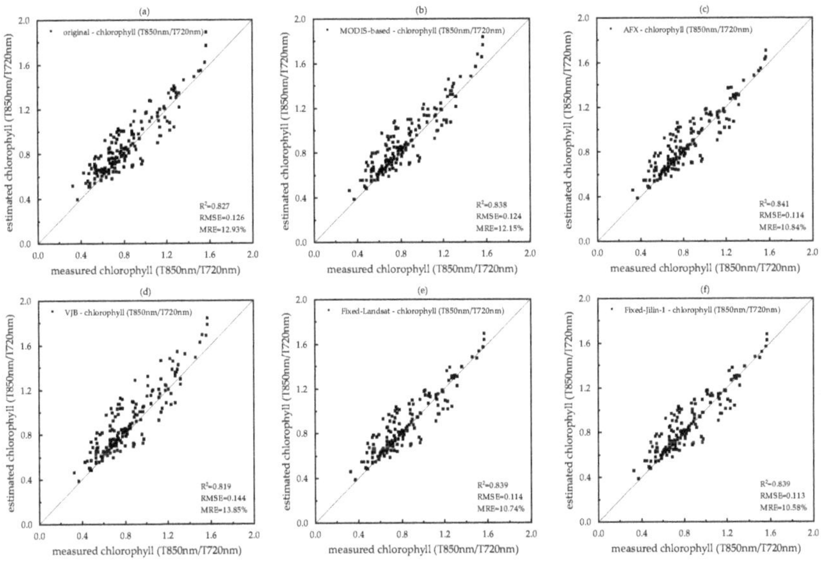

Figure 10. Chlorophyll (T850 nm/T720 nm) validation based on field measured data. The X-axis is the measured reflectance, the Y-axis is the corrected reflectance, and the black solid line is the 1:1 line. (**a**) Original, (**b**) MODIS-based, (**c**) AFX, (**d**) VJB, (**e**) Fixed-Landsat, and (**f**) Fixed-Jilin-1.

In summary, aside from the VJB method, the remaining four methods for angular correction were shown to enhance the retrieval accuracy of the LAI, FV, and chlorophyll content. The MODIS-based, AFX, and Fixed-Landsat methods exhibited comparable performance in improving the retrieval accuracy of these crop parameters. Notably, the Fixed-Jilin-1 method demonstrated superior accuracy in retrieving the LAI, FVC, and CHL across the board.

Furthermore, the retrieval of the LAI, FVC, and CHL exhibited varying sensitivities to the angle effect. This effect also influenced the retrieval accuracy of crop parameters differently, depending on the specific growth stage of the crops.

5. Discussion

5.1. Novelty of the Fixed-Jilin-1 Angle Effect Correction Method

In this study, we introduced a novel method for correcting the angular effects inherent in high-resolution, side-view satellite imagery data. The accuracy of this proposed method was substantiated through validation against measured crop reflectance data, demonstrating superior performance compared to other prevalent methodologies. Extensive research, including the present study, has confirmed the applicability of MODIS BRDF parameters and the fixed BRDF parameters of Landsat satellites for angular correction across a spectrum of satellite sensors, thereby highlighting the universality of these parameters. However, variations in sensor responsiveness, imaging conditions, observation geometry, and spatial resolution necessitate further refinements to these ostensibly universal approaches. This is particularly pertinent for satellites capable of flexible side-swing imaging, which exhibit more dynamic observation geometries, thus necessitating more specialized angular correction methodologies.

Similar to the results of Jiang et al. [20], the VJB method does not show better accuracy than other methods, which is different from the results of Claverie et al. [22]. Both this paper and Jiang et al. [20] focus on crops, while Claverie et al. [22] focus on semi-desert biome, so it can be inferred that the VJB method is greatly affected by the type of land cover. Additionally, the fixed BRDF parameters of Landsat satellites, when compared to other methods, exhibits marked discrepancies in the red band. This discrepancy is primarily attributed to the broader bandwidth of the Jilin-1 GP01/02 red band in contrast to that of the Landsat TM red band (50 nm versus 30 nm), coupled with a wider viewing angle (±45° side-view imaging versus sub-satellite point imaging).

The AFX method achieves superior accuracy in correcting sensor angle effects compared to the MODIS parameter-based approach. This enhanced precision is attributed to AFX's capacity to limit scale differences in BRDF profiles to a significant degree. At low viewing angles, the stability of the Landsat-based fixed parameters surpasses that of the MODIS method in angle correction for Jilin-1 GP01/02 PMS data, due to their reduced sensitivity to land cover type, condition, or surface disturbance. However, the fixed BRDF parameters falter in accurately depicting BRDF variations at high viewing angles. The AFX method compensates by incorporating surface variation data, suggesting a complementary application with the fixed parameter method under large observation angles. Consequently, the proposed Fixed-Jilin-1 method integrates the strengths of both the AFX and Landsat fixed parameter methods, employing fixed BRDF parameters that are more universally applicable and leveraging AFX to diminish the impact of spatial scale discrepancies. Overall, the proposed method corrects angular effects in Jilin-1 GP01/02 PMS data more effectively, offering a distinct advantage particularly at larger zenith angles.

5.2. Impact of Angle Effect Correction on Vegetation Parameters Retrieval

Given the high spatial heterogeneity of agricultural fields and the rapid fluctuations in crop growth, crop parameters vary considerably across different growth phases, leading to a differential impact of the angular effect on various crop parameters. During the middle and late stages of crop growth, when plant density is elevated, the inversion accuracy for the LAI is markedly lower compared to that for the FVC and CHL. This is partly due to the differential response of the LAI to red band saturation as opposed to the FVC and CHL, a phenomenon also observed by Jiang et al. [47]. Furthermore, the angular effect differentially influences the reflectance of distinct spectral bands [50], resulting in a pronounced variation in the impact of the angular effect on various vegetation indices. Consequently, the inversion algorithms for crop parameters based on these indices will also exhibit variations.

The observational data from the Jilin-1 GP01/02 satellites, which were synchronized with the measured data in this study, are predominantly backscattered, leading to a lower reflectivity post-correction compared to pre-correction. The binomial reflection from vegetation is a complex process involving the transmission of canopy and leaf radiation, which is intricately linked to the type of vegetation, growth phase, and incident light conditions. The data collected in this study pertain to the middle and late stages of crop growth, thus not delving deeply into the more complex, mixed radiation characteristics of the vegetation–soil interface observed during the early stages of crop development.

While the angular effect correction outcomes for the Jilin-1 GP01/02 PMS presented herein are promising, there are inherent limitations. The correction method introduced in this study has been tested solely within the mid-latitudes of the Northern Hemisphere, indicating a need for future research to evaluate the method's efficacy across a broader range of geographical and environmental contexts. Furthermore, the Jilin-1 GP01/02 PMS encompasses a multitude of narrowband visible light, red edge, and near-infrared bands in addition to the green, red, and near-infrared bands. Future studies will also explore BRDF correction methodologies for these bands to fully leverage the high-resolution and multispectral capabilities of this satellite imagery.

6. Conclusions

This paper assesses the impact of various angular correction methods on the data acquired from the high-resolution, flexible side-view imaging satellite, Jilin-1 GP01/02. We have developed a correction method specifically tailored for the Jilin-1 GP01/02 PMS to enhance the precision of crop parameter extraction. Moreover, this study employed ground-measured data to validate the accuracy of LAI, FV, and CHL retrieval using both pre- and post-correction reflectance. Compared with commonly used angular effect correction methods, the method introduced in this study, termed the Fixed-Jilin-1 method, effectively mitigates the influence of observation geometry on data accuracy. The findings indicate that the Fixed-Jilin-1 method markedly enhances the precision of crop parameter extraction, followed by the AFX method, the MODIS-based method, the Fixed-Landsat method, and the VJB method, respectively. Notably, the retrieval accuracy for the LAI and FVC saw an improvement of over 25%.

Thus, by addressing the angular effect in Jilin-1 GP01/02 PMS data through the Fixed-Jilin-1 method, significant strides are made towards bolstering the inversion accuracy of agricultural vegetation parameters. This approach holds substantial promise for the monitoring of crops using CubeSat constellations and may pave the way for a variety of novel applications within the realm of precision agriculture.

Author Contributions: Conceptualization, J.G. and X.Z.; methodology, J.G.; validation, J.G. and Z.X.; formal analysis, J.G. and R.Z.; investigation, Z.X.; resources, D.W.; data curation, Z.X.; writing—original draft preparation, J.G.; writing—review and editing, R.Z. and J.Y.; visualization, J.Y.; supervision, D.W.; project administration, R.Z.; funding acquisition, D.W. All authors have read and agreed to the published version of the manuscript.

Funding: This research was funded by the China Postdoctoral Science Foundation, grant number 2022M720538, and the Major Science and Technology Program of Jilin Province, grant number YDZJ202303CGZH023.

Data Availability Statement: Restrictions apply to the availability of these data.

Acknowledgments: We thank the Inner Mongolia Autonomous Region Surveying and Mapping Geographic Information Center for supporting the field survey work. We are also grateful to the MODIS team for providing BRDF products that are available online. We also would like to thank the reviewers for their helpful comments.

Conflicts of Interest: Authors Jialong Gong, Xing Zhong, Ruifei Zhu, Zhaoxin Xu, Dong Wang were employed by the company Chang Guang Satellite Technology Co., Ltd. The remaining authors declare that the research was conducted in the absence of any commercial or financial relationships that could be construed as a potential conflict of interest.

References

1. Moran, M.S.; Inoue, Y.; Barnes, E.M. Opportunities and limitations for image-based remote sensing in precision crop management. *Remote Sens. Environ.* **1997**, *61*, 319–346. [CrossRef]
2. Zhang, N.; Wang, M.; Wang, N. Precision agriculture—A worldwide overview. *Comput. Electron. Agric.* **2002**, *36*, 113–132. [CrossRef]
3. Sishodia, R.P.; Ray, R.L.; Singh, S.K. Applications of Remote Sensing in Precision Agriculture: A Review. *Remote Sens.* **2020**, *12*, 3136. [CrossRef]
4. Atzberger, C. Advances in Remote Sensing of Agriculture: Context Description, Existing Operational Monitoring Systems and Major Information Needs. *Remote Sens.* **2013**, *5*, 949–981. [CrossRef]
5. Revill, A.; Myrgiotis, V.; Florence, A.; Hoad, S.; Williams, M. Combining Process Modelling and LAI Observations to Diagnose Winter Wheat Nitrogen Status and Forecast Yield. *Agronomy* **2021**, *11*, 314. [CrossRef]
6. Xie, Q.; Dash, J.; Huete, A.; Jiang, A.; Yin, G.; Ding, Y.; Peng, D.; Hall, C.C.; Brown, L.; Shi, Y.; et al. Retrieval of crop biophysical parameters from Sentinel-2 remote sensing imagery. *Int. J. Appl. Earth Obs. Geoinf.* **2019**, *80*, 187–195. [CrossRef]
7. Jiménez-Berni, J.A.; Zarco-Tejada, P.J.; Suárez, L.; Fereres, E. Thermal and Narrowband Multispectral Remote Sensing for Vegetation Monitoring From an Unmanned Aerial Vehicle. *IEEE Trans. Geosci. Remote Sens.* **2009**, *47*, 722–738. [CrossRef]
8. Houborg, R.; McCabe, M. High-Resolution NDVI from Planet's Constellation of Earth Observing Nano-Satellites: A New Data Source for Precision Agriculture. *Remote Sens.* **2016**, *8*, 768. [CrossRef]
9. Revel, C.; Lonjou, V.; Marcq, S.; Desjardins, C.; Fougnie, B.; Luche, C.C.-D.; Guilleminot, N.; Lacamp, A.-S.; Lourme, E.; Miquel, C.; et al. Sentinel-2A and 2B absolute calibration monitoring. *Eur. J. Remote Sens.* **2019**, *52*, 122–137. [CrossRef]
10. Irons, J.R.; Dwyer, J.L.; Barsi, J.A. The next Landsat satellite: The Landsat data continuity mission. *Remote Sens. Environ.* **2012**, *122*, 11–21. [CrossRef]
11. Ranson, K.J.; Biehl, L.L.; Bauer, M.E. Sun-view angle studies of corn and soybean canopies in support of vegetation canopy reflectance modeling. In *AGRISTARS Annual Technical Summary, Report Prepared Undercontract NAS9-16528*; Purdue University: West Lafayette, IN, USA, 1983; pp. 95–116.

12. Roujean, J.L.; Leroy, M.; Deschanps, P.Y. A bidirectional reflectance model of the Earth's surface for the correction of remote sensing data. *J. Geophys. Res. Atmos.* **1992**, *972*, 20455–20468. [CrossRef]
13. Jiao, Z.; Zhang, H.; Dong, Y.; Liu, Q.; Xiao, Q.; Li, X. An Algorithm for Retrieval of Surface Albedo From Small View-Angle Airborne Observations Through the Use of BRDF Archetypes as Prior Knowledge. *IEEE J. Sel. Top. Appl. Earth Obs. Remote Sens.* **2015**, *8*, 3279–3293. [CrossRef]
14. Leigh, L.M.; Mishra, N.; Haque, M.O.; Aaron, D.; Markham, B. Radiomeric cross calibration of Landsat 8 Operatinal Land Imager. *Remote Sens.* **2014**, *6*, 12619–12638.
15. He, T.; Liang, S.; Wang, D.; Cao, Y.; Gao, F.; Yu, Y.; Feng, M. Evaluating land surface albedo estimation from Landsat MSS, TM, ETM +, and OLI data based on the unified direct estimation approach. *Remote Sens. Environ.* **2018**, *204*, 181–196. [CrossRef]
16. Sun, Y.; Wang, B.; Teng, S.; Liu, B.; Zhang, Z.; Li, Y. Continuity of Top-of-Atmosphere, Surface, and Nadir BRDF-Adjusted Reflectance and NDVI between Landsat-8 and Landsat-9 OLI over China Landscape. *Remote Sens.* **2023**, *15*, 4948. [CrossRef]
17. Roy, D.P.; Li, J.; Zhang, H.K.; Yan, L.; Huang, H.; Li, Z. Examination of Sentinel-2A multi-spectral instrument (MSI) reflectance anisotropy and the suitability of a general method to normalize MSI reflectance to nadir BRDF adjusted reflectance. *Remote Sens. Environ.* **2017**, *199*, 25–38. [CrossRef]
18. Roy, D.; Li, Z.; Zhang, H. Adjustment of Sentinel-2 Multi-Spectral Instrument (MSI) Red-Edge Band Reflectance to Nadir BRDF Adjusted Reflectance (NBAR) and Quantification of Red-Edge Band BRDF Effects. *Remote Sens.* **2017**, *9*, 1325. [CrossRef]
19. Liu, L.; Gao, H.; Pan, Z.; Qian, Y.; Han, Q.; Zhang, X.; Shi, T. Satellite-based time series calibration of GF-1 WFV sensors for large view zenith angle observations. *Int. J. Remote Sens.* **2018**, *39*, 8293–8316. [CrossRef]
20. Jiang, H.; Jia, K.; Wang, Q.; Shang, J.; Liu, J.; Xie, X.; Dong, T. Angular Effect Correction for Improved LAI and FVC Retrieval Using GF-1 Wide Field View Data. *IEEE Trans. Geosci. Remote Sens.* **2023**, *61*, 1–14. [CrossRef]
21. Lu, J.; He, T.; Liang, S.; Zhang, Y. An Automatic Radiometric Cross-Calibration Method for Wide-Angle Medium-Resolution Multispectral Satellite Senso Using Landsat Data. *IEEE Trans. Geosci. Remote Sens.* **2022**, *60*, 5604011. [CrossRef]
22. Claverie, M.; Vermote, E.; Franch, B.; He, T.; Hagolle, O.; Kadiri, M.; Masek, J. Evaluation of Medium Spatial Resolution BRDF-Adjustment Techniques Using Multi-Angular SPOT4 (Take5) Acquisitions. *Remote Sens.* **2015**, *7*, 12057–12075. [CrossRef]
23. Roy, D.P.; Zhang, H.K.; Ju, J.; Gomez-Dans, J.L.; Lewis, P.E.; Schaaf, C.B.; Sun, Q.; Li, J.; Huang, H.; Kovalskyy, V. A general method to normalize Landsat reflectance data to nadir BRDF adjusted reflectance. *Remote Sens. Environ.* **2016**, *176*, 255–271. [CrossRef]
24. Vermote, E.; Justice, C.O.; Breon, F.M. Towards a Generalized Approach for Correction of the BRDF Effect in MODIS Directional Reflectances. *IEEE Trans. Geosci. Remote Sens.* **2009**, *47*, 898–908. [CrossRef]
25. Bréon, F.-M.; Vermote, E. Correction of MODIS surface reflectance time series for BRDF effects. *Remote Sens. Environ.* **2012**, *125*, 1–9. [CrossRef]
26. Jiao, Z.; Hill, M.J.; Schaaf, C.B.; Zhang, H.; Wang, Z.; Li, X. An Anisotropic Flat Index (AFX) to derive BRDF archetypes from MODIS. *Remote Sens. Environ.* **2014**, *141*, 168–187. [CrossRef]
27. Delegido, J.; Verrelst, J.; Rivera, J.P.; Ruiz-Verdú, A.; Moreno, J. Brown and green LAI mapping through spectral indices. *Int. J. Appl. Earth Obs. Geoinf.* **2015**, *35*, 350–358. [CrossRef]
28. Fang, H.; Baret, F.; Plummer, S.; Schaepman-Strub, G. An Overview of Global Leaf Area Index (LAI): Methods, Products, Validation, and Applications. *Rev. Geophys.* **2019**, *57*, 739–799. [CrossRef]
29. Kimm, H.; Guan, K.; Jiang, C.; Peng, B.; Luo, Y. Deriving high-spatiotemporal-resolution leaf area index for agroecosystems in the U.S. Corn Belt using Planet Labs CubeSat and STAIR fusion data. *Remote Sens. Environ.* **2020**, *239*, 111615. [CrossRef]
30. Villa, P.; Vecchia, A.D.; Piaser, E.; Bolpagni, R. Assessing PROSPECT performance on aquatic plant leaves. *Remote Sens. Environ.* **2024**, *301*, 113926. [CrossRef]
31. Gitelson, A.A.; Viña, A.; Arkebauer, T.J.; Rundquist, D.C.; Keydan, G.; Leavitt, B. Remote estimation of leaf area index and green leaf biomass in maize canopies. *Geophys. Res. Lett.* **2003**, *30*, 1248. [CrossRef]
32. Liu, S.; Zeng, W.; Wu, L.; Lei, G.; Chen, H.; Gaiser, T.; Srivastava, A.K. Simulating the Leaf Area Index of Rice from Multispectral Images. *Remote Sens.* **2021**, *13*, 3663. [CrossRef]
33. Glenn, E.P.; Huete, A.R.; Nagler, P.L.; Nelson, S.G. Relationship between remotely-sensed vegetation indices, canopy attributes and plant physiological processes What vegetation indices can and cannot tell us about the landscape. *Sensors* **2008**, *8*, 2136–2160. [CrossRef] [PubMed]
34. Ma, Y.; Zhang, Q.; Yi, X.; Ma, L.; Zhang, L.; Huang, C.; Zhang, Z.; Lv, X. Estimation of Cotton Leaf Area Index (LAI) Based on Spectral Transformation and Vegetation Index. *Remote Sens.* **2021**, *14*, 136. [CrossRef]
35. Yang, S.; Huang, S.; Bai, Y.; Jia, Y.; Ba, Q.; Tian, S.; Zhong, X. On-orbit absolute radiometric calibration for the Multi-Spectral Imager of Jilin-1/GP02 based on multiple stable targets. *Remote Sens. Technol. Appl.* **2023**, *38*, 803–815. [CrossRef]
36. Yibo, D.; Ruifei, Z.; Jialong, G.; Dong, W.; Zhong, X. Retrieval of crop Leaf Area Index based on Jilin-1GP image. *Remote Sens. Technol. Appl.* **2023**, *38*, 816–826.
37. Cerovic, Z.G.; Moise, N.; Agati, G.; Latouche, G.; Ghozlen, N.B.; Meyer, S. New portable optical sensors for the assessment of winegrape phenolic maturity based on berry fluorescence. *J. Food Compos. Anal.* **2008**, *21*, 650–654. [CrossRef]
38. Jia, K.; Liang, S.; Gu, X.; Baret, F.; Wei, X.; Wang, X.; Yao, Y.; Yang, L.; Li, Y. Fractional vegetation cover estimation algorithm for Chinese GF-1 wide field view data. *Remote Sens. Environ.* **2016**, *177*, 184–191. [CrossRef]

39. Nicodemus, F.E.; Richmond, J.C.; Hsia, J.J.; Ginsberg, I.W.; Limperis, T. *Geometrical Considerations and Nomenclature for Reflectance;* National Bureau of Standards, US Department of Commerce: Washington, DC, USA, 1977.
40. Schläpfer, D.; Richter, R.; Feingersh, T. Operational BRDF Effects Correction for Wide-Field-of-View Optical Scanners (BREFCOR). *IEEE Trans. Geosci. Remote Sens.* **2014**, *53*, 1855–1964. [CrossRef]
41. Schaaf, C.B.; Gao, F.; Strahler, A.H.; Lucht, W.; Li, X.; Tsang, T.; Strugnell, N.C.; Zhang, X.; Jin, Y.; Muller, J.P. First operational BRDF, albedo nadir reflectance products from MODIS. *Remote Sens. Environ.* **2002**, *83*, 135–148. [CrossRef]
42. Wanner, W.; Ah, S.; Li, X. On the derivation of kernels for kernel-driven models of bidirectional reflectance. *J. Geophys. Res. Biogeosci.* **1995**, *100*, 21077–21089. [CrossRef]
43. Zhang, H.; Jiao, Z.; Dong, Y.; Li, X. Evaluation of BRDF Archetypes for Representing Surface Reflectance Anisotropy Using MODIS BRDF Data. *Remote Sens.* **2015**, *7*, 7826–7845. [CrossRef]
44. Jordan, C.F. Derivation of Leaf-Area Index from Quality of Light on the Forest Floor. *Ecology* **1969**, *50*, 663–666. [CrossRef]
45. Rouse, J.; Haas, R.; Schell, J.; Deering, D. Monitoring vegetation systems in the great plains with ERTS. In *R. Monitoring Vegetation Systems in the Great Plains with ERTS[C]//Third NASA Earth Resources Technology Satellite Symposium;* NASA Goddard Space Flight Center 3d ERTS-1 Symp., Vol. 1, Sect. A; NASA Goddard Space Flight Center: Greenbelt, MD, USA, 1973; pp. 309–317. Available online: https://ntrs.nasa.gov/citations/19740022614 (accessed on 14 May 2024).
46. Gitelson, A.; Merzlyak, M.N. Spectral reflectance changes associated with autumn senescence of Aesculus hippocastanum L. and Acer platanoides L. Leaves. Spectral features and relation to chlorophyll estimation. *J. Plant Physiol.* **1994**, *143*, 286–292. [CrossRef]
47. Jiang, Z.; Huete, A.; Didan, K.; Miura, T. Development of a two-band enhanced vegetation index without a blue band. *Remote Sens. Environ.* **2008**, *112*, 3833–3845. [CrossRef]
48. Qi, J.; Chehbouni, A.; Huete, A.R.; Kerr, Y.H.; Sorooshian, S. A Modified Soil Adjusted Vegetation Index. *Remote Sens. Environ.* **1994**, *48*, 119–126. [CrossRef]
49. Haboudane, D.; Miller, J.R.; Pattey, E.; Zarco-Tejada, P.J.; Strachan, I.B. Hyperspectral vegetation indices and novel algorithms for predicting green LAI of crop canopies: Modeling and validation in the context of precision agriculture. *Remote Sens. Environ.* **2004**, *90*, 337–352. [CrossRef]
50. Gao, F.; He, T.; Masek, J.G.; Shuai, Y.; Schaaf, C.B.; Wang, Z. Angular Effects and Correction for Medium Resolution Sensors to Support Crop Monitoring. *IEEE J. Sel. Top. Appl. Earth Obs. Remote Sens.* **2014**, *7*, 4480–4489. [CrossRef]

remote sensing

MDPI

Article

Comprehensive Assessment of NDVI Products Derived from Fengyun Satellites across China

Lei Wang [1], Xiuzhen Han [2,3,4,*], Shibo Fang [5] and Fengjin Xiao [1]

1 National Climate Center, Beijing 100081, China; wang_lei@cma.gov.cn (L.W.); xiaofj@cma.gov.cn (F.X.)
2 National Satellite Meteorological Center (National Center for Space Weather), Beijing 100081, China
3 Innovation Center for FengYun Meteorological Satellite (FYSIC), Beijing 100081, China
4 Key Laboratory of Radiometric Calibration and Validation for Environmental Satellites/Key Laboratory of Space Weather, China Meteorological Administration, Beijing 100081, China
5 Chinese Academy of Meteorological Sciences, Beijing 100081, China; fangshibo@cma.gov.cn
* Correspondence: hanxz@cma.gov.cn

Abstract: NDVI data are crucial for agricultural and environmental research. The Fengyun-3 (FY-3) series satellites are recognized as primary sources for retrieving NDVI products on a global scale. To apply FY-3 NDVI data for long-term studies, such as climate change, this study conducted a thorough evaluation to detect the potentials of the FY-3B and FY-3D satellites for generating a long time series NDVI dataset. For this purpose, the spatiotemporal consistency between the FY-3B and FY-3D satellites was evaluated, and their performances were compared. Then, a grey relational analysis (GRA) method was applied to detect the factors influencing the consistency among the different satellites, and a gradient boosting regression (GBR) model was constructed to create a long-term FY-3 NDVI product. The results indicate an overall high consistency between the FY-3B and FY-3D NDVIs, suggesting that they could be used as complementary datasets for generating a long-term NDVI dataset. The correlations between the FY-3D NDVI and the MODIS NDVI, as well as the leaf area index (LAI) measurements, were both higher than those of FY-3B, which indicates a better performance of FY-3D in retrieving NDVI data. The grey correlation degrees between the NDVI differences and four parameters, which were land cover (LC), DEM, latitude (LAT) and longitude (LON), were calculated, revealing that the LC was the most related to the NDVI differences. Finally, a GBR model with FY-3B NDVI, LC, DEM, LAT and LON as the input variables and FY-3D NDVI as the target variable was established and achieved a robust performance. The R values between the GBR-estimated NDVI and FY-3D NDVI reached 0.947, 0.867 and 0.829 in the training, testing and validation datasets, respectively, indicating the feasibility of the established model for generating long time series NDVI data by combining data from the FY-3B and FY-3D satellites.

Keywords: Fengyun satellites; NDVI product; assessment; long term; grey relational analysis; gradient boosting regression

1. Introduction

Vegetation coverage and growth conditions are crucial information for estimating global carbon sequestration, investigating vegetation–climate interactions and managing natural resources [1–3]. Over the past decades, a large number of remote sensing vegetation indicators, such as the enhanced vegetation index (EVI), normalized difference vegetation index (NDVI), fractional vegetation cover (FVC) and leaf area index (LAI), have been developed for mapping vegetation dynamics on a regional scale. As one of the most important and useful vegetation parameters, the remote-sensed NDVI has been extensively adopted for forecasting agricultural production, detecting drought intensities and exploring the climate effects on vegetation conditions in many previous studies [4–6]. In general, NDVI products can be derived from surface reflectance in the near-infrared and red wavelengths through several primary sensors, including the advanced very high resolution radiometer (AVHRR) on board the National Oceanic and Atmospheric Administration (NOAA) satellites, the moderate resolution imaging spectroradiometer (MODIS) on board Terra and Aqua, the enhanced thematic mapper plus (ETM+) on board Landsat, the medium resolution spectral imager (MERSI) and the second-generation MERSI (MERSI-II) on board the Fengyun-3 (FY-3) series satellites [7–10].

The FY-3 series satellites, including the FY-3A, FY-3B, FY-3C, FY-3D, FY-3E, FY-3G, FY-3F and additional satellites to be launched in the next few years, are China's second-generation polar-orbiting meteorological satellites. These satellites are equipped with more than ten sets of advanced remote sensing instruments, such as the visible and infrared radiometer (VIRR), global navigation occultation sounder (GNOS), microwave radiometric imager (MWRI), high spectral infrared atmospheric Sounder (HIRAS), microwave temperature sounder (MWTS) and microwave humidity sounder (MWHS), thus enabling them to collect the quantitative, multispectral, three-dimensional and all-weather earth surface characteristic parameters on a global scale [11,12]. Among these instruments, the MERSI and MERSI-II sensors were designed as key visible and infrared spectral imaging instruments for detecting the atmospheric, terrestrial and oceanic features, and providing important operational products such as the outgoing longwave radiation (OLR), precipitable water vapor (PWV) and NDVI. Compared to MERSI onboard FY-3A/B/C satellites, the performance of MERSI-II onboard FY-3D has been greatly upgraded and improved, with more imaging channels, a higher radiometric calibration accuracy and an enhanced infrared detection capability [12]. In many previous studies, FY-3 MERSI and MERSI-II data have been widely used to monitor the vegetation coverage and conditions, detect natural disaster events and support weather forecasting sand environmental studies [13–16].

Cross-comparisons among vegetation indices (VIs) derived from different satellites and sensors is of great significance for their application, calibration and cooperative inversion using multisource remote sensing data [11]. Many previous studies have utilized methods such as regression analysis to recognize differences between VIs extracted from different satellite systems and sensors [17–19]. For example, seven VIs, including NDVI, EVI, normalized difference water index (NDWI), soil adjusted vegetation index (SAVI), green vegetation index (GVI), land surface water index (LSWI) and normalized burn ratio (NBR), provided by two sensors, which are the ETM+ on board Landsat 7 and the operational land imager (OLI) on board Landsat 8, were compared by researchers. The results showed that the correlation coefficients between different indices from ETM+ and OLI were high, indicating that those two sensors can be used to collect complementary data [17]. Additionally, the NDVI values across different land cover types from Landsat 7 and Landsat 8 were compared, revealing that differences between NDVI images from the two satellites were larger in low-vegetation-covered areas than in high-vegetation-covered areas, which indicates the need for calibration in low-vegetation-covered areas to achieve higher consistency between Landsat 7 and Landsat 8 NDVI images [20]. Despite the large number of early studies regarding cross-comparison analyses between vegetation parameters from different sensors and satellites, few studies have focused on the NDVI product from different generations of the FY-3 series satellites.

The FY-3D satellite equipped with the upgraded MERSI-II instrument was launched in November 2017 and started to provide an operational NDVI product in 2019, which indicates the short time series of the FY-3D NDVI data. Thus, the application of FY-3D NDVI in fields such as climate change was limited.

To solve this issue, this work proposes a novel scheme, as follows: Firstly, the consistency between FY-3B and FY-3D NDVI data was assessed to detect the possibility of combining different FY-3 satellites' data into complementary datasets for generating a long-term NDVI dataset. Secondly, a correlation analysis between FY-3 products and two reference datasets was conducted to determine the satellite with the better performance. Thirdly, a grey correlation method was applied to detect the possible factors that could affect the consistency degree among different satellites. Finally, an ensemble regression model was established for generating long time series NDVI data by combining different FY-3 satellites. The paper is organized as follows: Section 2 introduces the study area, data, and methods. In Section 3, the results are described. Sections 4–6 illustrate the discussion, practical applications and conclusions, respectively.

2. Data and Methods

2.1. Study Area

China is the world's third largest country with an approximately 9.6 million km^2 territory area. It is located in the east of Asia and on the west coast of the Pacific Ocean with coordinates of 18°N to 54°N and 73°E to 135°E (Figure 1). The terrain of China gradually increases from lower than 100 m in the east to more than 8000 m in the west. The Qinghai Tibet Plateau, known as the roof of the world, is located in the southwest of the study area, with an average elevation of 4000 m (Figure 1a). There are mainly six land cover types across China, including forest, grassland, farmland, barren, urban lands and water bodies, as shown in Figure 1b. Most parts of China are covered by various vegetation types. The forests are mainly distributed in southeastern and northeastern China, showing a regular zonal distribution from north to south. Grasslands are widely distributed in the northwestern, southeastern and northeastern regions of the study area. The distribution of farmland in China is quite unbalanced due to the effects of topography and climate environments, especially the water resources. Most cultivated lands are densely located in the central and eastern regions with flat terrain, mild climate and abundant precipitation.

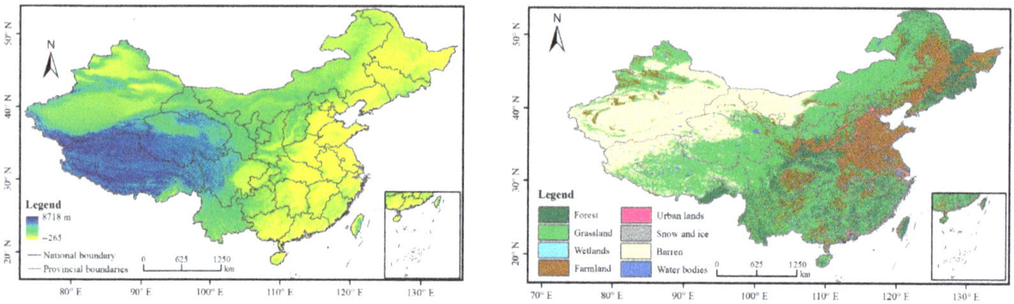

(a) (b)

Figure 1. The DEM (**a**) and land cover types (**b**) in China.

Over the past several decades, China has experienced a rapid increase in extreme climate event occurrences in the context of global climate change, which have led to spatio-temporal changes in vegetation coverage and other vegetation activities [21,22]. Therefore, it is urgent and of great importance to achieve accurate and continuous NDVI dataset on a regional scale for China's vegetation monitoring and carbon sequestration detection.

2.2. Data

2.2.1. RCH-CEOS NDVI Product Derived from Fengyun Satellites

To produce a highly homogenous and long time series NDVI product for vegetation applications in fields such as climate change, the National Satellite Meteorological Center (NSMC) adopted the unified calibration method to achieve the Retrospective Calibration of Historical Chinese Earth Observation Satellites (RCH-CEOS) NDVI dataset. The RCH-CEOS NDVI product was derived from the calibrated Level 1 reflectance of FY-3A, FY-3B MERSI and FY-3D MERSI-II, with spatial resolutions of 1 km and 250 m. The maximum value composite (MVC) approach was applied to generate the composite NDVI maps from 2010 to 2023 at ten-day and monthly intervals [10,23]. Among them, the NDVI products for 2010, 2011–2020 and 2019–2023 were derived from the FY-3A, FY-3B and FY-3D satellites, respectively. In this study, pixel values were extracted from FY-3B/MERSI and

FY-3D/MERSI-II NDVI products for the months from January to December 2020. The averaged NDVI values for forest, grassland, farmland, barren land and the entire country were then calculated and compared to assess the consistency among the RCH-CEOS NDVI products derived from different Fengyun satellites.

2.2.2. In Situ LAI Measurements from CMA

The website for the China meteorological data (https://data.cma.cn/, accessed on 18 October 2023), which is operated by the National Meteorological Information Center (NMIC) of the China Meteorological Administration (CMA), collects and releases an abundance of weather-related data, including meteorological station-based measurements, radar and satellite remote sensing images, numerical forecasting products, reanalysis data, atmospheric composition data and climate forecasting products. Here, 16 in situ LAI measurements collected from agrometeorological stations were downloaded from the China meteorological data website, and the latitudes and longitudes of each station were applied to calculate the rows and columns of station pixels located at the FY-3 NDVI images. Then the correlation between NDVI values at those pixels and the LAI measurements was analyzed to identify the ability of FY-3B and FY-3D for monitoring the vegetation growth on a site scale.

2.2.3. Land Cover and NDVI Products Derived from MODIS

The MODIS acquires the features of the land, ocean and atmosphere every 1–2 days and provides important time series datasets for quantifying changes in the global environment [24]. Many of the MODIS's data and products, such as the MODIS land cover (MODIS LC) and MODIS NDVI products, have been widely used for land cover type classification and crop yield estimation in many early studies [25,26]. The MODIS LC products (MOD12Q1) contain 17 land cover types based on the international geosphere biosphere programme (IGBP) scheme. Here, these land cover types were merged into 8 types, including forest, grassland, wetland, farmland, urban and built-up land, barren, snow and ice, and water bodies, for assessing the consistency among NDVI values of FY-3 satellites across various land types. The MODIS NDVI products (MOD13A3, v006) for the months from January to December 2020 were downloaded from the NASA website (https://ladsweb.modaps.eosdis.nasa.gov/, accessed on 11 October 2023) as the reference dataset. Then, the correlation coefficients between MODIS NDVI and NDVI values derived from the different FY-3 satellites were calculated to evaluate the ability of FY-3B and FY-3D for monitoring vegetation conditions on a regional scale.

2.2.4. DEM Data from GMTED2010

The digital elevation model (DEM) is one of the branches of the digital terrain model (DTM), which is a discrete mathematical representation of the Earth's surface topography, and is frequently selected as auxiliary data for use in many areas of research [27,28]. The global multiresolution terrain elevation data in the 2010 (GMTED2010) dataset contain multisource elevation data, including shuttle radar topography mission (SRTM) digital terrain elevation data from US National Geospatial-Intelligence Agency (NGR), Canadian digital elevation data, US continental and Alaska elevation data, the second DEM for Australia and the DEM from Antarctic and Greenland radar and laser altimeters [29]. The GMTED2010 dataset has proven to be more accurate than the GTOPO30 model and was adopted to map the topography across the study area in this study.

2.3. *Methods*

2.3.1. Grey Relational Analysis Method

The grey relational analysis (GRA) method, which is an important part of the theory of grey systems, has been extensively applied in conducting relational analyses, performing predictions, and solving decision-making problems in various sectors [30,31]. In general, the degree of similarity or diversity of the developing trend among factors are measured using the GRA method for achieving the relevance of multiple factors and ranking the alternatives [32–34].

The GRA method involves the following steps: Step 1: Define the reference sequence (X^*) and m comparison sequences (X^i) (i = 1, 2,..., m) as:

$$X^* = [X_1{}^*, X_2{}^*, \ldots, X_k{}^*, \ldots, X_n{}^*] \tag{1}$$

$$X^i = [X_1{}^i, X_2{}^i, \ldots, X_k{}^i, \ldots, X_n{}^i] \tag{2}$$

where n is the number of samples in each sequence, k (k = 1, 2, ..., n) is the index of samples, $X_k{}^*$ represents the reference sequence for the kth sample, and $X_k{}^i$ represents the comparison sequence for the kth sample of the ith alternative. Step 2: Transform the original data into nondimensional sequences ($C_k{}^i$) ranging from 0 to 1 using the following equations:

$$C_k{}^* = \frac{X_k{}^* - X_{\min}{}^*}{X_{\max}{}^* - X_{\min}{}^*} \tag{3}$$

$$C_k{}^j = \frac{X_k{}^j - X_{\min}{}^j}{X_{\max}{}^j - X_{\min}{}^j} \tag{4}$$

where $X_{\min}{}^*$ and $X_{\max}{}^*$ are the minimum and maximum values for the reference sequence, respectively. $X_{\min}{}^j$ and $X_{\max}{}^j$ are the minimum and maximum values for the jth comparison sequence, respectively. Step 3: Calculate the grey relational coefficient ($\xi_i{}^k$) and grey relational degree (D_i) using the following equations:

$$\xi_i{}^k = \frac{\min\limits_{1\le i\le m}\min\limits_{1\le k\le n}\left|C_k{}^* - C_k{}^i\right| + \rho \max\limits_{1\le i\le m}\max\limits_{1\le k\le n}\left|C_k{}^* - C_k{}^i\right|}{\left|C_k{}^* - C_k{}^i\right| + \rho \max\limits_{1\le i\le m}\max\limits_{1\le k\le n}\left|C_k{}^* - C_k{}^i\right|} \tag{5}$$

$$D_i = \frac{1}{n}\sum_{k=1}^{n} \xi_i{}^k \tag{6}$$

where ρ is the identification coefficient ranging from 0 to 1, which was set as ρ = 0.5 in this study. $\xi_i{}^k$ represents the grey relational coefficient between $C_k{}^i$ and $C_k{}^*$. D_i represents the grey relational degree between the reference sequence and the ith comparison sequence. A greater value of D_i indicates a higher proximity of sequence X^i to the reference sequence [35].

In this study, the GRA method was adopted to analyze the importance of factors influencing the NDVI bias from different Fengyun satellites. Here, the difference between the FY-3B and FY-3D NDVI values in 2020 was set as the reference sequence, and several potential influencing factors, such as the land cover types, during the same period were set as the sequences to be compared. Then, the grey relational degrees between the reference and comparison sequences were calculated to rank the importance of each factor influencing the NDVI difference.

2.3.2. Gradient Boosting Regression Model

Various versions of the boosting algorithm, such as AdaBoost and gradient boosting (GB), have proven to be very competitive in a wide range of applications [36,37]. Among them, the GB model, which was introduced by Friedman, in 2001, has been extensively applied for solving classification and regression problems as one of the most powerful ensemble learning techniques [38–41].

In general, the performance of the GB model is improved by training a series of basic learners and minimizing the expectations of the loss function [41]. There are several most frequently used loss functions, including the Gaussian L2 loss function, Laplace L1 loss function, Huber loss function and Quantile loss function [39]. Here, the Huber loss function was selected considering its robustness to outliers. The Huber loss function comprises two parts and is designed as follows:

$$L(y, f(x))_{Huber,\delta} = \begin{cases} (y - f(x))^2/2 & |y - f(x)| \leq \delta \\ \delta(|y - f(x)| - \delta/2) & |y - f(x)| > \delta \end{cases} \quad (7)$$

where y is the true value, $f(x)$ is the predicted value and δ is the threshold parameter, which specifies the robustness of the Huber loss function. To achieve a more accurate global optimum and reduce the time needed to update those parameters, the mean squared error (MSE) is applied when the prediction error is less than or equal to δ, and a linear error indicator such as mean absolute error (MAE) is applied when the prediction error is greater than δ.

The GB model is built in a stepwise fashion. For each step, the pseudo-residual of each weak learner decreases in the direction of the negative gradient by fitting the negative gradient of the loss function of the previously cumulative model, so that the cumulative model loss decreases after the addition of the weak learner. Assume that the number of training samples is n. The approximate solution is initialized as:

$$F_0(x) = \arg\min_{f \in F} \sum_{i=1}^{m} L(y_i, f(x_i)) \quad (8)$$

where f is one of the basic learners, and F is the set of basic learners.

Assume that the number of applied basic learners is m. For step j ($j = 1, 2, \ldots, m$), the negative gradient G and learner f for the current step are computed as follows:

$$G_j(x_i) = \left[-\frac{\partial L(y_i, f(x_i))}{\partial f(x_i)} \right]_{f(x_i) = f_{j-1}(x_i)} \quad (9)$$

$$f_j(x_i) = v_j G_j(x_i) \quad (10)$$

where the parameter v is the multiplier and is computed based on the following equation:

$$v_j = \arg\min_{\rho} \sum_{i=1}^{n} L(y_i, f(x_i) + v_j G_j(x_i)) \quad (11)$$

Then, the GB model is updated as follows:

$$F_j(x_i) = F_{j-1}(x_i) + f_j \quad (12)$$

In this study, to further improve the consistency among the NDVI products derived from different Fengyun satellites, the gradient boosting regression (GBR) model was built for the NDVI estimations, pixel by pixel, using multiple inputs.

2.4. Evaluation Criteria

To evaluate the consistency between the FY-3D NDVI product and the NDVI derived from FY-3B and the estimated NDVI using the multivariate GBR model, five statistical indicators, including the correlation coefficient (R), root mean square error (RMSE), mean absolute error (MAE) and rate of change (RoC), were adopted in this study.

The R represents the relative accuracy between the original FY-3B NDVI retrievals or the estimated NDVI values using the GBR model and the FY-3D NDVI retrievals, respectively, and can be defined by Equation (13).

$$\mathrm{R} = \frac{\sum_{i-1}^{N} (CNDVI_i - \overline{CNDVI})(RNDVI_i - \overline{RNDVI})}{\sqrt{\sum_{i-1}^{N} (CNDVI_i - \overline{CNDVI})^2} \sqrt{\sum_{i-1}^{N} (RNDVI_i - \overline{RNDVI})^2}} \quad (13)$$

where $CNDVI_i$ represents the NDVI values to be compared, which refer to the FY-3B NDVI and the GBR-estimated NDVI values in this study. $RNDVI_i$ represents the reference NDVI values, i.e., the FY-3D NDVI retrievals. $\overline{CNDVI}$ represents the average of the FY-3B NDVI retrievals or the GBR NDVI estimations across all pixels in the study area, $\overline{RNDVI}$ indicates the average of the FY-3D NDVI retrievals, and N represents the number of valid pixels in each dataset.

The RMSE and MAE evaluate the absolute difference between the FY-3B NDVI product or NDVI estimations using the GBR model and the FY-3D NDVI data, which can be calculated by Equations (14) and (15).

$$\text{RMSE} = \sqrt{\frac{\sum_{i=1}^{N}(CNDVI_i - RNDVI_i)^2}{N}} \tag{14}$$

$$\text{MAE} = \frac{1}{N}\sum_{i-1}^{N}|CNDVI_i - RNDVI_i| \tag{15}$$

The RoC evaluates the degree of the GBR model for improving the consistency between NDVI products from FY-3B and FY-3D, and it can be defined as the following equation:

$$\text{RoC} = \frac{E_g - E_b}{E_b} \times 100\% \tag{16}$$

where E_g represents the error metrics between the GBR-estimated NDVI and the FY-3D NDVI data, and E_b represents the error metrics between the FY-3B NDVI and the FY-3D NDVI data.

3. Results and Analysis

3.1. Consistency Assessment of NDVI Products Derived from Different Fengyun Satellites

The variation curves of FY-3B NDVI (solid line) and FY-3D NDVI (dash line) for different land cover types in 2020 are displayed in Figure 2. Generally, the FY-3B and FY-3D NDVI data showed an overall high consistency, with the averaging NDVI difference values across the study area being lower than 0.01. However, the NDVI differences varied in different time periods and for different land cover types. In particular, the deviations between FY-3B and FY-3D from October to December were larger than in other months. For the forest and farmland areas, the NDVI value of FY-3D was slightly higher than that of FY-3B. The FY-3D NDVI was slightly lower than FY-3B NDVI in most months for the grassland.

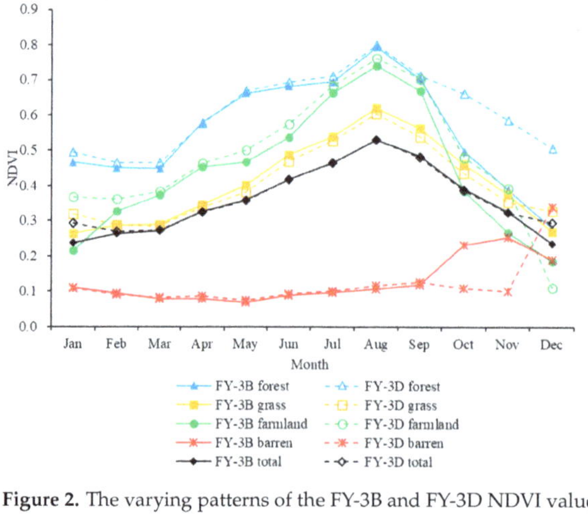

Figure 2. The varying patterns of the FY-3B and FY-3D NDVI values for different land cover types from January 2020 to December 2020.

The spatial distributions of the NDVI difference values between FY-3B and FY-3D in January, April, July, and October 2020 are mapped in Figure 3. Generally, the FY-3B NDVI values in the southeast region of the study area were lower than those of FY-3D, and the FY-3B NDVI values in southwest China were greater than those of FY-3D. The NDVI differences vary in different months. Compared with April and July, the FY-3B and FY-3D NDVIs' difference in January and October was significantly higher.

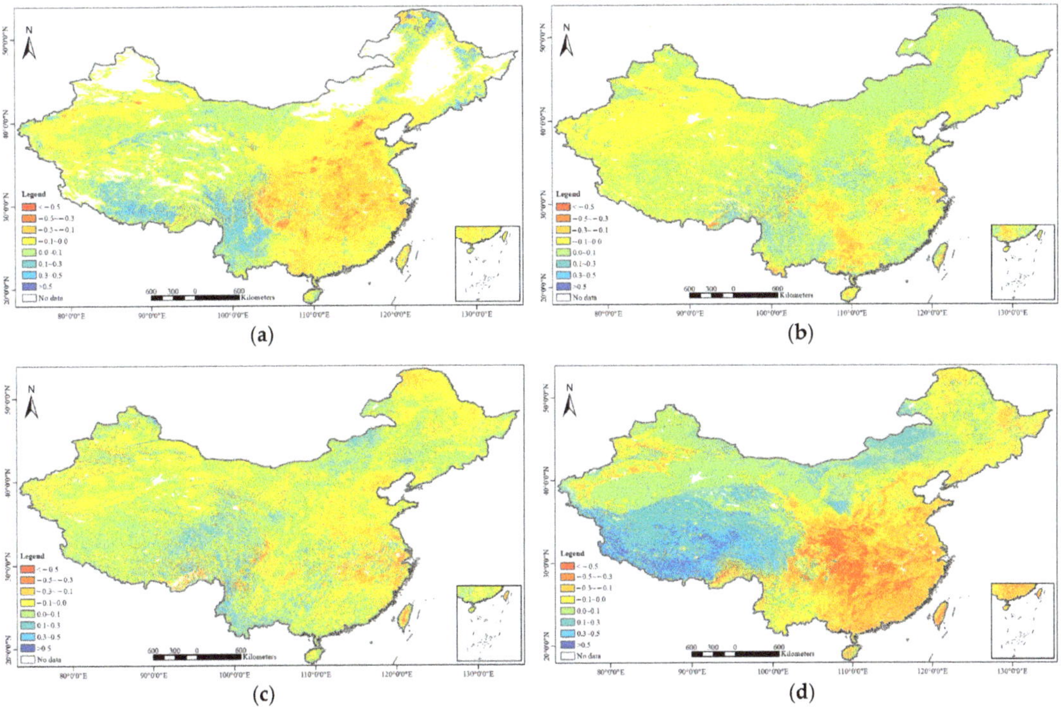

Figure 3. The spatial distribution of the difference in the NDVI values between FY-3B and FY-3D: (**a**) January; (**b**) April; (**c**) July; (**d**) October.

3.2. Comparing the Performances of Different Fengyun Satellites in Monitoring Vegetation Conditions

To further evaluate the potentials of the FY-3B and FY-3D NDVIs for monitoring vegetation growth conditions, a correlation analysis between the measured LAI data and the NDVI values extracted from the FY-3B and FY-3D satellites was conducted. The scatterplots and the coefficients of determination (R^2) are shown as Figure 4. The results show that the R^2 value between the FY-3D NDVI and the station-based LAI (0.349) was higher than that between the FY-3B NDVI and the in situ LAI measurements (0.231), which indicates that the FY-3D NDVI is more reliable in reflecting vegetation growth and conditions on a site scale.

The MODIS NDVI is recognized as one of the most popular vegetation indices; thus, it was selected to evaluate the abilities of FY-3B and FY-3D for monitoring vegetation conditions on a grid scale. The R and MAE values between MODIS NDVI and NDVI products extracted from the FY-3B and FY-3D satellites were calculated, as shown in Figure 5. The results show that the R value between the MODIS NDVI and the NDVI values calculated by FY-3D and FY-3B across all pixels in the study area reached 0.983 and 0.973, respectively, indicating that the overall consistency between the MODIS NDVI and the FY-3D NDVI was slightly higher than that of the FY-3B NDVI. In addition, the averaged

MAE values of FY-3B and FY-3D were both less than 0.01, indicating that the FY-3 NDVI and the MODIS NDVI were generally close.

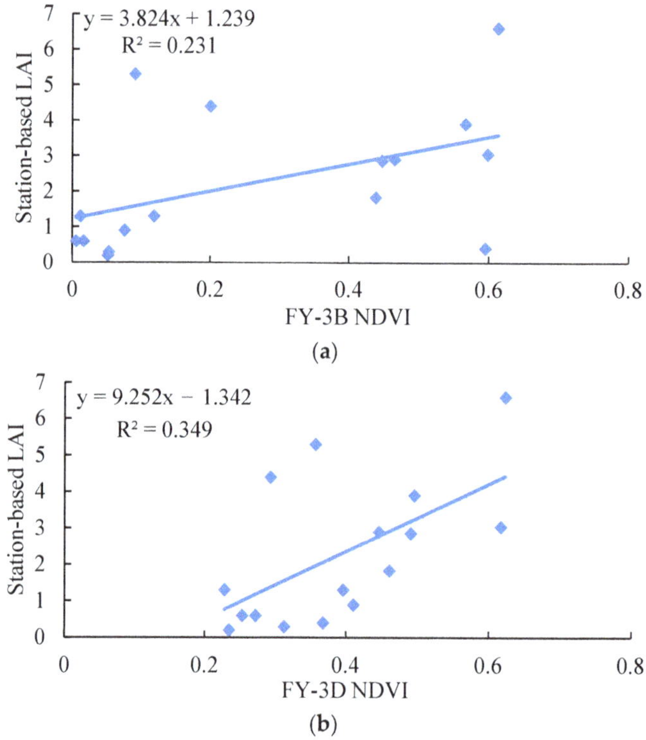

Figure 4. Scatterplots between the station-based LAI measurements and the NDVI values derived from (**a**) FY-3B and (**b**) FY-3D.

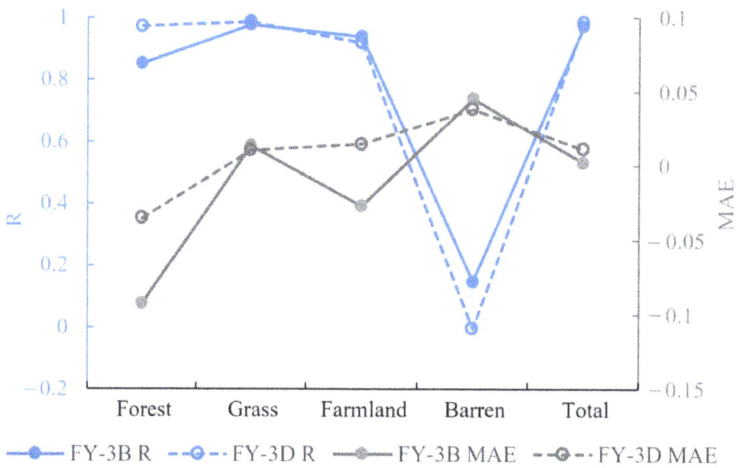

Figure 5. The correlation coefficients and mean absolute errors between the MODIS NDVI and the NDVI derived from FY-3B and FY-3D for different land cover types.

The consistency between the FY-3 and MODIS NDVI varied for the different land cover types. The FY-3 NDVI and MODIS NDVI for grassland achieved the highest consistency, with R values between the MODIS NDVI and the NDVI derived from FY-3B and FY-3D, reaching 0.976 and 0.985, and the MAE values reached 0.013 and 0.010, respectively. The FY-3B and FY-3D NDVI values for forest were lower than that of the MODIS NDVI, and the underestimation degree of FY-3B (MAE = −0.092) was higher than that of FY-3D (MAE = −0.035). For farmland, the FY-3B NDVI was lower than the MODIS NDVI, with an MAE value equivalent to −0.027, while the FY-3D NDVI was much closer to the MODIS NDVI with an MAE value equivalent to 0.014. Among those land cover types, the correlations with barren areas were the worst, with average R values lower than 0.2.

3.3. Detecting the Factors Affecting the NDVI Difference Using the GRA Method

In this study, the GRA method was applied to analyze the factors affecting the NDVI difference in January, April, July and October. Firstly, the FY-3B and FY-3D NDVIs' difference values at each pixel were extracted to construct the reference sequence. The DEM, LC, LAT and LON values at pixels were used to construct four sequences to be compared. Then, the correlation degree values between the reference sequence and those comparison sequences were calculated according to Equations (3)–(6), as shown in Figure 6. The grey correlation between the NDVI difference sequence and the LC sequence was the highest in each month, with correlation degree values reaching 0.578, 0.601, 0.603 and 0.603 in January, April, July and October, respectively, which illustrated that the land cover condition was closely related to the FY-3B and FY-3D NDVIs' difference. By ranking the grey correlation degrees in decreasing order, the importance of those affecting factors were ranked as follows: LC > DEM > LAT > LON in January; LC > LAT > DEM > LON in April; LC > LAT > LON > DEM in July; and LC > LON > LAT > DEM in October.

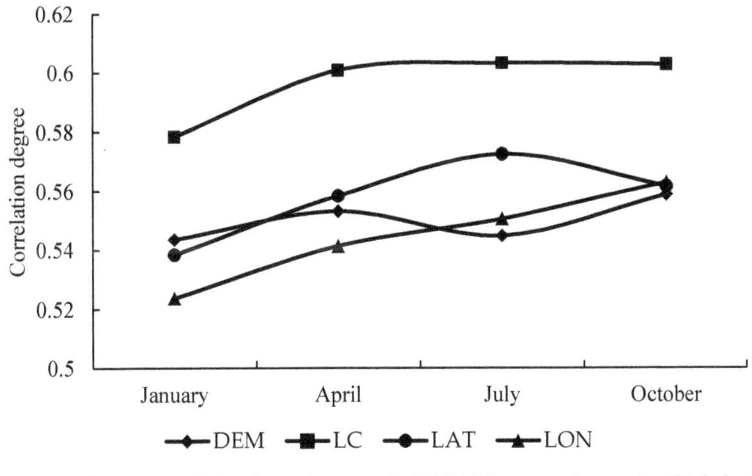

Figure 6. The grey correlation degree between the NDVI difference values and multiple factors using the GRA method.

3.4. Establishing a GBR Model for Forming Long Time Series NDVI Data

In this study, the monthly FY-3B and FY-3D NDVI products in 2019 and 2020, as well as several auxiliary data, including the DEM, LC, LAT and LON, were selected to establish and validate the GBR model for improving the consistency among NDVI products extracted from different FY-3 satellites. To be specific, 90% and 10% of the samples in 2019 were adopted as the training and testing datasets, respectively, and the samples in 2020 were used as the validation dataset. Considering that the inconsistency issue was relatively more serious in October, as demonstrated in Section 3.1, a GBR model with the FY-3B NDVI, LC, DEM, LAT and LON as the input variables and FY-3D NDVI as the target variable was built to correct the FY-3B NDVI for the formation of a long-term FY-3 NDVI product in October.

The error metrics, including R, RMSE, MAE and RoC, between the reference variable (FY-3D NDVI) and the NDVI values extracted from the FY-3B satellite, as well as the estimated NDVI using the GBR model (GBR NDVI), were calculated as shown in Table 1. The results showed that the R values between the GBR NDVI and FY-3D NDVI reached 0.947, 0.867 and 0.829 in the training, testing and validation datasets, respectively, which greatly increased by 5.0%, 17.2% and 51.6% compared with that between the FY-3B and FY-3D NDVIs. In addition, the RMSE values decreased by 27.2%, 15.8% and 34.3%, and the MAE values decreased by 29.2%, 13.3% and 32.7% during the training, testing and validation processes, respectively, which indicated a robust performance of the GBR for generating long time series NDVI data by combining data from the FY-3B and FY-3D satellites.

Table 1. The error metrics between the FY-3D NDVI retrievals and the NDVI values extracted from the FY-3B and the GBR model.

Error / Dataset	FY-3B NDVI				GBR NDVI						
	R	***p***	**RMSE**	**MAE**	**R**	***p***	**RoC**	**RMSE**	**RoC**	**MAE**	**RoC**
Training	0.902	<0.001	0.092	0.065	0.947	<0.001	+5.0%	0.067	−27.2%	0.046	−29.2%
Testing	0.774	<0.001	0.120	0.090	0.867	<0.001	+17.2%	0.101	−15.8%	0.078	−13.3%
Validation	0.547	<0.001	0.213	0.159	0.829	<0.001	+51.6%	0.140	−34.3%	0.107	−32.7%

The importance coefficients of each input factor were calculated by the GBR model and are shown in Figure 7. Among those variables, the FY-3B NDVI contributed the most, with the coefficient of importance (CI) reaching 0.529, followed by the latitude (CI = 0.150) and land cover (CI = 0.143). The longitude and DEM contributed less than the other variables, with the CI values equivalent to 0.097 and 0.081, respectively.

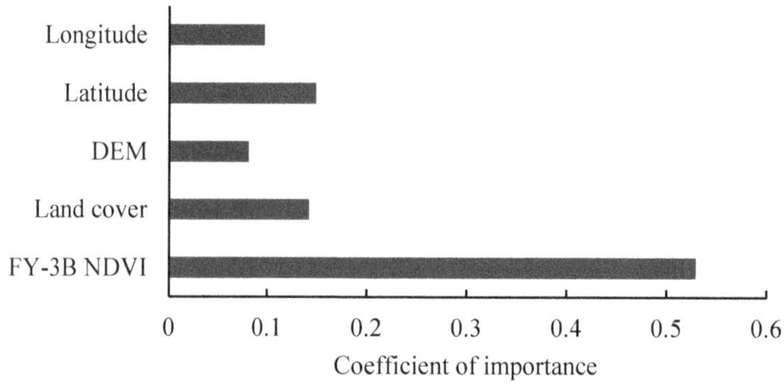

Figure 7. Features' importance generated for the GBR model.

The established GBR model was applied to perform the NDVI estimations across the study area, pixel by pixel. The FY-3D, FY-3B and GBR model-based NDVI images were mapped and are displayed in Figure 8. Compared to the FY-3D NDVI (Figure 8a), the FY-3B NDVI (Figure 8b) significantly overestimated the NDVI values in the southwest (mainly the Qinghai–Tibet Plateau) and northeast parts of the study area. The overestimation issue was largely improved in the estimated NDVI product based on the GBR model (Figure 8c), which indicates the feasibility of the use of the established GBR model for generating long time series Fengyun NDVI data with higher consistency.

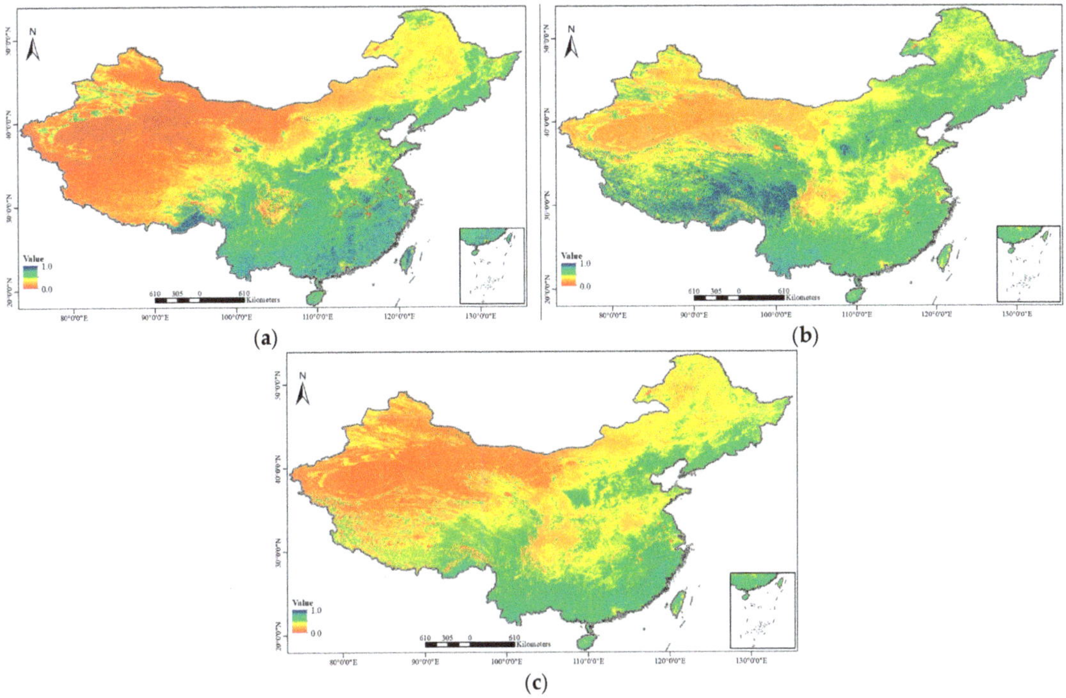

Figure 8. The spatial patterns of the NDVI values: (**a**) FY-3D; (**b**) FY-3B; (**c**) GBR model.

4. Discussion

A large number of China's Fengyun meteorological satellites were launched in recent years to support weather forecasting and extreme events detection. However, the products derived from the Fengyun satellites are limited for long-term applications in fields such as climate change due to their short time series. This study explored the possibility of generating a long-term NDVI dataset by combining different Fengyun satellite NDVI products. The results are promising and can be used as a reference idea and method for evaluating and improving Chinese Fengyun satellite products for weather and climate applications.

Previous studies have pointed out that the observed NDVI signal could be greatly affected by the aerosol optical depth (AOD), which could significantly underestimate the amount of vegetation [42,43]. Apart from the AOD, atmospheric water vapor, clouds, and differences in spectral wavelengths could also cause differences in NDVI values [44,45]. The results in this work have shown a relatively lower consistency between the FY-3B and FY-3D NDVIs during the months from October to December than in other months. This can be attributed to the relatively higher aerosol loads in winter, as well as the differences in spectral wavelengths of FY-3B/MERSI and FY-3D/MERSI-II. Compared with the MERSI, the number of MERSI-II channels increased from 20 to 25, with more infrared channels and enhanced infrared detection capability [23]. Moreover, the FY-3D NDVI applied the 6SV method for atmospheric correction and greatly decreased the atmospheric effects [10], thus achieving more accurate NDVI data in periods with higher aerosol optical thickness.

Field campaigns and ground truth could greatly contribute to the evaluation and validation processes of satellite remote sensing products [46]. In this study, the in situ LAI measurements were selected as reference data to evaluate and compare the capacity of the FY-3B and FY-3D for monitoring vegetation growth conditions because of the unavailability of ground-based NDVI or spectral measurements. The results indicate a higher correlation between the FY-3D and LAI measurements than that of the FY-3B, which could be explained by the fact that the FY-3D carries more advanced instruments than the FY-3B. Although a reasonable result was achieved, there are several issues with the current LAI data, which could cause a biased evaluation of the performance of remote sensing products. The in situ LAI data contain only two land cover types, which are grassland and farmland; thus, the representativeness of the LAI samples needs to be further improved in future studies for a more reliable evaluation. In additions, the mismatch of scales between meteorological stations and satellites could also potentially influence the reliability of the assessment [47].

5. Practical Applications

This work provides a reliable method for evaluating and correcting NDVI products derived from different Chinese meteorological satellites for forming long time series NDVI data. The generated FY-3 long-term NDVI dataset could contribute to research in fields such as climate change, ecological environment detection, crop growth monitoring, phenological extraction and carbon sequestration trend analysis. For example, long-term NDVI time series data, together with other climatic data, can be applied to detect the effect of extreme events on vegetation activities [21]. This information is quite crucial for local ecosystem conservation, especially in climate-sensitive regions. Apart from these potential application fields of long-term NDVI data, the idea proposed by this study could be adopted as a reference method to process various remote sensing products, such as the land surface temperature (LST), outgoing longwave radiation (OLR) and surface soil moisture (SSM), from different Fengyun satellites for weather and climate applications in the future.

6. Conclusions

This study introduced an overall strategy that could be extended for evaluating and improving the Chinese FY-3 meteorological satellite systems for long-term environmental research on a regional scale. There are four main components contained in this strategy: (1) consistency assessment among different satellites to provide information for supplementary dataset selection and further improvement; (2) performance comparison among different satellites to recognize the target data for the correction model; (3) correlation analysis to detect the factors affecting the differences among satellites; and (4) an ensemble machine learning model to form long time series data by combining different satellites.

In this study, the variation trends of the FY-3B/MERSI and FY-3D/MERSI-II NDVIs for different land cover types in 2020 were compared, and the results show an overall high consistency between the FY-3B and FY-3D NDVIs, indicating the potential of the FY-3 satellites for providing long-term NDVI datasets as complementary data sources. The correlations between the FY-3 NDVI and two reference data, which were the MODIS NDVI and in situ LAI measurements, were analyzed to determine the satellite with the better performance. The correlation coefficients indicate that FY-3D/MERSI-II has a better capacity for retrieving an NDVI product on both the regional and site scales. This study applied the GRA method to detect the factors related to the NDVI differences among different satellites. The results show that the grey correlation degree between the LC and the NDVI difference sequences was the highest in each month, implying that the LC is an important parameter for the FY-3 NDVI product correction.

This study established a GBR model with the FY-3B NDVI, LC, DEM and geographical locations as input variables and the FY-3D NDVI as the target variable to generate long-term NDVI data using the FY-3B and FY-3D satellites. The R values between the model-based NDVI estimates and the FY-3D NDVI reached 0.947, 0.867 and 0.829 in the training, testing and validation datasets, respectively, which significantly improved compared with the R values between the FY-3B and FY-3D. Therefore, the GBR model has the feasibility to correct the FY-3B NDVI for generating a long-term FY-3 NDVI dataset, with the historical data more consistent with the upgraded data. In addition, this study calculated the feature importance of the GBR model. The results showed that the FY-3B NDVI contributed the most (0.529) to the model, followed by LAT (0.150), LC (0.143), LON (0.097) and DEM (0.081). Note that although the GBR model is one of the most widely used ensemble machine learning models for solving classification and regression problems, it has some limitations, such as a high sensitivity to outliers and complex parameter optimization, which suggests that it will be necessary to further optimize the parameter analysis in future studies to minimize its drawbacks and increase its performance.

Author Contributions: Conceptualization, L.W. and X.H.; methodology, L.W., X.H. and S.F.; software, L.W.; validation, S.F. and F.X.; data curation, L.W. and F.X.; writing—original draft preparation, L.W.; writing—review and editing, X.H., S.F. and F.X.; visualization, L.W.; supervision, X.H. and F.X.; funding acquisition, L.W. and X.H. All authors have read and agreed to the published version of the manuscript.

Funding: This research was funded by the National Natural Science Foundation of China (grant numbers 42205190, U2142212 and U2242211), Fengyun Application Pioneering Project (FY-APP-ZX-2023.02) and the Youth Innovation Team of China Meteorological Administration (CMA2023QN15).

Data Availability Statement: The FY-3 NDVI data can be downloaded from the website of NSMC. Available online: https://satellite.nsmc.org.cn/PortalSite/Data/Satellite.aspx (accessed on 9 April 2024). The MODIS land cover and NDVI products can be downloaded from the website of NASA. Available online: https://ladsweb.modaps.eosdis.nasa.gov/ (accessed on 9 April 2024). The in-situ LAI data are available at https://data.cma.cn/ with the permission of NMIC.

Acknowledgments: We thank all of the data centers for providing essential help in obtaining the datasets. Sincere gratitude also goes to Hongliang Fang from the Institute of Geographic Sciences and Natural Resources Research, CAS, for the valuable discussion.

Conflicts of Interest: The authors declare no conflicts of interest.

References

1. Dai, X.; Yang, G.; Liu, D.; Wan, R. Vegetation Carbon Sequestration Mapping in Herbaceous Wetlands by Using a MODIS EVI Time-Series Data Set: A Case in Poyang Lake Wetland, China. *Remote Sens.* **2020**, *12*, 3000. [CrossRef]
2. Zhang, X.; Cao, Q.; Chen, H.; Quan, Q.; Li, C.; Dong, J.; Chang, M.; Yan, S.; Liu, J. Effect of Vegetation Carryover and Climate Variability on the Seasonal Growth of Vegetation in the Upper and Middle Reaches of the Yellow River Basin. *Remote Sens.* **2022**, *14*, 5011. [CrossRef]
3. Thackway, R.; Lee, A.; Donohue, R.; Keenan, R.; Wood, M. Vegetation information for improved natural resource management in Australia. *Landsc. Urban Plan.* **2007**, *79*, 127–136. [CrossRef]
4. Becker-Reshef, I.; Vermote, E.; Lindeman, M.; Justice, C. A generalized regression-based model for forecasting winter wheat yields in Kansas and Ukraine using MODIS data. *Remote Sens. Environ.* **2010**, *114*, 1312–1323. [CrossRef]
5. Khan, J.; Wang, P.; Xie, Y.; Wang, L.; Li, L. Mapping MODIS LST NDVI imagery for drought monitoring in Punjab Pakistan. *IEEE Access* **2018**, *6*, 19898–19911. [CrossRef]
6. Bhuyan, U.; Zang, C.; Vicente-Serrano, S.M.; Menzel, A. Exploring Relationships among Tree-Ring Growth, Climate Variability, and Seasonal Leaf Activity on Varying Timescales and Spatial Resolutions. *Remote Sens.* **2017**, *9*, 526. [CrossRef]
7. Xiao, Z.; Liang, S.; Tian, X.; Jia, K.; Yao, Y.; Jiang, B. Reconstruction of long-term temporally continuous NDVI and surface reflectance from AVHRR data. *IEEE J. STARS* **2017**, *14*, 5551–5568. [CrossRef]
8. Jin, Z.; Xu, B. A Novel Compound Smoother-RMMEH to Reconstruct MODIS NDVI Time Series. *IEEE Trans. Geosci. Remote Sens.* **2013**, *10*, 942–946. [CrossRef]
9. Ke, Y.; Im, J.; Lee, J.; Gong, H.; Ryu, Y. Characteristics of Landsat 8 OLI-derived NDVI by comparison with multiple satellite sensors and in-situ observations. *Remote Sens. Environ.* **2015**, *164*, 298–313. [CrossRef]
10. Han, X.; Weng, F.; Han, Y.; Huang, H.; Li, S. Vegetation indices derived from Fengyun-3D MERSI-II data. In Proceedings of the IGARSS 2020—2020 IEEE International Geoscience and Remote Sensing Symposium, Waikoloa, HI, USA, 26 September–2 October 2020.
11. Xiao, F.; Liu, Q.; Li, S.; Qin, Y.; Huang, D.; Wang, Y.; Wang, L. A Study of the Method for Retrieving the Vegetation Index from FY-3D MERSI-II Data. *Remote Sens.* **2023**, *15*, 491. [CrossRef]
12. Yang, Z.D.; Zhang, P.; Gu, S.Y. Capability of Fengyun-3D satellite in earth system observation. *J. Meteorol. Res.* **2019**, *33*, 1113–1130. [CrossRef]
13. Liu, Y.; Han, X.; Weng, F.; Xu, Y.; Zhang, Y.; Tang, S. Estimation of Terrestrial Net Primary Productivity in China from Fengyun-3D Satellite Data. *J. Meteorol. Res.* **2022**, *36*, 401–416. [CrossRef]
14. Jin, S.; Zhang, M.; Ma, Y.; Gong, W.; Chen, C.; Yang, L.; Hu, X.; Liu, B.; Chen, N.; Du, B.; et al. Adapting the Dark Target Algorithm to Advanced MERSI Sensor on the FengYun-3-D Satellite: Retrieval and Validation of Aerosol Optical Depth Over Land. *IEEE Trans. Geosci. Remote Sens.* **2021**, *59*, 8781–8797. [CrossRef]

15. Ma, X.; Yao, Y.; Zhang, B.; Du, Z. FY-3A/MERSI precipitable water vapor reconstruction and calibration using multi-source observation data based on a generalized regression neural network. *Atmos. Res.* **2022**, *265*, 105893. [CrossRef]
16. Han, X.; Weng, F.; Han, Y. Fengyun-3D MERSI True Color Imagery Developed for Environmental Applications. *J. Meteorol. Res.* **2019**, *33*, 914–924. [CrossRef]
17. Mancino, G.; Ferrara, A.; Padula, A.; Nolè, A. Cross-Comparison between Landsat 8 (OLI) and Landsat 7 (ETM+) Derived Vegetation Indices in a Mediterranean Environment. *Remote Sens.* **2020**, *12*, 291. [CrossRef]
18. Wei, X.; Gu, X.; Meng, Q.; Yu, T.; Jia, K.; Zhan, Y.; Wang, C. Cross-Comparative Analysis of GF-1 Wide Field View and Landsat-7 Enhanced Thematic Mapper Plus Data. *J. Appl. Spectrosc.* **2017**, *84*, 829–836. [CrossRef]
19. Hao, C.; Wu, S.; Xu, C. Comparison of some vegetation indices in seasonal information. *Chin. Geogr. Sci.* **2008**, *18*, 242–248. [CrossRef]
20. Xu, D.; Guo, X. Compare NDVI Extracted from Landsat 8 Imagery with that from Landsat 7 Imagery. *Am. J. Remote Sens.* **2014**, *2*, 10–14. [CrossRef]
21. Yang, Q.; Jiang, C.; Ding, T. Impacts of Extreme-High-Temperature Events on Vegetation in North China. *Remote Sens.* **2023**, *15*, 4542. [CrossRef]
22. Xu, X.; Piao, S.; Wang, X.; Chen, A.; Ciais, P.; Myneni, R.B. Spatio-temporal patterns of the area experiencing negative vegetation growth anomalies in China over the last three decades. *Environ. Res. Lett.* **2012**, *7*, 035701. [CrossRef]
23. Han, X.; Gao, H.; Yang, J.; Li, Y.; Geng, W. Advances in ecological applications of Fengyun satellite data. *J. Meteorol. Res.* **2021**, *35*, 743–758. [CrossRef]
24. Vermote, E.F.; Saleous, N.Z.E.; Justice, C.O. Atmospheric correction of MODIS data in the visible to middle infrared: First results. *Remote Sens. Environ.* **2002**, *83*, 97–111. [CrossRef]
25. Song, X.; Huang, C.; Feng, M.; Sexton, J.O.; Channan, S.; Townshend, J.R. Integrating global land cover products for improved forest cover characterization: An application in North America. *Int. J. Digit. Earth* **2014**, *7*, 709–724. [CrossRef]
26. Garcia-Mora, T.J.; Mas, J.; Hinkley, E.A. Land cover mapping applications with MODIS: A literature review. *Int. J. Digit. Earth* **2012**, *5*, 63–87. [CrossRef]
27. Meng, L.; Liu, H.; Zhang, X.; Ren, C.; Ustin, S.; Qiu, Z.; Xu, M.; Guo, D. Assessment of the effectiveness of spatiotemporal fusion of multi-source satellite images for cotton yield estimation. *Comput. Electron. Agric.* **2019**, *162*, 44–52. [CrossRef]
28. Wang, L.; Fang, S.; Pei, Z.; Zhu, Y. Using FengYun-3C VSM Data and Multivariate Models to Estimate Land Surface Soil Moisture. *Remote Sens.* **2020**, *12*, 1038. [CrossRef]
29. Danielson, J.; Gesch, D. *Global Multi-Resolution Terrain Elevation Data 2010 (GMTED2010)*; U.S. Geological Survey Open-File Report 2011-1073; USGS: Reston, VA, USA, 2011; 26p.
30. Jiang, H.; Sun, Z.; Guo, H.; Weng, Q.; Du, W.; Xing, Q.; Cai, G. An assessment of urbanization sustainability in China between 1990 and 2015 using land use efficiency indicators. *npj Urban Sustain.* **2021**, *1*, 34. [CrossRef]
31. Xie, Y.; Liu, S. Research on evaluations of several grey relational models adapt to grey relational axioms. *J. Syst. Eng. Electron.* **2009**, *20*, 304–309.
32. Liu, H.; Wang, H.; Yuan, Y.; Zhang, C. Models for multiple attribute decision making with picture fuzzy information. *J. Intell. Fuzzy Syst.* **2019**, *37*, 1973–1980. [CrossRef]
33. Zhang, J.; Zhang, Q.; Zhang, J. The result greyness problem of the grey relational analysis and its solution. *J. Intell. Fuzzy Syst.* **2023**, *44*, 6079–6088. [CrossRef]
34. Wang, L.; Wang, P.; Li, L.; Xun, L.; Kong, Q.; Liang, S. Developing an integrated indicator for monitoring maize growth condition using remotely sensed vegetation temperature condition index and leaf area index. *Comput. Electron. Agric.* **2018**, *152*, 240–349. [CrossRef]
35. Kuo, Y.; Yang, T.; Huang, G. The use of a grey-based Taguchi method for optimizing multi-response simulation problems. *Eng. Optim.* **2008**, *40*, 517–528. [CrossRef]
36. Bühlmann, P.; Hothorn, T. Boosting Algorithms: Regularization, Prediction and Model Fitting. *Stat. Sci.* **2007**, *22*, 477–505.
37. Natekin, A.; Knoll, A. Gradient boosting machines, a tutorial. *Front. Neurorobot.* **2012**, *7*, 21. [CrossRef] [PubMed]
38. Friedman, J.H. Greedy function approximation: A gradient boosting machine. *Ann. Stat.* **2001**, *29*, 1189–1232. [CrossRef]
39. Xu, N.; Wang, Z.; Dai, Y.; Li, Q.; Zhu, W.; Wang, R.; Finkelman, R.B. Prediction of higher heating value of coal based on gradient boosting regression tree model. *Int. J. Coal Geol.* **2023**, *274*, 104293. [CrossRef]
40. Yurttakal, A.H. Extreme gradient boosting regression model for soil thermal conductivity. *Therm. Sci.* **2021**, *25*, S1–S7. [CrossRef]
41. Zhang, F.; Zhu, X.; Hu, T.; Guo, W.; Chen, C.; Liu, L. Urban Link Travel Time Prediction Based on a Gradient Boosting Method Considering Spatiotemporal Correlations. *ISPRS Int. J. Geoinf.* **2016**, *5*, 201. [CrossRef]
42. Huete, A.; Justice, C.; van Leeuwen, W. Modis Vegetation Index (MOD13). *Algorithm Theor. Basis Doc.* **1999**, *3*, 213.
43. Wang, D.; Chen, Y.; Wang, M.; Quan, J.; Jiang, T. A New Neighboring Pixels Method for Reducing Aerosol Effects on the NDVI Images. *Remote Sens.* **2016**, *8*, 489. [CrossRef]
44. Hideki, K.; Dennis, G.D. Atmospheric conditions for monitoring the long-term vegetation dynamics in the Amazon using normalized difference vegetation index. *Remote Sens. Environ.* **2005**, *97*, 519–525.

45. Teillet, P.M.; Ren, X. Spectral band difference effects on vegetation indices derived from multiple satellite sensor data. *Can. J. Remote Sens.* **2008**, *34*, 159–173. [CrossRef]
46. Wang, S.; Li, X.; Ge, Y.; Jin, R.; Ma, M.; Liu, Q.; Wen, J.; Liu, S. Validation of Regional-Scale Remote Sensing Products in China: From Site to Network. *Remote Sens.* **2016**, *8*, 980. [CrossRef]
47. Jackson, T.J.; Bindlish, R.; Cosh, M.H.; Zhao, T.J.; Starks, P.J.; Bosch, D.D.; Seyfried, M.; Moran, M.S.; Goodrich, D.C.; Kerr, Y.H.; et al. Validation of soil moisture and ocean salinity (SMOS) soil moisture over watershed networks in the U.S. *IEEE Trans. Geosci. Remote Sens.* **2012**, *50*, 1530–1543. [CrossRef]

remote sensing

MDPI

Review

Remote Sensing Application in Chinese Medicinal Plant Identification and Acreage Estimation—A Review

Jihua Meng [1,2,*], Xinyan You [1,2,3], Xiaobo Zhang [4], Tingting Shi [4,*], Lei Zhang [1,2], Xingfeng Chen [5], Hailan Zhao [1,2,3] and Meng Xu [1,2,6]

1 Key Laboratory of Digital Earth Science, Aerospace Information Research Institute, Chinese Academy of Sciences, Beijing 100094, China; youxinyan21@mails.ucas.ac.cn (X.Y.); zhanglei@radi.ac.cn (L.Z.); zhaohailan20@mails.ucas.ac.cn (H.Z.); xumeng10273@navinfo.com (M.X.)
2 International Research Center of Big Data for Sustainable Development Goals, Beijing 100094, China
3 University of Chinese Academy of Sciences, Beijing 100049, China
4 National Resource Center for Chinese Materia Medica, China Academy of Chinese Medical Sciences, Beijing 100700, China; jack110007@163.com
5 State Environmental Protection Key Laboratory of Satellite Remote Sensing, Aerospace Information Research Institute, Chinese Academy of Sciences, Beijing 100094, China; chenxf@aircas.ac.cn
6 School of Earth Sciences and Resources, China University of Geosciences, Beijing 100083, China
* Correspondence: mengjh@aircas.ac.cn (J.M.); shi_tingt@163.com (T.S.); Tel.: +86-13911012239 (J.M.); +86-15901089605 (T.S.)

Abstract: Chinese Materia Medica Resources (CMMRs) are crucial for developing the tradition and industry of traditional Chinese medicine. Given the increasing demand for CMMRs, an accurate and effective understanding of CMMRs is urgently needed. Chinese medicinal plants (CMPs) are the most important sources of CMMRs. Traditional methods of investigating medicinal plant resources have limitations, including severe subjectivity and poor timeliness, which make it difficult to meet the demand for real-time monitoring of large-scale medicinal plant resources. In recent years, remote sensing technology has become an important means of obtaining information on medicinal plants, and the application of this technology has made up for the shortcomings of traditional methods. This paper first discusses the development of investigation methods of CMMRs; points out the importance of remote sensing technology in the application of spatial distribution and information identification and extraction of Chinese medicinal plant resources (CMPRs); analyzes the characteristics of CMPs in different planting patterns, different habitats, and different regions from the perspective of remote sensing information extraction; and explores the selection of suitable data sources, providing a reference for medicinal plant identification and information extraction. Secondly, according to the existing classification and identification methods, previous studies are summarized from the perspectives of classification scales, classification features, and classification accuracy, and the advantages and disadvantages of the commonly used remote sensing classification methods in the investigation of CMPRs are summarized and compared. Finally, the development trend of remote sensing technology in the identification and information extraction of CMPs is examined, and the key technical problems to be solved in the identification and classification of CMPs and the extraction of area information are summarized so as to provide technical support and experience references for the application of remote sensing in the investigation of CMPRs.

Keywords: Chinese medicinal plants; remote sensing; recognition extraction; classification

1. Introduction

The protection and sustainable utilization of Chinese Materia Medica Resources (CMMRs) are highly valued by the government. Four national surveys of CMMRs were conducted, respectively, in the 1960s, 1970s, 1980s, and 2010s [1]. Due to the increasing demand for traditional Chinese medicine (TCM) health care services around the world, the vigorous development of TCM-related industries, the transformation of medical models, the growth of people's health needs, and the speedy development of the TCM cause and TCM-related industries led to a significant increase in the demand for CMMRs. From 2011 to 2020, the National Administration of TCM organized the fourth national survey of CMMRs; compared with the third national survey of CMMRs, great changes have broadly occurred in the CMMR industry. The type, distribution, and application of CMMRs are constantly changing in China, so it is urgent to grasp the type and distribution of CMMRs in real time accurately and quickly in order to further determine the landscape of CMMRs in China and provide a reliable scientific basis for the protection measures of CMMRs and the development policy of related industries.

The sources of CMMRs are complex and diverse, including plants, bacteria, animals, and minerals, among which Chinese medicinal plants (CMPs) are the most important type of CMMR. Traditional surveys of CMMRs have mainly focused on field surveys [2] and historical data collection, combined with subjective estimation and ground surveys, which are greatly influenced by subjective factors, complicated, and lack scientific rigor. There are also phenomena of random cutting and felling of CMPs, and scattered planting can lead to significant differences in the spatial distribution of Chinese medicinal plant resources (CMPRs). Therefore, accurate spatial distribution, information identification, and extraction of CMPRs are of great significance for their protection and utilization. At the same time, compared with agricultural and forestry resources, CMPRs have many special characteristics, such as diverse species, diverse habitats, dispersed growth distribution, and non-population-common species, which make research problems more complex [3].

Remote sensing has the advantages of large coverage, high efficiency, timeliness, and labor reduction, providing timely and accurate farmland information for the agricultural sector. Compared with traditional survey methods, it has incomparable advantages and is increasingly used in agricultural production and management. As early as the 1960s, the Remote Sensing Agricultural Applications Laboratory at Purdue University in the United States was the first to use remote sensing data for crop planting area monitoring research, successfully monitoring a single corn crop and demonstrating the feasibility of satellite remote sensing data for crop monitoring [4]. In China, in 1979, Academician Chen Shupeng began to advocate using remote sensing technology for crop yield estimation. During Chian's "Eighth Five-Year Plan" period, the remote sensing estimation of crop yield became a national science and technology task, and experimental research on remote sensing estimation of large-scale wheat, maize, and rice was conducted; as well as dynamic monitoring research on grass–cattle equilibrium in northern grasslands; monitoring and forecasting the planting area, growth, and yield of winter wheat in Hebei, Shandong, Henan, northern Anhui, Beijing, and Tianjin; rice in Hubei, Jiangsu, and Shanghai and corn in Jilin Province [5]. This indicates that remote sensing technology is continuously maturing in crop monitoring in China, and many experts and scholars have conducted research on remote sensing monitoring of medicinal plant resources, with some progress. Sun et al. [6] proposed the idea and method of applying remote sensing to the investigation of Chinese medicinal materials resources. Sun et al. [7] studied the remote sensing monitoring methods of medicinal plant resources in different ecological environments, and, respectively, expounded the monitoring methods of wild and cultivated medicinal plant resources, providing ideas and methods for remote sensing monitoring of different types of medicinal plants. Sun et al. [8] conducted the first survey and monitoring of wild *Atractylodes lancea* (Thunb.) DC. medicinal resources in China using remote sensing technology, showing the feasibility of using remote sensing technology for surveying and monitoring individual medicinal plants. With the development and progress of technology, remote sensing has been integrated with many disciplines, effectively realizing the estimation of wild CMP reserves [9], the CMP planting areas in artificial plantations, dynamic monitoring [7], etc. Remote sensing technology is indispensable in the real-time dynamic monitoring and sustainable development of CMPRs. However, due to the complexity of remote sensing data, diversity of classification methods, and sensitivity to scale, research on remote sensing identification and planting information extraction of CMPs has not been systematically sorted out, making it difficult to conduct universal research and wide-ranging applications. Figure 1 shows the trend of paper publication of remote sensing applied to CMP monitoring since 2000 (a) and keyword co-occurrence analysis.

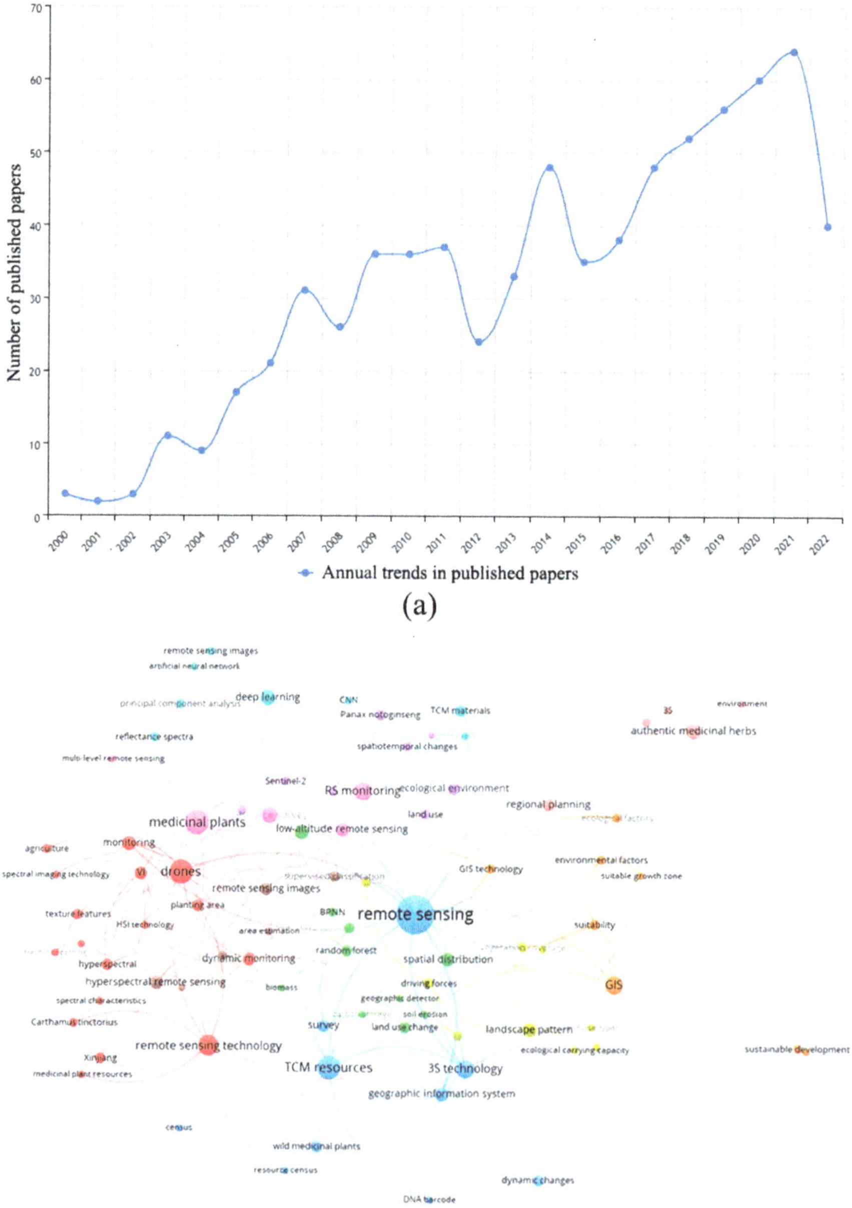

Figure 1. The trend of paper publication of remote sensing applied to CMP monitoring since 2000 (**a**) and keyword co-occurrence analysis based on VOSviewer (**b**).

Therefore, the planting patterns and habitat characteristics of CMPs are explored from the perspective of remote sensing application in this paper, providing references for data source selection, and the application of remote sensing in the identification and area extraction of CMPs is discussed. The identification and classification technology highlighted by remote sensing in the identification and area extraction of CMPs is emphasized in this paper, and the research progress of scholars is summarized and discussed, taking into account the classification scales, classification features, and classification methods. The issues and technical difficulties encountered in the research are also summarized and discussed. Then, the relevant research progress of CMP identification and information extraction is separately discussed in this paper according to different classification methods, and their advantages are compared and applicable objects described. Remote sensing data sources and classification methods can be preliminarily selected by researchers based on their own research content, methods, and objectives, combined with the types, planting patterns, and precision requirements of the CMP being studied, with reference to this paper. Finally, the key technical problems that still need to be solved in remote sensing identification and information extraction of CMPs are summarized based on previous literature reviews, and the future development direction of research is hypothesized, aiming at providing some references for further promoting remote sensing identification and information extraction research on CMPs.

2. Planting Characteristics and Applicable Satellite Remote Sensing Data for CMPs

The application of remote sensing technology in the investigation and dynamic monitoring of CMPRs is the focus of the research on the sustainable development of CMMRs, and the selection of appropriate data sources is the primary task of remote sensing monitoring of medicinal plant resources. How to select remote sensing data according to the types of CMPs to participate in the research has become a problem that many scholars need to consider first. The responses of different planting patterns and habitat characteristics to the monitoring of CMPRs should be explored in combination with actual planting characteristics, and more appropriate data sources should be selected according to the distribution and research value of medicinal plant resources in the study area. On the one hand, the accuracy of remote sensing monitoring of medicinal plant resources should be ensured; on the other hand, resource waste caused by purchasing high-resolution data sources should be avoided. Therefore, it is necessary to select appropriate data sources through research to provide reliable information for researchers of CMPRs.

2.1. Planting Patterns and Habitat Characteristics

In terms of planting patterns, CMPRs can be divided into wild and cultivated categories [10]. The former refers to non-artificially planted CMPRs distributed in a certain geographical area. There are various kinds of wild medicinal plant resources and different habitats in China. According to their living types, they can be categorized into woody plants (trees, shrubs, bamboo, vines, epiphytes, and parasitic trees), semi-woody plants (semi-shrubs and dwarf shrubs), herbaceous plants (perennial herbaceous plants, annual herbaceous plants, parasitic herbaceous plants, saprophytic herbaceous plants, and aquatic herbaceous plants), and thallophytes (mosses and lichens, algae, and fungi) [11]. Wild medicinal resources can also be further divided into abundant and rare plant species. Tree-based medicinal resources are generally densely distributed and have large scales, while shrubs are not as tall and independently distributed as trees and are more common in mountain grasslands and understories at higher elevations, making remote sensing identification and extraction of shrub-based medicinal plants more complex.

For rare medicinal plant resources, firstly, the distribution of wild rare medicinal plant resources is not concentrated in a single or dominant population but dispersed across different plant communities and geographically dispersed [7], which also creates certain difficulties in selecting remote sensing monitoring methods. Secondly, multi-spectral remote sensing data generally cannot directly obtain information on rare medicinal plant resources but can monitor vegetation cover in the medicinal plant habitat. Therefore, research on remote sensing monitoring methods for rare medicinal plant resources can be used to establish the inherent connection between the number of rare medicinal plant species and remote sensing monitoring methods to investigate the amount of rare medicinal plant resources. The selection of remote sensing images should be based on the characteristics of the rare medicinal resources themselves. When using remote sensing images for data information extraction, the spatial resolution, spectral resolution, and radiation resolution of the data should be considered from the perspective of monitoring rare medicinal plant resources.

At present, the natural quantity of rare medicinal plant resources in China is decreasing year by year [12] due to people's pursuit of wild rare medicinal plant resources in recent years and their destruction of the ecological environment system, which has led to a continuous decline in the number of rare wild medicinal plant resources. In this situation, cultivated medicinal plant resources have also received increasing attention [13]. On the one hand, cultivated medicinal plant resources can further alleviate the market demand. In this social context, remote sensing monitoring methods can effectively estimate the total amount, output, and demand of cultivated medicinal plant resources overall. Cultivated medicinal plants are mostly centrally distributed and recorded for filing, making it easier to analyze their resource distribution and information extraction than for wild medicinal plants. The investigation of cultivated medicinal plant resources, the calculation of the planting area, and the estimation of reserves can prevent the blind introduction of cultivation and provide a theoretical basis for guiding scientific and reasonable site selection for cultivation bases. Therefore, remote sensing identification and area extraction estimation of cultivated medicinal plant resources are important. More than 200 kinds of medicinal plants commonly used in China are mainly cultivated artificially. Due to the influence of growth years, planting areas, and management methods, the area and yield of medicinal plants obtained by conventional methods have a large error. Remote sensing technology has important application value and guiding significance in estimating and predicting the area and yield of cultivated medicinal plants [14]. Cultivated medicinal plants are mainly categorized based on their planting patterns: field cultivation, arable land cultivation, understory cultivation (intercropping), facility agriculture (greenhouse) cultivation, and special habitat cultivation. The research results of remote sensing in the field cultivation of CMPs can refer to the more mature remote sensing monitoring methods of some crops, and they are easier to carry out than other planting modes. Figure 2 shows classification of planting patterns of CMPs and examples of CMPs in different planting patterns.

Nowadays, in the research on remote sensing in the field of identification and classification of CMP arable-land-based cultivation and understory cultivation (intercropping), the lack of a large number of annotated data in the corresponding field and the subjectivity of manually annotated data have generally been mentioned by scholars [15]. In the case of dispersed planting of multiple kinds of traditional CMPRs within the same administrative region, if precise discrimination of the distribution area of each CMPR is required, an unmanned aerial vehicle (UAV) low-altitude remote sensing system can be used to obtain high-resolution image data of the ground, combined with texture and spectral features and ground surveys for further accurate and quantitative analysis, and secondary verification can be performed with ground sample data. Using this method, the assessment accuracy of the UAV remote sensing system in identifying, classifying, and performing area extraction for CMPRs can be confirmed to distinguish different CMPR types and distribution areas with a certain level of accuracy. Some CMPs do require intercropping with other crops to improve quality and yield. Since the spectrum characteristics of each plant and the modes of interplanting and single species are different, different classification models need to be constructed for different situations when using remote sensing image recognition to extract planting distribution areas. The typical CMP in a facility agriculture (greenhouse) model is greenhouse-cultivated ginseng, which is grown in a common warm greenhouse. The obstacle of continuous cropping of greenhouse-cultivated ginseng is an important factor restricting the planting of greenhouse-cultivated ginseng. The available ginseng resources are an important weakness that troubles the development of the ginseng industry. The dynamic monitoring of plantation distribution area and area changes using remote sensing technology is of guiding significance for the production and planning of ginseng. The roof of the ginseng shed is made of plastic film. Because of this planting mode, the garden ginseng has a particularly prominent advantage in satellite remote monitoring. Compared with the surrounding forest vegetation, the characteristics of the ginseng shed in remote sensing images are more prominent.

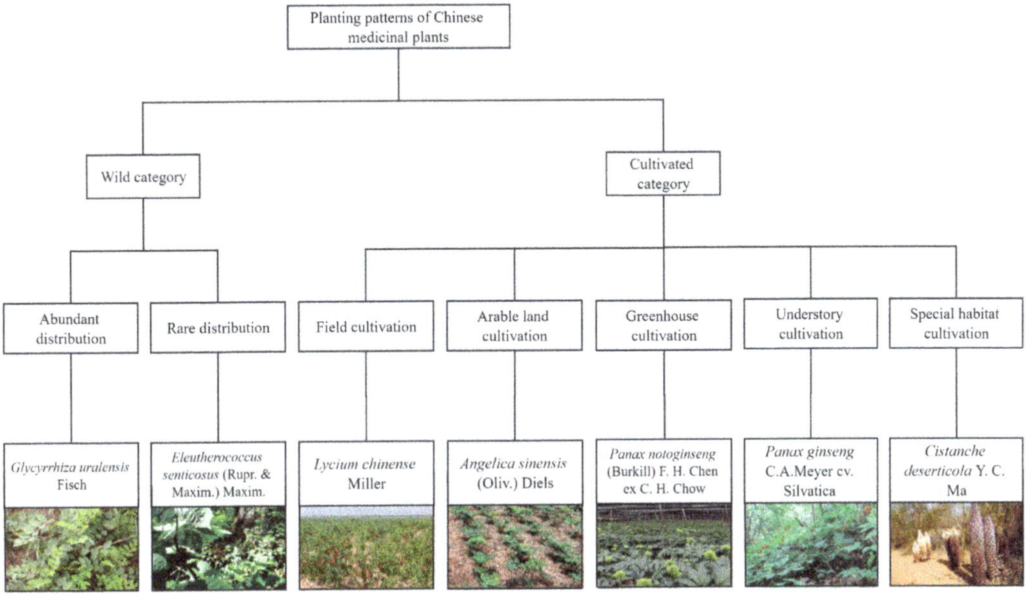

Figure 2. Classification of planting patterns of CMPs and examples of CMPs in different planting patterns.

In addition to diverse planting patterns, China has a rich and complex ecosystem, and the distribution of CMPRs varies greatly depending on the ecosystem. CMPRs have significant regional characteristics [16], and spatial characteristics and geographical environments play a decisive role in shaping the quality of CMPs, i.e., the "stress response" of medicinal plants [17]. Medicinal plants in different habitats have different spectral reflectance characteristics, which can be monitored using remote sensing techniques. In addition, some scholars use habitat factor interaction and correlation analysis with effective components of CMPs, combined with spatial overlapping analysis, to extract suitable areas for CMPs. On the basis of literature review and analysis of ecological factors such as elevation, annual precipitation and average annual temperature, Dong et al. [18] studied the distribution area of *Swertia mussotii* Franch. in Sichuan province by obtaining ecological factors, superposition analysis and GPS field verification data based on remote sensing and GIS technology. Remote sensing techniques are also involved in medicinal plants in some special habitats. Medicinal plants in certain special aquatic ecological environments can be monitored using different remote sensing methods based on their unique ecological environment information. For example, radar remote sensing can be used to estimate the resource quantity of aquatic TCM lotus, and remote sensing technology can be used to extract the special ecological environment of the parasitic plant *Cistanche deserticola* Y. C. Ma and some symbiotic plants in the desert for resource quantity estimation due to their special ecological environment types. Different types of medicinal plants have different special characteristics. Therefore, how to extract specific medicinal plant cultivation information should be analyzed according to its characteristics, and reasonable and effective remote sensing survey methods should be adopted. It is impossible to use a universal method for all cases.

2.2. Data Sources

The planting patterns and habitat characteristics of CMPs were introduced in the previous article. The selection of remote sensing data sources for monitoring CMPs should take these characteristics into account. Although the spatial resolution of early high-resolution satellites can be as high as meters, the number of bands is small, and image recognition needs to cooperate with the human eye. Chen et al. [9] used Landsat-7 ETM data from the United States, SPOT 5 data from France, and QuickBird image data from the United States to identify Jilin ginseng, mainly using manual interpretation and automatic interpretation methods. The results showed that the manual interpretation accuracy of high-resolution satellite images can reach 100%, the combination of medium-high and medium-resolution images can reach 97%, and medium-resolution satellite images can reach 90%. Although the former two have high accuracy, the cost is also high, and the manual interpretation cannot be widely promoted. Meanwhile, the workload of manual interpretation is large and time-consuming. This study explored a method for applying remote sensing technology to investigate medicinal plants. Zhang et al. [19] measured the area of wild *Glycyrrhiza uralensis* Fisch in Yanchi County, Ningxia Hui Autonomous Region, located on the Ordos Plateau, using medium-resolution ETM (15 m) satellite images combined with field data and the visual interpretation method. The comparative analysis of field verification points and image interpretation results showed that the manual interpretation accuracy of the survey sample area reached 90% when the ground resolution was 15 m, 97% when the ground resolution was 5 m, and 100% when the ground resolution was 0.61 m. This indicates that the higher the resolution of remote sensing data, the better the recognition accuracy in the survey of large area wild *G. uralensis* Fisch resources. This study discussed the influence of the spatial resolution of remote sensing data on extraction accuracy, provided a technical reference for the area estimation of large areas of wild Chinese medicinal materials, and provided a great technical scheme for wild resources investigation. Bahdu [20] studied the planting area of *Carthamus tinctorius* L. using data sources with different resolutions, such as the Worldview-2 satellite, ZY-3 satellite, and ZY-1 02C satellite, and found that high-resolution data could greatly improve the interpretation accuracy of *C. tinctorius* L., but it was difficult for automatic interpretation of *C. tinctorius* L. In order to analyze and compare which data source is more suitable for remote sensing identification of wild medicinal vegetation with more complex growth conditions, Na et al. [21,22] took wild *Apocynum venetum* L. nature reserve in Altai Alahak Township as the research object and adopted the medium-low resolution remote sensing image of domestic satellite ZY-3 and the high-resolution remote sensing image of foreign satellite Worldview-2. In this study, the same classification system was adopted for the two data sources, and the primary component analysis and texture feature were added to the traditional classification method to improve the classification accuracy. It was found that the high-resolution image of Worldview-2 was more suitable for the remote sensing identification of wild *A. venetum* L. with more complex growth conditions, while the domestic medium- and low-resolution ZY-3 was only suitable for the identification of densely distributed *A. venetum* L. and could not identify sporadic *A. venetum* L. Therefore, selecting remote sensing data sources should be based on the distribution of target objects in the specific work area. If the target objects in the study area are densely distributed and large in scale, such as artificial cultivation or single growing medicinal plant, the target objects can be identified well using remote sensing data of medium and low resolution, and there is no need to purchase high-resolution data, thus avoiding resource waste. For objects scattered and accompanied by other wild vegetation, it is considered that high-resolution satellite image data can be used to identify the objects well. Scattered wild medicinal plants are mostly planted in mountainous areas with suitable natural environmental conditions and complex terrain, and the planting plots are scattered. It is vital to provide decision support for the protection and utilization of CMMRs by studying the extraction method of the CMP planting area based on remote sensing. With the development of space remote sensing data in the direction of high spatial and temporal resolution, rich spectrum information, and wide band coverage, it has increasingly met the needs of large-scale surveys of CMMRs. Some scholars have applied Sentinel-2 [23] data and GF-6 data [24] to identify medicinal plants and achieved good results.

In addition, the unmanned aerial vehicle low-altitude remote sensing system (UAVRS) has the advantages of low cost, flexible and convenient operation and control, low-altitude flight under clouds, high timescales, etc. [14]. The spatial resolution has reached the centimeter level [25], which meets the sampling accuracy requirements, greatly reduces working time and human and material resources, and can also be used for the survey of CMMRs collected. It is especially suitable for rare medicinal plants with dispersed distribution and complex habitats. According to the characteristics of different remote sensing platforms and the distribution of medicinal plant resources, Xie [26] proposed the technical scheme of using low-altitude remote sensing such as the UAVRS to investigate medicinal plant resources, introduced the application characteristics and key technologies of unmanned aerial vehicles, and discussed the advantages and feasibility of unmanned aerial vehicles to investigate medicinal plant resources. She pointed out that the UAVRS can be used to select some representative areas based on low-resolution remote sensing images for high-resolution image supplementary acquisition, which can improve the multiple requirements of CMMR investigation. Xie [27] also conducted a survey of low-altitude remote sensing resources of endangered F. K. M. Shen in Xinjiang, and the research results clearly identified the geographical distribution area of F. K. M. Shen in Xinjiang for the first time, and the error between monitoring results and field data was only 14%. The research indicated that low-altitude remote sensing technology had the advantage of quickly obtaining the distribution status of wild F. K. M. Shen in Xinjiang and could effectively evaluate the dynamic change in wild F. K. M. Shen resources in Xinjiang. The author's research confirms that the UAVRS has a great application prospect in the investigation of CMMRs.

3. Research Status of Remote Sensing Identification of CMPs

According to different data sources and existing technical means of CMP information extraction research using satellite remote sensing technology, experts and scholars have proposed different methods of CMP identification and area information extraction, covering various sensor types, time/space/spectral resolution remote sensing data, and involving a variety of principles and technologies. The final classification effect and accuracy can be summarized by three levels of classification scales, classification features, and classification methods. Therefore, this paper summarizes and analyzes the existing research methods and progress in remote sensing in CMP identification and area information extraction from the perspective of classification scales, features, and methods. Figure 3 shows research flow chart of CMP monitoring based on remote sensing.

3.1. Classification Scales

3.1.1. Pixels

The pixel-based classification method is a classical method for remote sensing image analysis. In 2001, the Institute of Medicinal Plants, the Chinese Academy of Chinese Medical Sciences, used remote sensing technology for the first time to study the survey methodology of ginseng and other CMPRs, which created a precedent for the application of remote sensing technology in the survey of CMMRs. Sun et al. [8] used remote sensing technology for the first time to investigate the resources of the wild CMP *A. Lancea* (Thunb.) DC., they screened and classified the main influencing factors of the growth of *A. Lancea* (Thunb.) DC. According to the differences in reflectance spectral characteristics of different ground objects, the satellite remote sensing images of the Maoshan area in Jiangsu Province were interpreted, translated and inverted, and combined with field investigation data. The relationship between field data, remote sensing data, and habitat data of *A. Lancea*

(Thunb.) DC. was established, and the resources of *A. Lancea* (Thunb.) DC. in Maoshan were estimated. The results were not much different from about 10,000 plants provided by local farmers engaged in *A. Lancea* (Thunb.) DC. observation. The spatial resolution of remote sensing images will affect the extraction accuracy of vegetation information. For example, Chen et al. [9] used remote sensing data sources with different spatial resolutions to realize the investigation of ginseng resources and discussed the influence of spatial resolution of remote sensing data on extraction accuracy, which provided a useful reference for the selection of data for the investigation of medicinal species resources.

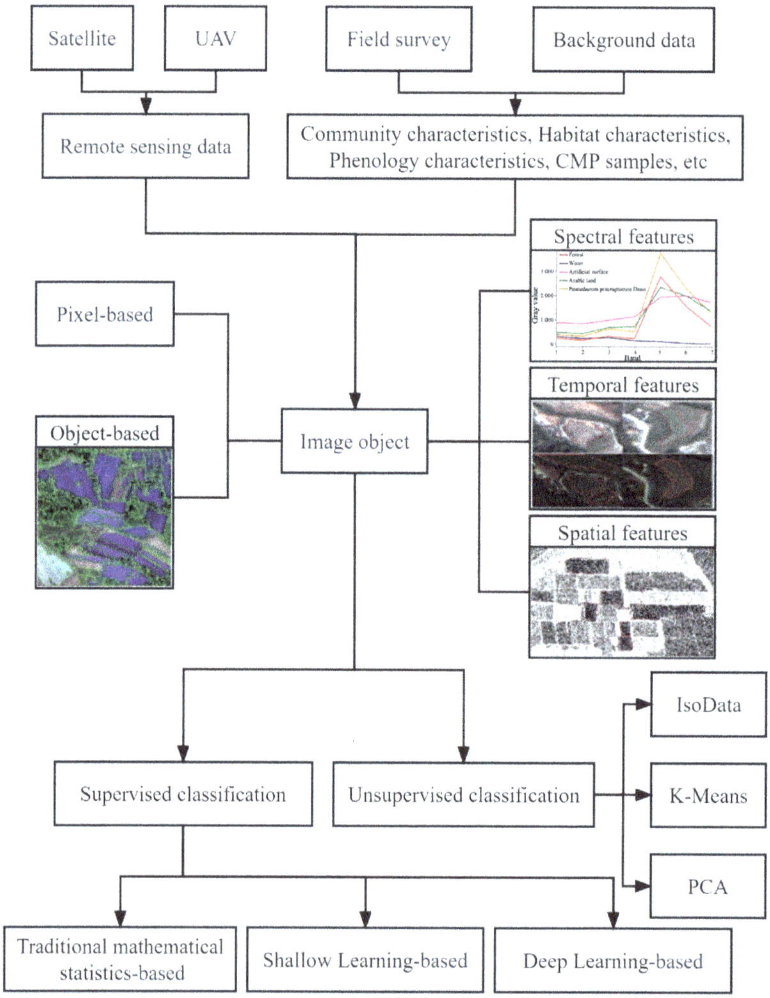

Figure 3. Research flow chart of CMP monitoring based on remote sensing.

Most remote sensing recognition of CMPs is still based on image elements, at present. The pixel-based classification method is mainly based on the spectral information characteristics of ground objects and is simple, intuitive, and easy to understand. For a large number of wild and continuously distributed medicinal plants and artificial large-scale cultivation, remote sensing data with low spatial resolution can be selected on the basis of ensuring classification accuracy, and the pixel is used as the unit for classification. This not only reduces the salt-and-pepper noise easily generated by pixel-based classification, but more importantly, the data cost is low, the steps are simple, and the time is short. However, it cannot process spatial information and is susceptible to the influence and interference of the high local heterogeneity of high-resolution images, with low classification accuracy; for the medicinal plants with complex habitats and fragmented distribution, it is difficult to obtain the ideal classification effect.

3.1.2. Objects

The traditional classification methods are difficult to apply to the CMP information extraction of high-spatial-resolution remote sensing images. Aiming at the above problems, an object-oriented remote sensing image classification method can also be used for remote sensing recognition of CMPs. An object-based CMP remote sensing image monitoring method refers to the segmentation of images into multiple homogeneous units with consistent attributes based on multi-scale segmentation technology and then the introduction of certain texture features or mathematical rules to improve the accuracy of the layout structure and distribution of medicinal plants. The key point is that the object-based classification method is not based on a single pixel but a homogeneous unit, which makes full use of the spectral statistical characteristics of the image and a series of information such as the size, shape, texture, and adjacent relationship of the ground object [28]. This method plays an important role in extracting CMP planting information and solves the problem of the fragmentation of the extraction plot of TCM. Shi et al. [15] took Fusong County, Jilin Province, as the research area and used ZY-3 satellite data as the data source to explore a method of ginseng planting area extraction based on a rule set object-based method. The research showed that, in terms of the results of multi-scale segmentation, the optimal segmentation scale of ginseng planting plots could be determined, and the fragmented and distributed ginseng planting plots could be accurately extracted based on a certain spectral feature rule set, which was beyond the reach of traditional pixel-based classification methods. Among them, the multi-scale segmentation algorithm used in this study is common in image-segmentation technology.

The object-based classification method can overcome the limitations of traditional methods and make full use of the spectral and spatial information of high-resolution images so as to extract the necessary thematic information quickly and accurately. However, the accuracy is much lower for low-resolution images, and the operation is more complicated. Compared with the classification of medicinal plants based on pixels, the object-oriented classification of medicinal plants first utilizes the different characteristics of CMPs and other vegetation and non-vegetation ground objects to establish different rules and segment the images. Therefore, the classification effect of the object-oriented classification of CMPs in complex habitats with diverse vegetation types and relatively fragmented distribution is better. And to a certain extent, the phenomenon of mismarking and missing marks is avoided. Based on the classification of objects, how to accurately and effectively classify the CMPs and other vegetation to be monitored into different objects is very important.

3.2. *Categorical Features*

3.2.1. Spectral Features

As the physical basis of crop remote sensing classification, spectral features result from the joint action of crop physiological and biochemical parameters and environmental factors [29]. The focus of this method is to find the "key phenological period" of different crops—that is, to find the specific period when the spectral features of the CMP to be separated have the largest difference from the background spectral features. For example, Liu et al. [30] took the medicinal plant safflower in the Changji region of Xinjiang as the research object and found that the safflower canopy and the remaining safflower canopy after picking had characteristic absorption peaks near 770 nm through the analysis of its canopy spectral features. The method proposed aimed to study the special hyperspectral features of *C. tinctorius* L. under a specific phenological stage. It provided an early theoretical basis for monitoring the growth state and physiological parameters of *C. tinctorius* L. and also laid the groundwork for the estimation of distribution area and dynamic fine monitoring of *C. tinctorius* L. in the future. Based on GF-6 remote sensing data, Li et al. [24] conducted a comparative study on the growth cycle of CMPs and winter wheat and found a special stage in the growth cycle of winter wheat. The research in this article showed that the reflectance of the spectral features of winter wheat in the grain filling period and milk maturity period in May decreased in the green band and near-red band, while the CMP showed signs of luxuriant branches and leaves in May, and the reflection of the green band and the near-red band was higher. According to the difference and magnification, the difference in spectral curve between *Angelica dahurica* (Fisch. ex Hoffm.) Benth. and Hook. f. ex Franch. and Sav., *Taraxacum mongolicum* Hand.-Mazz., *A. Lancea* (Thunb.) DC., *Saposhnikovia divaricata* (Trucz.) Schischk., and winter wheat could be distinguished, and the planting area of CMPs could be extracted. However, this study needed a parameter to distinguish whether it was a CMP or winter wheat, but this parameter depended on the empirical model and field investigation, and the selection of this parameter had a great impact on the classification results, so it needed to be carefully selected and screened. As an alternative index to study vegetation, the quantitative measurement of the vegetation index can indicate vegetation vitality, etc. [31,32]. Among them, the NDVI is the most widely used, and the NDVI is defined as the ratio of the difference between the values of the NIR band and the red band and the sum of the values of the NIR band and the red band. Zhang et al. [33] used *Paeonia lactiflora* Pall. as the data source of CMPs and analyzed the NDVI values of wild *P. lactiflora* Pall. growing points in different periods. Combined with the data of sampling points, the authors monitored the distribution area of *P. lactiflora* Pall. for multiple cycles. The result showed that the change trend of the distribution of peony resources could be retrieved and monitored based on the change in the vegetation index in different periods.

All the above studies have proved the feasibility of using spectral features for remote sensing identification and information extraction of CMPs, but the classification method based on spectral information will inevitably cause the phenomenon of "the same object with different spectrum" and "the different object with same spectrum" in the classification results [34], resulting in low classification accuracy. Because of the wide variety of CMP planting, it is often difficult to capture the image of the "best recognition period" of planting. Only relying on spectral feature classification to extract planting information of CMPs cannot achieve ideal results.

3.2.2. Temporal Features

Temporal features refer to the changing rules of vegetation in different temporal remote sensing images, which mainly reflect the physiological and biochemical differences in medicinal plants in different phenological stages. The growth and development of vegetation have phenological characteristics, and its spectral features will change with the change in seasons [35], which leads to the fact that the image classification effect of a single period is often not ideal. Therefore, the advantages of vegetation classification based on multi-temporal remote sensing images are more obvious [36]. The planting extraction method based on multi-temporal image data sources can make full use of the seasonal rhythm characteristics of medicinal plants, which has gradually become the mainstream method for remote sensing extraction of medicinal plant planting information. Liu et al. [23] studied the classification accuracy of four random forest (RF) classification methods using the multi-temporal Sentinel-2 (MSI) as the remote sensing data source, respectively, using spectral bands alone or in combination, vegetation indices (VIs), and topographic variables, and took *Zanthoxylum bungeanum* Maxim. from the Linxia Hui Autonomous Prefecture of Gansu Province as the research object. The result showed that the classification accuracy of only using spectral bands (40 input features) can reach 95.43%, and the classification accuracy can also approach 95% when only using the 15 most important features. This study demonstrated the potential of multi-temporal remote sensing data in improving the accuracy of planting area extraction of *Z. bungeanum* Maxim, which possibly also allowed for detailed remote sensing identification and classification of other special crops or tree species in other regions.

The vegetation classification method using multi-temporal information can be used in the remote sensing identification and classification extraction of CMPs. Because it follows the growth law of CMPs and conforms to the current vegetation remote sensing classification trend, the classification accuracy is greatly improved. Especially for medicinal plants grown in the field, we can start with the regional bulk crop calendar to determine the images of the different periods between the growth of medicinal plants and the crop calendar, which can be well distinguished. However, the classification method is complicated and requires time series data, and sometimes the available data cannot meet the needs.

3.2.3. Spatial Features

Rational use of the spatial features of CMPs can realize the distinction between CMPs and other non-CMPs. With the rapid development of image processing technology, spatial features have become an important means to assist spectral features and temporal features for remote sensing identification of CMPs and extracting planting information [37], especially for suppressing the phenomenon of "the same object with different spectrum". In the identification of medicinal plants and extraction of planting areas, only using the spectral information of remote sensing images cannot effectively extract the information of medicinal plants. For medicinal plants, the reflection spectrum of surrounding vegetation is similar to them, which is easily confused. If texture features can be reasonably used to find the special texture features of medicinal plants that are different from other vegetation, combined with their specific growth period, it will greatly improve the classification accuracy of remote sensing images. Jia et al. [16] took the medicinal plant Ledebouriellae Radix planted in a large area of Naiman Banner as an example, used the fused 2 m resolution Resource No. 3 (ZY-3) remote sensing image as the data source, and used the texture and spectral information of the remote sensing image to extract and understand the distribution range and planting area of Ledebouriellae Radix in Naiman Banner. The research result showed that combining spectral information and texture information could realize the discrimination of Ledebouriellae Radix. This study gradually applied spectral information and texture information to the identification of CMPs and provided a new research idea and method for the further identification and extraction of CMPs via remote sensing. Wu et al. [38] first extracted the distribution data of the *Panax notoginseng* (Burkill) F. H. Chen ex C. H. Chow planting area based on TM/OLI images using a supervised classification method. On this basis, combined with the topographic distribution characteristics of the image in the planting area of *P. notoginseng* (Burkill) F. H. Chen ex C. H. Chow, a decision tree was constructed to correct the rough extraction results of supervised classification so as to achieve the goal of rapid extraction of the distribution data of the *P. notoginseng* (Burkill) F. H. Chen ex C. H. Chow planting area and realize the rapid extraction of the spatial distribution information of the *P. notoginseng* (Burkill) F. H. Chen ex C. H. Chow planting area. The accuracy of the results was high, which could meet the requirements of research and application.

The coverage of CMPs is usually low, and the planted area of cultivated medicinal plants is relatively fragmented compared to other large crops, while wild medicinal plants may be more fragmented. Using spatial features for the remote sensing identification and distribution area extraction of medicinal plants can improve the interpretability and visualization effect of classification and can provide more comprehensive information, which is of great help to improve accuracy. However, it is usually necessary to consider the relationship between pixels and the complex influence of the surrounding environment, so its extraction and calculation are difficult and require a lot of computing resources and time.

3.3. Classification Methods

3.3.1. Supervised Classification

- Based on traditional mathematical statistics

Remote sensing image classification is divided into two categories according to whether it has prior knowledge: supervised classification and unsupervised classification [39]. Supervised classification usually selects the region of interest (ROI) of the corresponding known object from the remote sensing image in advance and uses the image features of the sample area to train the classification algorithm to complete the classification of the whole study area. At present, the supervised classification method based on traditional mathematical statistics is mainly used in the identification of cultivated medicinal plants planted in cultivated land and fields, because the identification of medicinal plants in this planting mode is relatively easy. Yi et al. [40] used the domestic Satellite Resource No. 3 (ZY-3) image to perform remote sensing identification of *Paeonia ostii* T. Hong et J. X. Zhang using the maximum likelihood method and classified and extracted the planting area of *P. ostii* T. Hong et J. X. Zhang. The overall classification accuracy was 97.81%, and the KAPPAA coefficient was 0.96. The result showed that it was feasible to estimate the *P. ostii* T. Hong et J. X. Zhang planting area based on remote sensing image classification using the maximum likelihood method. Yang et al. [41] took a ZY-3 remote sensing image as the data source and used the minimum distance method, maximum likelihood method, parallelepiped method, and Markov distance method to identify *Cynanchum otophyllum* Schneid. in Dangshan County in 2018 via remote sensing and compared and analyzed the results, providing a reference for remote sensing monitoring of the Chinese *C. otophyllum* Schneid. planting area.

The supervised classification method based on traditional mathematical statistics is suitable for situations that require interpretation, small sample data, or clear problem modeling, but it has certain requirements on the distribution and characteristics of data. Although it is often used in cultivated medicinal plants that are more easily identified, it is also very limited. If the plots are scattered, the accuracy of remote sensing image extraction is not high enough. It is easy to be restricted by the phenomena of "same thing different spectrum", "foreign body same spectrum", and "mixed pixel".

- Based on Shallow Learning

Machine learning will shift the focus of the research to how to better improve the prediction effect of the model, concerned with the problem directly related to improving the prediction effect. The non-parametric model, which tends to be automated, does not need to assume normal distribution of data and has gradually become a hot spot in the research of extracting CMP information. Dai et al. [42] used Landsat data to extract *P. notoginseng* (Burkill) F. H. Chen ex C. H. Chow planting areas in Wenshan Prefecture and Honghe Prefecture in 2010, 2012, 2014, and 2015 through a support vector machine. The classification accuracy could meet the requirements of spatial analysis and practical applications and proved the feasibility of real-time dynamic monitoring of the *P. notoginseng* (Burkill) F. H. Chen ex C. H. Chow planting area by using classification technology based on a support vector machine. As an information extraction method based on spatial data mining and knowledge discovery, a decision tree is an efficient spectral classification method. The basic idea is that the remote sensing data set is graded downward according to specific preset rules, and finally, subcategories with single or different attributes are obtained [43]. Some scholars have applied this method to the research of remote sensing identification and information extraction of CMPs. Zhu et al. [44] improved the expert decision tree model based on the domestic open-source image of GF-1 with a 16 m resolution and extracted shade image spots of *Panax notoginseng* (Burkill) F. H. Chen ex C. H. Chow using decision tree classification. In evaluating the reliability of identification and the accuracy of area extraction of *Panax notoginseng* (Burkill) F. H. Chen ex C. H. Chow, Google Images was used instead of field drawing, and visual interpretation results were used as the benchmark of spot area. The result showed that the accuracy of decision tree classification and visual interpretation is 87% and 99%, respectively, and the overall area accuracy of the results extracted using expert decision tree classification was 80%. This study showed that compared with the traditional method of artificial marking based on high-resolution commercial images and visual interpretation, the method used in the area extraction of *Panax notoginseng* (Burkill) F. H. Chen ex C. H. Chow based on a decision tree model could investigate *Panax notoginseng* (Burkill) F. H. Chen ex C. H. Chow planting resources in Wenshan at a faster speed under the condition of ensuring a certain accuracy. Traditional shallow learning is an effective, fast, and high-precision data mining method; requires a relatively small amount of data; does not require a high-quality data set; and creates a good interpretation of the model. However, it relies on artificial feature extraction, and due to the simple structure of the model, its generalization ability is limited, which reduces the classification accuracy of this method to a certain extent. The planting pattern of CMPs is closely related to the terrain, humidity, climate, soil, farming system, and other factors. Therefore, in the extraction of the planting structure of CMPs, in addition to the pixel spectral band features, other non-remote sensing information such as statistical data of CMPs, landforms, phenological characteristics, and prior knowledge of experts participate in the extraction of the planting structure of CMPs as important auxiliary information to improve the classification accuracy, but the effect of extracting complex features in the image is poor.

- Based on Deep Learning

The concept of deep learning (DL) was first proposed in 2006 [45]. Compared with traditional methods, DL algorithms can adaptively learn various features from a large number of samples and can directly input remote sensing images into the model network. At present, DL algorithms have been applied in vegetation remote sensing recognition and information extraction. However, there is little research on the application of CMPs, especially research on the fine classification of CMPs in high-resolution remote sensing images. Shi et al. [46] applied the deep convolutional neural network model to the recognition of CMPs mimicking wild planting. The method of deep learning was used to solve the problems of insufficient utilization of high-resolution image information and "salt-and-pepper noise" caused by intra-class pixel differences in traditional classification. Remote sensing data were obtained using an unmanned aerial vehicle (UAV), and the convolutional neural network was applied to the fine classification of *Lonicera japonica* Thunb. wild planting. The result showed that the overall classification accuracy of the optimized model after training could reach 97.5%, and the total accuracy of the area was 94.6%. The deep convolutional neural network based on GoogLeNet could effectively extract the information of *L. japonica* Thunb. with a broken distribution in the image and realized the fine classification of *L. japonica* Thunb. Deep learning was applied to the classification of medicinal plants from UAV high-resolution images, providing a reference for the application of a deep convolutional neural network method in the remote sensing classification of CMPs. It can be seen that deep learning algorithms have great development potential in the intelligent extraction of medicinal plant resource information in remote sensing big data.

Deep learning has the abilities of self-learning and high-speed searching for optimal solutions, and the accuracy of extracting medicinal plant types is high. However, this method needs to provide a large number of parameter training samples, the training time is long, it is difficult to provide the reasoning process and basis when classifying the types, and there are numerous deep learning models. The remote sensing recognition and classification methods of CMPRs still need further optimization.

3.3.2. Unsupervised Classification

Unsupervised classification is also known as the cluster analysis method. It can help uncover hidden structures, patterns, or relationships in the data and facilitate exploratory data analysis and insight and then complete the final classification with the classifier. Therefore, unsupervised classification can identify the distribution pattern and habitat preference of CMPs without samples. Sun et al. [47] used LANDSAT-5 TM as the data source to extract and analyze *Ginkgo biloba* L. information from multi-period TM remote sensing images using the unsupervised clustering method and obtained the spatial distribution and multi-year dynamic change in *G. biloba* L. in Pizhou City, Jiangsu Province, with the support of field investigation data. It provides an idea and method for monitoring the dynamic change in medicinal plant resources.

Since unsupervised classification does not require pre-labeled samples, it is difficult to distinguish medicinal plants from general plants, and it is more suitable for monitoring medicinal plant resources where there are basically no other similar types of vegetation in the distribution area and the species result is single, so it is rarely used in remote sensing recognition of CMPs.

4. Key Problems and Prospects

Remote sensing has made significant progress in the dynamic monitoring and identification classification of CMPRs, which provides scientific planning guidance for the sustainable development of the CMPR industry and has a good prospect in extracting the planting area of CMPRs. However, there are still some technical deficiencies that restrict the accuracy and efficiency of remote sensing extraction results. For example, cultivated medicinal plants have low vegetation coverage and are more scattered, while wild ones are more fragmented and diverse. How to accurately extract the planting area by remote sensing images will always be a problem that needs to be studied. For a long period of time in the future, the research on remote sensing identification and extraction of CMPRs should aim at reducing mixed pixels and improving extraction accuracy and should combine the characteristics of CMP types, give full play to the advantages of combined utilization of multi-source data, and carry out research on the extraction of planting information around the diversification of CMP types and the multi-scale coverage area. The research on key technologies such as data preprocessing, feature parameter extraction, and efficient classifier selection was strengthened to improve the spatial scale of the remote sensing extraction of plant CMMRs so as to meet the needs of a variety of plant CMMR investigations.

4.1. Key Problems

- There are many kinds of Chinese medicinal materials, great differences in planting patterns, and complex planting landscape patterns

There are many kinds of plant Chinese medicinal materials, and different Chinese medicinal materials have different spectral and image characteristics [48]. Due to their different characteristics, there are a variety of planting patterns. Taking field planting as an example, there are also widespread rotation, intercropping, and intercropping phenomena [49], such as bupleur–maize intercropping, rice–Coptis rice–water and drought rotation, Pinellia–maize intercropping, and so on. There were differences in spatial pattern, vertical distribution, and time variation among different planting patterns of CMPs.

At the same time, unlike the large-scale cultivation of traditional food crops, the cultivation of Chinese medicinal materials is relatively scattered, with small plots and fragmented landscapes [50]. In this case, spectral feature recognition using satellite remote sensing data is prone to mixed-pixel problems. These characteristics of CMP planting have created new requirements for remote sensing technology in terms of spatial resolution, time resolution, spectral recognition degree, recognition features, and recognition methods. It is therefore necessary to explore new remote sensing technology.

- The selection of data sources

How to make full use of multi-source data is a key issue that needs to be considered. Remote sensing identification and area extraction of CMPs involve many types, which require high spatial, temporal, and spectral resolutions of remote sensing images. How to deal with the contradictory relationship between the spectral, temporal, and spatial resolution to realize the identification of CMPRs and the accurate extraction of the planting area is a problem that needs to be considered. In addition, it has also been mentioned above that non-remote sensing data sources, such as meteorology, topography, and phenology, are often introduced into the extraction of auxiliary plant planting information of CMPs to improve the accuracy of remote sensing extraction. However, the quality of data from different sources and the difference in spatial and temporal scales will affect the accuracy of remote sensing extraction in turn. Therefore, future research needs to strengthen the research on the substitution and integration of multi-source data to achieve the "complementary advantages" between different data sources.

4.2. Prospects

- Establishment of planting pattern, habitat, and phenological characteristics database of medicinal plants

There are a variety of cultivation patterns for medicinal plants. In addition, the distribution of CMPs in different ecosystems is varied and has significant regional characteristics [51], leading to complex habitat characteristics. A sample database with reasonable classification and a sufficient quantity of information such as planting patterns, habitat characteristics, and phenological characteristics of medicinal plants should be built and applied to remote sensing recognition of medicinal plants in combination with data assimilation technology. The classification algorithm can be guided by the information in the database to improve the accuracy of the identification of medicinal plants and to understand the distribution and growth of medicinal plants more accurately. Therefore, different monitoring schemes can be developed for different models of CMPs.

(1) For CMPs planted in cultivated fields and greenhouses, the planting mode is relatively simple, and the surface distribution and spectral characteristics are obvious. Based on the appropriate adjustment of the existing remote sensing identification technology methods for crop and land-cover types, the corresponding remote sensing identification and spatial distribution mapping can be carried out. After the improvement in monitoring technology and accuracy evaluation, this kind of CMP can be demonstrated and applied in a wide range to form the spatial distribution map and area data of the corresponding CMPs. Among them, based on the accurate identification of CMPs grown in the field, it can also carry out theme monitoring and application, including growth monitoring, maturity prediction, and unit yield estimation, which can support production management situation research and judgment.

(2) For the cultivation of CMPs under forest and other field conditions, the planting mode is complex, the planting plot is broken, and the spectral characteristics are difficult to obtain due to the influence of vertical structure. At the same time, spectral and mixed pixel interference are serious due to the influence of adjacent vegetation and surrounding high vegetation. Given this situation, it is necessary to fully discover the characteristics of multi-spectral and multi-feature remote sensing technology and combine optical and microwave data, microwave data of different bands (C\X\S band), and microwave data of different polarizations (VV\VH\HH\HV) to explore the technology and method of identification and monitoring by combining multi-source remote sensing data.

(3) For the cultivation of CMPs under complex landscape patterns and CMPs under complex terrain and special habitats, the existing identification cannot be directly applied due to the complexity of their species patterns. It is necessary to carry out a targeted analysis of its spatial distribution landscape pattern and optimize the satellite remote sensing data with a corresponding resolution based on the monitoring scale. Then, the spectral and spatial characteristics were analyzed to form the recognition feature system. Finally, the application effect of different classification methods in crop recognition was discussed, and the suitable recognition method was optimized. Through the breakthrough of recognition scale, recognition feature, and recognition method, a suitable recognition technology and process were formed and applied.

- Establishment of spectral sample library of medicinal plants

Spectral data have important reference significance for remote sensing signals to identify ground objects and extract land surface information. Research on the spectral features of ground objects has attracted attention since the 1940s, and establishing a spectral database of ground objects plays an important role in quantitative remote sensing modeling, land surface parameter inversion, and environmental ecological monitoring. In recent years, various research institutions at home and abroad have built part of the ground object spectrum database [52–54], which better supports the research in ground object classification and target recognition. There is no spectral sample database for CMPs, with few sample data and scattered management. At the same time, there is a lack of a data-driven large-scale sample database covering different planting patterns and habitat characteristics, as well as multi-type sample data of CMPs such as multi-spectral, hyperspectral, and infrared [55].

Based on the spectral data of ground objects, the spectral characteristics of CMPs in visible light, near-infrared, and short-wave infrared can be studied. The spectral changes in plants under different climates, terrains, and nutrient conditions can be analyzed. The amplitude and rate of spectral variation in different growth stages of TCM plants can be analyzed, and the observation window and potential identification bands of TCM plants can be proposed. The difference in spectral characteristics between different TCM plants and between TCM plants and other plants can be analyzed, and the spectral characteristic parameters of TCM plant recognition can be established. The establishment of a CMP spectral database, the study of the relationship between CMP spectral features and remote sensing information, and the matching and recognition of non-imaging spectral data and multi-spectral imaging data can provide a theoretical basis for the remote sensing identification of CMPs and play an important role in the construction of CMP information. In particular, the main research object of species identification of wild medicinal plants should be the individual wild medicinal plants. The key technical links need to use ground-based spectrometers to measure and obtain the spectral data of medicinal plants, analyze and study them, clarify the spectral characteristics of each medicinal plant species, and establish the standard spectral characteristics database of medicinal plant species. It should be noted that in the process of establishing the database, in addition to making it have a complete analysis function, strict data specification standards should be established. The establishment of a spectral database of medicinal plants is a complex and continuous work, which requires many efforts and resources. At the same time, the spectral characteristics of different medicinal plants may vary greatly, so it is necessary to take this diversity into account and develop corresponding data collection and analysis strategies. In addition, even for the same variety, under different growth environmental conditions its spectral data are significantly different, and samples of various medicinal plants should be collected from different geographical locations to ensure that a variety of species and growth environments are covered. At the time of sample collection, data such as detailed geographic information, growth conditions, and growth stage are recorded, which will facilitate subsequent data analysis. Over time, as new data and technologies become available, spectral databases need to be constantly updated and improved.

- Fusion of multi-source and multi-temporal data

From the perspective of data source selection, the data sources applied to extract CMP information using remote sensing technology should be diversified. At present, most studies focus on the use of a single data source; however, the use of a single data source makes it difficult to meet the needs of high-precision extraction of medicinal plant information. Satellite remote sensing data are also sometimes affected by clouds, bad weather, orbital period, resolution, and other factors, resulting in data pollution, reducing remote sensing observation, and increasing the spatio-temporal discontinuity of remote sensing observation [56]. UAVRS makes up for this shortage with its flexibility, convenience, low cost of data acquisition, and the ability to fly at low altitudes under clouds [57,58]. The development of UAVRS technology has improved the accuracy and real-time performance of data collection and area extraction and expanded the application scope of medicinal plant resources research based on satellite remote sensing. It has accumulated valuable experience and provided reference value for UAVRS in the future extraction of large-area wild medicinal resources information. Microwave remote sensing has the advantages of working all day in all weather and strong penetration [59]. If it is applied to the dynamic monitoring of CMMRs, it can make the data acquisition independent of the local weather, and for some short varieties of CMPs shielded by the canopy, they can also be successfully monitored. Using multi-source and multi-temporal data fusion technology, multi-source data improve the spatial resolution of remote sensing images, and multi-temporal data improve the time resolution of remote sensing images, which is beneficial to obtain the best time phase of medicinal plant remote sensing information, better characterize the whole process of growth and development of medicinal plants, and improve the accuracy of CMP identification and information extraction.

With the development of science and technology in remote sensing data sources, the relationship between spectral resolution, spatial resolution, and temporal resolution of remote sensing data should be fully considered in view of different planting patterns and habitat characteristics of medicinal plants, combined with space remote sensing and aerial remote sensing data, as well as optical, thermal infrared, and microwave remote sensing data. Therefore, it is also necessary to strengthen the research on the scale conversion of multi-source data. It is important to improve the flexibility and accuracy of planting information extraction of plant CMMRs, meet the needs of dynamic and real-time monitoring under the broken agricultural landscape pattern, and improve the algorithm to solve or alleviate the problem of significantly reduced spatial independence of forecast data in view of the current various spatial–temporal data fusion technologies, which will reduce the spatial independence of data. Using the fused multi-source remote sensing data, the ability of feature extraction and data mining is improved so as to further improve the accuracy of medicinal plant resource information extraction. If the data of a single period do not meet the accuracy requirements for extracting information, establishing a technical system for comprehensive identification of medicinal plants with multi-source remote sensing data of the whole growth period should also be considered.

- The application of deep learning and classification automation

The continuous development of deep learning technology has also shown great potential in the automatic remote sensing monitoring and classification of CMPs. Compared with traditional methods, deep learning technology can automatically learn and extract the characteristics of medicinal plants in remote sensing monitoring and identification classification of medicinal plants, which greatly reduces the cost of manual intervention. This automatic feature extraction method can not only significantly improve the accuracy of monitoring and recognition classification but also shorten the start-up cycle and theoretical threshold. In addition, deep learning technology can deal with more complex data features and has higher flexibility and adaptability [60], as well as very good compatibility and integration capabilities. These advantages jointly promote the continuous exploration and progress of automatic monitoring, identification, and classification of medicinal plants in the field of remote sensing. Therefore, the quality and quantity of data used for deep learning model training, the appropriate network architecture, and the selection and extraction of optimal features need to be focused on. In particular, the selection of feature variables for the classification of medicinal plants has great limitations, and more attention is paid to the characteristics of medicinal plants in experimental areas, and its planting and phenological characteristics are rarely considered. It is important to make full use of the advantages of deep learning in information mining and extraction and apply it as a tool in remote sensing identification of CMP planting types and remote sensing inversion of physiological/biochemical parameters so as to improve the ability and accuracy of CMP monitoring. Evaluating the contribution rate of different features to crop classification and identification and determining the key set of identifying features for different medicinal plant types and their combinations is a concept that can be explored in future research. After improving the feature extraction method of the existing deep convolutional neural network, the method test and accuracy evaluation are carried out through the changes in samples. According to different planting patterns, differentiated target extraction schemes can be adopted:

(1) Field planting pattern: Starting with the regional bulk crop calendar, the images of the different periods of TCM growth and crop calendar were determined. The method of spectral characteristic parameter identification and classification is used to extract the target.
(2) Greenhouse planting pattern: Methods such as brightness index or building index and object form index are used to extract greenhouse information, such as the form, size, and spatial combination of the characteristics of regional traditional Chinese medicine planting greenhouses. Then combined with ground sampling survey to further confirm.
(3) Mountain planting pattern: Regional environmental information (topography, geomorphic features) was used to classify potential planting targets for medicinal plants. Based on the luminance index, morphology index, and other parameters, the "skylight" block in the forest is separated. The classification method of the vegetation characteristic index was used to distinguish medicinal plants from grassland, cultivated land, and other bare land.
(4) Underwood planting pattern: With radar and multi-spectral data, multi-source data fusion was carried out to identify and classify the medicinal plants under the forest.

5. Conclusions

This paper mainly analyzes the selection of remote sensing monitoring methods for CMPRs in different remote sensing data scales and remote sensing classification scales. It introduces the research progress of the investigation work of researchers on medicinal plant resources in China and discusses the difficulties and hot issues in the identification and classification of CMPs using remote sensing. This paper compares and analyzes the common techniques and methods of remote sensing recognition and classification at the present stage and discusses the shortcomings of related research and suggestions for improvement. In general, multi-source remote sensing technology will continue to be used in the identification and classification of medicinal plant resources, and the establishment of medicinal plant planting characteristics and a spectral database in advance can guide identification and classification and improve its accuracy. Not only that, but with the rapid development of artificial intelligence, remote sensing extraction of CMP area information has begun to develop from redundant manual interpretation to deep learning technology, which will greatly reduce the consumption of human and material resources, provide a new and fast method for the identification and extraction of planting information of CMPRs, and is expected to develop in the direction of automation. This study summarized the latest frontier results at home and abroad; based on the new needs of the development of medicinal plant resources in China, it provides a reference for improving remote sensing technology in further research on the identification and area extraction of CMPRs.

Author Contributions: Conceptualization, J.M. and X.Y.; methodology, J.M. and X.Y.; software, J.M. and X.Y.; validation, J.M. and X.Y.; formal analysis, J.M. and X.Y.; investigation, J.M. and X.Y.; resources, J.M., X.Y., X.Z., T.S., L.Z., X.C. and H.Z.; data curation, J.M. and X.Y.; writing—original draft preparation, J.M., X.Y. and M.X.; writing—review and editing, J.M., X.Y., X.Z., T.S. and X.C.; visualization, J.M. and X.Y.; supervision, J.M. and X.Y.; project administration, J.M. and X.Y.; funding acquisition, J.M., X.Z., T.S., L.Z. and X.C. All authors have read and agreed to the published version of the manuscript.

Funding: This research was funded by the Scientific and Technological Innovation Project of the China Academy of Chinese Medical Sciences (No. CI2021A03902, CI2021A03901, and CICI2021B013); the Guangxi Science and Technology Development Plan (No. 2022AA01030); the National Key Research and Development Program of China (No. 2022YFD2001102).

Data Availability Statement: No new data were created for this article.

Acknowledgments: We would like to thank China Academy of Chinese Medical Sciences for providing helpful feedback on the original draft of the manuscript.

Conflicts of Interest: The authors declare no conflict of interest.

References

1. Ru, M.; Dai, Z.; Li, Y.; Lu, H.; Feng, J.; Lu, Q.; Guo, M. Investigation and research on medicinal plant resources in Yongning District, Guangxi. *J. Chin. Med. Mater.* **2023**, *6*, 8. [CrossRef]
2. Ding, R.; Luo, J.; Wang, C.; Yu, L.; Yang, J.; Wang, M.; Zhong, S.; Gu, R. Identifying and mapping individual medicinal plant *Lamiophlomis rotata* at high elevations by using unmanned aerial vehicles and deep learning. *Plant Methods* **2023**, *19*, 38. [CrossRef] [PubMed]
3. Sun, Y. *Remote Sensing Monitoring of Resources of Medicinal Plants* Atractylodes lancea *(Thunb.) DC. and* Ginkgo biloba *L.*; China Academy of Chinese Medical Sciences: Beijing, China, 2008.
4. Xiao, P.; Qian, P.; Xu, J.; Lu, M. A bibliometric analysis of the application of remote sensing in crop spatial patterns: Current status, progress and future directions. *Sustainability* **2022**, *14*, 4104. [CrossRef]
5. Chen, S.; Sun, J. Key technical links and solution ways of setting up aworking system for yield estimations of the maincrops of china by satellite remote sensing. *J. Nat. Resour.* **1997**, *12*, 363–369. [CrossRef]
6. Sun, Y.; Huang, L.; Guo, L.; Zhu, W.; Pan, Y. Application of remote sensing technology in the investigation of traditional Chinese medicine resources. *Mod. Chin. Med.* **2006**, *8*, 7–10. [CrossRef]
7. Sun, Y.; Guo, L.; Zhu, W.; Huang, L.; Pan, Y. Remote sensing monitoring methods for medicinal plants resource in different types of ecological environment. *China J. Chin. Mater. Med.* **2007**, *32*, 1490–1492.
8. Sun, Y.; Guo, L.; Zhu, W.; Huang, L.; Gu, X.; Han, L.; Pan, Y. Remote sensing resource monitoring on *Atractylodes lancea*. *China J. Chin. Mater. Med.* **2008**, *33*, 353–356.
9. Chen, S.; Zhang, B.; Zhang, J.; Zhou, Y.; Su, G.; Xiao, S.; Yang, Z.; Xiao, P. Study of method for application of remote-sensing technology to investigation of reserves of *Ginseng* resources. *World Sci. Technol.* **2005**, *7*, 37–43. [CrossRef]
10. Jia, X.; Sun, X. Significance and development prospect of medicinal plant cultivation in our area. *Rural. Sci. Technol.* **2001**, *4*, 18. [CrossRef]
11. Huang, L.; Wang, Y. *Technical Specification for National Survey of Chinese Medicine Resources*, 1st ed; Shanghai Scientific & Technical Publishers: Shanghai, China, 2015; pp. 95–96. ISBN 978-7-5478-2138-1.
12. Shan, Z.; Ye, J.; Hao, D.; Xiao, P.; Chen, Z.; Lu, A. Distribution patterns and industry planning of commonly used traditional Chinese medicinal plants in China. *Plant Divers.* **2022**, *44*, 255–261. [CrossRef] [PubMed]
13. Tang, X.; Yuan, Y.; Wang, L.; Chen, S.; Liu, X.; Zhang, J. Identifying prioritized planting areas for medicinal plant *Thesium chinense* Turcz. under climate change in China. *Ecol. Inform.* **2021**, *66*, 101459. [CrossRef]
14. Lan, J.; Zhang, F.; Lian, C.; Li, C. Application of remote sensing technology in medicinal plant resources. *Tech. Innov. Appl.* **2021**, *29*, 167–170.
15. Shi, T.; Zhang, X.; Qu, X.; Qiu, Z.; Zhang, Q.; Guo, L.; Huang, L.; Fu, Q.; Lv, Z.; Gong, Y. Research on extraction of *Ginseng* planting distribution information based on object-oriented classification—By case study of Fusong country in Jilin. *China J. Chin. Mater. Med.* **2017**, *42*, 4353–4357. [CrossRef]
16. Li, Q. From the perspective of biodiversity: Sustainable development of forest medicinal plant resources. *China For. Ind.* **2023**, *6*, 54–56.
17. Huang, L.; Guo, L. Secondary metabolites accumulating and geoherbs formation under environmental stress. *China J. Chin. Mater. Med.* **2007**, *32*, 277–280.
18. Dong, Y.; Luo, Y.; Zhu, C.; Peng, W.; Xu, X.; Fang, Q. Application of remote sensing and GIS in study of suitability distribution of *Swertia mussotii*, a Tibetan medicine in Sichuan province. *China J. Chin. Mater. Med.* **2017**, *22*, 4387–4394. [CrossRef]
19. Zhang, B.; Chen, S.; Zhang, J.; Su, G.; Zhou, Y.; Yang, Z.; Xiao, S. Resources survey of *Glycyrrhiza uralensis* based on remote sensing technique. *Chin. Tradit. Herb. Drug.* **2005**, *36*, 112–115. [CrossRef]
20. Bahdu, S.T. *Xinjiang Specialty Herbal Resources Survey Research Methods to Interpret with Remote Sensing—Case of Type Herbal Plantation*; Xinjiang University: Xinjiang, China, 2014.
21. Na, R.; Zheng, J.; Guo, B.; Shi, M.; Sen, B.; Sun, Z.; Sun, C.; Li, X.; Jia, X. Evaluation of high-resolution images application for wild medicinal plants macro monitoring: A case of *Apocynum*. *China J. Chin. Mater. Med.* **2014**, *39*, 1777–1781. [CrossRef]
22. Na, R.; Zheng, J.; Guo, B.; Shi, M.; Sen, B.; Sun, Z.; Sun, C.; Li, X.; Jia, X. The application and evaluation of China ZY3 satellite on wild *Apocynum* monitoring. *Xinjiang Agric. Sci.* **2014**, *51*, 497–503.
23. Liu, M.; Liu, J.; Atzberger, C.; Jiang, Y.; Ma, M.; Wang, X. *Zanthoxylum bungeanum* Maxim mapping with multi-temporalSentinel-2 images: The importance of different features and consistency of results. *ISPRS J. Photogram. Remote Sens.* **2021**, *174*, 68–86. [CrossRef]
24. Li, Y. Research on information extraction of Chinese medicinal materials based on GF-6 remote sensing data. *South China Agric.* **2019**, *13*, 111–113. [CrossRef]
25. Li, H.; Xia, C.; Zhang, X.; Wang, Y.; Zhang, Y. Extraction of maize lodging information at mature stage based on UAV multispectral images. *Agric. Res. Arid. Areas* **2023**, *5*, 198–206.

26. Xie, C.; Chen, S.; Lin, Z.; Zhou, Y.; Li, Y. The application of UAV remote sensing technology in the investigation of medicinal plant resources. *Mod. Chin. Med.* **2007**, *6*, 4–6. [CrossRef]
27. Xie, C.; Shi, M.; Guo, B.; Shi, L.; Zeng, F.; Fu, D.; Li, X.; Jia, X. Resource investigation for endangered wild *Ferulasinkiangensis* based on low altitude remote sensing. *World Sci. Technol.* **2014**, *11*, 2480–2486. [CrossRef]
28. Guo, Q.; Kelly, M.; Gong, P.; Liu, D. An object-based classification approach in mapping tree mortality using high spatial resolution imagery. *GISci. Remote Sens.* **2007**, *44*, 24–47. [CrossRef]
29. Fortes, C.; Demattê, J.A.M. Discrimination of sugarcane varieties using Landsat 7 ETM+ spectral data. *Int. J. Remote Sens.* **2006**, *27*, 1395–1412. [CrossRef]
30. Liu, J.; Guo, H.; Li, G.; Yue, J.; Lu, H.; Zhou, X.; Men, F.; Zhang, Y. Hyperspectral characteristics of *Carthamus tinctorius* in Xinjiang region. *China J. Chin. Mater. Med.* **2013**, *38*, 1335–1339. [CrossRef]
31. Xue, J.; Su, B. Significant Remote Sensing Vegetation Indices: A review of developments and applications. *J. Sens.* **2017**, *2017*, 1353691. [CrossRef]
32. Zeng, Y.; Hao, D.; Huete, A.; Dechant, B.; Berry, J.; Chen, J.M.; Joiner, J.; Frankenberg, C.; Bond-Lamberty, B.; Ryu, Y.; et al. Optical vegetation indices for monitoring terrestrial ecosystems globally. *Nat. Rev. Earth Environ.* **2022**, *3*, 477–493. [CrossRef]
33. Zhang, X.; Guo, L.; Ge, X.; Zhu, S.; Huang, L. The monitoring method for distribution change of *Paeonia lactiflora* pall. on normalized difference vegetation index. *Mod. Chin. Med.* **2014**, *16*, 96–99. [CrossRef]
34. Fan, D.; Su, X.; Weng, B.; Wang, T.; Yang, F. Research progress on remote sensing classification methods for farmland vegetation. *AgriEngineering* **2021**, *3*, 61. [CrossRef]
35. Cui, L.; Shi, J.; Du, H. Advances in remote sensing extraction of vegetation phenology and its driving factors. *Adv. Earth Sci.* **2021**, *1*, 9–16. [CrossRef]
36. Yang, C.; Wu, G.; Li, Q.; Wang, J.; Qu, L.; Ding, K. Research progress on remote sensing classification of vegetation. *Geogr. Geo-Inf. Sci.* **2018**, *4*, 24–32.
37. Liu, L.; Wang, J.; Zhao, C.; Song, X.; Li, C.; Zhang, C. Study on floating prior probability MLC based on spatial features and local spatial autocorrelation. *J. Remote Sens.* **2006**, *2*, 227–235. [CrossRef]
38. Wu, F.; Zheng, M.; Hu, W.; Hu, G.; Wang, L.; Dong, X.; Lu, Y. Remote sensing extraction and change analysis of *Panax notoginseng* planting area in Wenshan Prefecture of Yunnan Province in Past Decade Based on TM and OLI Data. *Acta Agric. Jiangxi* **2021**, *33*, 90–97. [CrossRef]
39. Xu, J.; Feng, G.; Zhao, T.; Sun, X.; Zhu, M. Remote sensing image classification based on semi-supervised adaptive interval type-2 fuzzy c-means algorithm. *Comput. Geosci.* **2019**, *131*, 132–143. [CrossRef]
40. Yi, M.; Xie, H.; Cheng, M.; Zhao, Y.; Shi, T.; Zhang, X.; Peng, H. Planting area estimation of *Paeonia ostii* based on RS—By case study in Dangshan of Anhui province. *China J. Chin. Mater. Med.* **2019**, *44*, 4107–4110. [CrossRef]
41. Yang, M.; Chen, L.; Zhang, X.; Zhao, Y.; Shi, T.; Cheng, M.; Peng, H. Extracting *Paeonia* lactiflora planted area in Dangshan of Anhui province based on ZY-3 remote sensing image. *China J. Chin. Mater. Med.* **2019**, *44*, 4101–4106. [CrossRef]
42. Dai, C.; Xie, X.; Xu, Z.; Du, P. Monitoring and analyzing herbal medicine plantation via remote sensing:A case study of pseudo-ginseng in Wenshan and Honghe Prefecture of Yunnan Province. *Remote Sens. Nat. Resour.* **2018**, *1*, 210–216. [CrossRef]
43. Pal, M.; Mather, P.M. An assessment of the effectiveness of decision tree methods for land cover classification. *Remote Sens. Environ.* **2003**, *86*, 554–565. [CrossRef]
44. Zhu, Z.; Yuan, X.; Gan, S.; Zhang, L.; Zhao, Q. GF-1 remote sensing data for *Panax notoginseng* planting information extraction in Wenshan, Yunnan Province. *J. Zhejiang A F Univ.* **2020**, *37*, 129–135. [CrossRef]
45. Hinton, G.E.; Salakhutdinov, R.R. Reducing the Dimensionality of Data with Neural Networks. *Science* **2006**, *313*, 504–507. [CrossRef] [PubMed]
46. Shi, T.; Zhang, X.; Guo, L.; Jing, Z.; Huang, L. Research on remote sensing recognition of wild planted *Lonicera japonica* based on deep convolutional neural network. *China J. Chin. Mater. Med.* **2020**, *45*, 5658–5662. [CrossRef]
47. Sun, Y.; Gu, X.; Yang, X.; Zhu, W.; Guo, L.; Huang, L.; Pan, Y. Study on dynamical changes of *Ginkgo bilobas* resources in Pizhou city of Jiangsu Province base on RS and GIS. *China J. Chin. Mater. Med.* **2007**, *32*, 1861–1864.
48. Chen, D. Application of remote sensing monitoring method in the monitoring of medicinal plant resources. *Sci. Technol. Inf.* **2013**, *8*, 41–43. [CrossRef]
49. Zhao, J.; Sun, Z.; Zhang, Z. Research progress of planting model of Chinese materia medica in China. *Mod. Chin. Med.* **2021**, *2*, 372–380. [CrossRef]
50. Jin, G.; Liu, S.; Li, Y.; Chen, H.; Liu, J. Some thoughts on the standardized planting of Chinese medicinal materials. *China J. Chin. Mater. Med.* **2006**, *2*, 175–176.
51. Liu, J.; Pan, M.; Li, G.; Ma, C.; Hou, J.; Meng, F.; Zhao, C. Application of 3S technology in resources survey of medicinal plants. *Chin. Tradit. Herb. Drug* **2016**, *4*, 695–700. [CrossRef]
52. Baldridge, A.M.; Hook, S.J.; Grove, C.I.; Rivera, G. The ASTER spectral library version 2.0. *Remote Sens. Environ.* **2019**, *113*, 711–715. [CrossRef]
53. Kokaly, R.F.; Clark, R.N.; Swayze, G.A.; Livo, K.E.; Hoefen, T.M.; Pearson, N.C.; Wise, R.A.; Benzel, W.M.; Lowers, H.A.; Driscoll, R.L.; et al. *USGS Spectral Library Version 7: U.S. Geological Survey Data Series 1035*; U.S. Geological Survey: Reston, VA, USA, 2017; p. 61. [CrossRef]

54. Zhang, Y.; Xiao, Q.; Wen, J.; You, D.; Dou, B.; Tang, Y.; Peng, S. Review on Spectral Libraries: Progress and Application. *J. Remote Sens.* **2017**, *21*, 12–26. [CrossRef]
55. Zhao, L.; Liu, S.; Chen, X.; Wu, Z.; Yang, R.; Shi, T.; Zhang, Y.; Zhou, K.; Li, J. Hyperspectral identification of ginseng growth years and spectral importance analysis based on random forest. *Appl. Sci.* **2022**, *12*, 5852. [CrossRef]
56. Long, D.; Yan, L.; Bai, L.; Zhang, C.; Li, X.; Lei, H.; Yang, H.; Tian, F.; Zeng, C.; Meng, X.; et al. Generation of MODIS-like land surface temperatures under all-weather conditions based on a data fusion approach. *Remote Sens. Environ.* **2020**, *246*, 111863. [CrossRef]
57. Araus, J.L.; Cairns, J.E. Field high-throughput phenotyping: The new crop breeding frontier. *Trends Plant Sci.* **2014**, *19*, 52–61. [CrossRef] [PubMed]
58. Zhang, H.; Wang, L.; Tian, T.; Yin, J. A review of unmanned aerial vehicle low-altitude remote sensing (UAV-LARS) use in agricultural monitoring in China. *Remote Sens.* **2021**, *13*, 1221. [CrossRef]
59. Wu, X.; Jin, S. Models and theoretical analysis of SoOp circular polarization bistatic scattering for random rough surface. *Remote Sens.* **2020**, *12*, 1506. [CrossRef]
60. Ball, J.; Anderson, D.; Chan, C. Comprehensive survey of deep learning in remote sensing: Theories, tools, and challenges for thecommunity. *J. Appl. Remote Sens.* **2017**, *11*, 42609. [CrossRef]

remote sensing

MDPI

Review

A Systematic Review of UAV Applications for Mapping Neglected and Underutilised Crop Species' Spatial Distribution and Health

Mishkah Abrahams [1,*], Mbulisi Sibanda [1], Timothy Dube [2], Vimbayi G. P. Chimonyo [3,4] and Tafadzwanashe Mabhaudhi [3,5]

1 Discipline of Geography, Environmental Studies & Tourism, Faculty of Arts, University of the Western Cape, Private Bag X17, Bellville 7535, South Africa; msibanda@uwc.ac.za
2 Institute of Water Studies, Faculty of Science, University of the Western Cape, Private Bag X17, Bellville 7535, South Africa; tidube@uwc.ac.za
3 Centre for Transformative Agricultural and Food Systems, School of Agricultural, Earth & Environmental Sciences, University of KwaZulu-Natal, P/Bag X01, Pietermaritzburg 3209, South Africa; v.chimonyo@cgiar.org (V.G.P.C.); t.mabhaudhi@cgiar.org (T.M.)
4 International Maize and Wheat Improvement Center (CIMMYT)-Zimbabwe, Mt Pleasant, Harare P.O. Box MP 163, Zimbabwe
5 International Water Management Institute (IWMI-SA), Southern Africa Office, Pretoria 0184, South Africa
* Correspondence: 3852812@myuwc.ac.za

Abstract: Timely, accurate spatial information on the health of neglected and underutilised crop species (NUS) is critical for optimising their production and food and nutrition in developing countries. Unmanned aerial vehicles (UAVs) equipped with multispectral sensors have significantly advanced remote sensing, enabling the provision of near-real-time data for crop analysis at the plot level in small, fragmented croplands where NUS are often grown. The objective of this study was to systematically review the literature on the remote sensing (RS) of the spatial distribution and health of NUS, evaluating the progress, opportunities, challenges, and associated research gaps. This study systematically reviewed 171 peer-reviewed articles from Google Scholar, Scopus, and Web of Science using the PRISMA approach. The findings of this study showed that the United States (n = 18) and China (n = 17) were the primary study locations, with some contributions from the Global South, including southern Africa. The observed NUS crop attributes included crop yield, growth, leaf area index (LAI), above-ground biomass (AGB), and chlorophyll content. Only 29% of studies explored stomatal conductance and the spatial distribution of NUS. Twenty-one studies employed satellite-borne sensors, while only eighteen utilised UAV-borne sensors in conjunction with machine learning (ML), multivariate, and generic GIS classification techniques for mapping the spatial extent and health of NUS. The use of UAVs in mapping NUS is progressing slowly, particularly in the Global South, due to exorbitant purchasing and operational costs, as well as restrictive regulations. Subsequently, research efforts must be directed toward combining ML techniques and UAV-acquired data to monitor NUS' spatial distribution and health to provide necessary information for optimising food production in smallholder croplands in the Global South.

Keywords: crop health; drones; food security; NUS; precision agriculture; spatial distribution; stomatal conductance; UAV

1. Introduction

Inherent water scarcity, which is exacerbated by factors such as climate change, population expansion, and changes in land use [1–3], has intensified the pressure on the agricultural sector, particularly in ensuring long-term food supply for the expanding populations [4,5]. Most of the agricultural production in developing regions is derived from rainfed farms, which occupy 97% of croplands [6]. However, 80% of these croplands are smallholder farms that contribute most of the food production in developing regions [1,2]. However, these croplands are marginal due to suppressing agronomic and climatological infrequencies. Therefore, climate variability and soil degradation have drastically reduced the production of staple cereal crops, such as maize, amplifying food and nutrition security issues in these regions. This has placed local food systems on the verge of catastrophe. Hence, establishing innovative methods to combat food and nutrition insecurity while optimising production is urgently needed.

A paradigm shift away from cultivating susceptible cereal crops towards diversifying climate-smart crops, such as underutilised crop species (NUS), is necessary [7]. NUS, also referred to as alternative, traditional, orphaned, or neglected crops, are adapted to flourish in fragile production systems where land degradation and drought are topical [3]. Examples include millet, quinoa, teff, Bambara ground nuts, sweet potato, and taro [3–5]. These crops are stress-tolerant, require fewer inputs, and are highly adaptable to a broad spectrum of ecological niches [8]. Including NUS in fragile production systems is one of many strategies for safeguarding the well-being of minority populations, meeting sustainable development goals (SDGs), leveraging indigenous knowledge systems, and facilitating traditional food and cultural heritages.

A major challenge to the mainstreaming of NUS is that basic information on genetic and phenotypic traits remains scanty and localised, and productivity is generally low. Generating spatially explicit information on NUS' biochemical and morphological characteristics to optimise their productivity could contribute to their development and promotion. In this regard, spatially explicit, high-throughput phenotyping technologies are required to augment the rapid advancements in phenotyping NUS. This will aid in maximising their productivity in fragmented smallholder croplands.

Generally, field surveys and other traditional methods have been widely used to measure the spatial extent, suitability, growth, and morphological attributes of NUS. However, such methods are time-consuming and expensive, making them unsuitable for continuous precision crop monitoring in areas with numerous crop varieties. Over the past few decades, satellite-based Earth-observation technologies, such as Landsat and MODIS, have become effective for monitoring plant growth and health changes [9]. Satellite-borne remote sensing technologies provide non-invasive, accurate, fast, and cost-effective data for estimating traits such as stomatal conductance and chlorophyll content as a proxy for crop productivity, hence their use [10]. However, the resolution of these freely available satellite sensors provides limited information for high-throughput phenotyping, particularly in the context of highly fragmented and diverse smallholder croplands [11]. In this regard, smallholder croplands require very high-spatial-resolution remote-sensing technologies that are affordable and highly efficient [11].

The advent of UAV-based phenotyping allows access to data with extremely high levels of detail and precision. This is ideal for the precise, consistent phenotyping of smallholder farms at the plot level [11]. However, the capacity of UAV-mounted sensors to differentiate crop types based on their spectral responses as a mechanism for plausible high-throughput field phenotyping is yet to be determined [12]. RS and ML techniques have recently greatly aided high-throughput phenotyping technologies [12]. For example Li, Burnside [6] combined three ML techniques, which included random forest (RFR), support vector regression (SVR), and artificial neural network (ANN), in combination with optimal VI's to predict the red-clover dry matter yields in various phenological growth periods.

Despite the usefulness of UAVs, their application in agriculture, rural development, and, more importantly, resource management remains limited [13]. Although some studies have attempted to assess the literature on the application of drone-acquired data [14,15], most of these studies did not systematically and quantitatively assess the literature on mapping the spatial distribution and health of NUS crops with a special interest in Global South trends. The objective of this study was to systematically review the literature on the application of UAV remotely sensed data for mapping the spatial distribution and health of NUS, with a particular focus on sweet potatoes and taro. The aim was to gain an understanding of the progress, challenges, opportunities, and gaps in this field. Gaining insights into the mapping of NUS' spatial extent, morphological features, and biochemical traits through remotely sensed data acquired by UAVs could pave the way for enhancing food production in South Africa's smallholder croplands.

2. Materials and Methods

2.1. Literature Search

This review followed the Preferred Reporting Items for Systematic Reviews and Meta-Analyses (PRISMA) checklist and guidelines for gathering and analysing the literature. In the initial literature search phase, keywords, terms, and phrases for searching the literature were generated from other literature reviews on NUS [16–18]. The methodology followed in this study was adopted from the literature [8–10,19]. The following keywords and their variants were used in this study to search for relevant literature: "neglected and underutilised crop species", "orphan crops", "traditional crops", "unmanned aerial vehicle(s)", "drone(s)", "remote sensing", "GIS", "crop health", "stomatal conductance" and "leaf area index", and "chlorophyll". Furthermore, the following PRISMA statement and its variants were used to search for research pertaining taro and sweet potatoes: "Taro", "sweet potato", "unmanned aerial vehicle(s)", "drone(s)", "remote sensing", "GIS", "crop health", "stomatal conductance" and "leaf area index", and "chlorophyll".

The SCOPUS, Web of Science, and Google Scholar databases were utilised to collect literature using the established key search terms. The PRISMA statement served as the framework for the literature search procedure. This search was not restricted in terms of time. The Google Scholar, Scopus, and Web of Science literature searches yielded 109, 1036, and 90 articles, respectively (n = 1235). In preparation for screening, all obtained material was organised in EndNote. The screening procedure considered in this study followed the PRISMA procedure, reported as a flowchart (Figure 1). For an article to be considered in the meta-analysis, it had to meet the following criteria;

(a) The study focuses on NUS crops, traditional, or orphaned crops, and no other vegetation types (e.g., forests or shrubs) were included, since they denoted different ecosystems;
(b) The study focuses on NUS productivity (i.e., LAI, chlorophyll, or stomatal conductance) or spatial distribution;
(c) The study was based on UAV or drone remotely sensed data, GIS, or remote-sensing techniques in NUS crop productivity and health mapping;
(d) The article was published in an accredited journal;
(e) The article was written in English;
(f) The article was accessible.

After eliminating all duplicates (n = 650), 585 remained. In this case, literature not written in English was excluded from the analysis (n = 20). The next step was to assess whether the retrieved literature covered the context of mapping the spatial distribution or assessed productivity elements of neglected and underutilised crops based on remotely sensed data by examining the abstracts. After the title and abstract screening, 332 studies were excluded and 253 remained. Of the remaining articles, 82 were not accessible as full texts and were excluded, with 171 articles remaining. The full-length articles of the selected abstracts were then sought and downloaded as PDF documents. After the screening procedure, 171 articles were retained (Figure 1). Then, the bibliographic information, including author names, year of publication, title, journal name, issue, volume number, keywords, and abstracts, was exported from Endnote as a text file to Microsoft Excel. The Excel database was then used to extract and store qualitative and quantitative data from each article, as indicated in the proceeding phase.

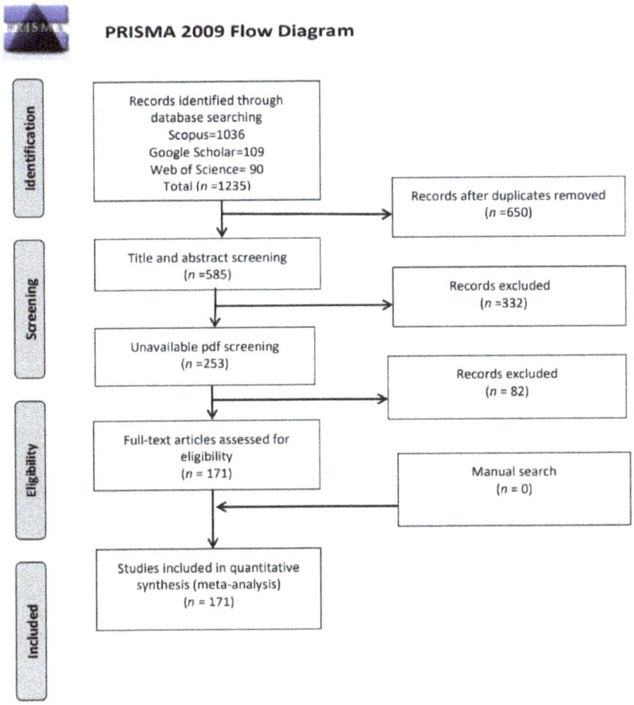

Figure 1. PRISMA flow diagram for selection of studies considered in the review.

2.2. Data Extraction

In the preceding phase, the extracted data in an Excel database were used to generate and extract information on progress, gaps, challenges, and opportunities of using UAV technologies to map the spatial distribution of different NUS and their health attributes. Specifically, information related to the study country, region, NUS type, the type of crop health attribute investigated, sensor type and platform type, vegetation indices, predictive or classification algorithms, and optimum spectral variables obtained, were all retrieved from the literature and documented in the spreadsheet. Some categorical variables were converted into numeric values to facilitate analysing and evaluating trends in the retrieved literature. During this step, relevant bibliometric information was also collected. As mentioned above, the qualitative and quantitative information extracted from each article were added to the Excel spreadsheet with the author names, region, year of publication, article title, journal name, and abstract, among the other bibliometric data gathered.

2.3. Data Analysis

The retrieved literature and extracted data were subjected to quantitative and qualitative analyses during this phase. Basic statistical frequencies were calculated for quantitative analysis [13]. In addition, exploratory trend analysis was carried out on the frequency of publications to assess the progress made in mapping NUS' spatial distribution and health attributes utilizing satellite and drone-borne sensors. A bibliometric analysis was also conducted to identify trends in co-occurring key terms from the retrieved literature. The trends were identified by quantitatively examining the occurrence and co-occurrence of key terms in the titles and abstracts using the VOS viewer software. Furthermore, the VOS viewer software used the titles and abstracts of all the retrieved literature (171 articles) and only those based on UAV-derived datasets (18 articles). This assisted in evaluating how concepts and topics evolved in mapping NUS using satellite-based remotely sensed data to drone-acquired data. VOS viewer provides network visualisation of key terms in linked clusters. Creating a map in VOS viewer includes four steps, which are:

(1) Selecting a counting method (binary counting or full counting);
(2) Selecting a minimum number of occurrences for a term (calculating similarity index);
(3) Calculating the relevance score for the co-occurrence terms and displaying the most relevant items based on this score;
(4) Displaying a map based on the selected terms.

Considering that only the occurrence, co-occurrence of key terms, and frequency distributions were computed, bias assessment was not conducted. As aforementioned, the PRISMA checklist (http://www.prisma-statement.org/, accessed on 1 July 2022) was used as a guideline to avoid biased reporting. In this regard, no further robust bias statistical assessments were conducted, since only exploratory data analysis was conducted. The review was divided into two main sections to address the research objectives. The first section investigated recent advances in mapping NUS crops' spatial distribution and health using remotely sensed data. This section presented and discussed quantitative literature trends in analysing NUS's spatial distribution and health. Throughout this phase, the crop health attributes, Earth-observation sensors (cameras), sensor platforms, algorithms, and optimal spectral variables used by the community of practice in the retrieved literature were assessed and presented. The last phase discussed the challenges, gaps, and opportunities for knowledge generation in mapping NUS's spatial distribution and health using drone-derived remotely sensed data.

3. Results

Figure 2 shows the co-occurrence of topical concepts derived from titles and abstracts from (a) studies based on satellite-borne data and (b) drone-acquired data. Figure 2a illustrates seven topical clusters in dark blue, light blue, red, green, purple, yellow, and orange for mapping crop spatial extent and health status. The key terms from the "red" cluster were "agriculture", "spad value", "low cost", "remote sensing data", "hyperspectral data", "multispectral data", "VIS", "processing", "size", and "crop field", which directly imply the utility of "low-cost remote sensing" systems for mapping and monitoring NUS crop productivity (spad-value/chlorophyll) with remotely sensed data in smallholder crop fields (Figure 2a). The second-largest cluster linked to UAVs was in yellow and contained "growth", "variety", "trait", "detection", "plant-height", "UAVs", "drone", "dsm", "rgb", and "msi". This cluster articulates the use of drone remotely sensed ("UAVs") data in detecting and mapping relevant NUS phenotypical attributes ("growth", "variety", "trait", "detection", and "plant height"). The third cluster in dark blue included 'growth stage', "lai", "crop height", "agdw", "canopy nitrogen", "weight", "fusion", "plsr", and "rmse" as the key terms in order of importance (Figure 2). This cluster relates to the estimation of NUS crop productivity attributes ("lai", "crop height", "agdw", "canopy nitrogen", "weight", "lai", "crop height", "agdw", "canopy nitrogen", and "weight") using remotely sensed data and regression techniques. The fourth cluster in green contained "population", "climate change", "water stress", "food security", "African leafy vegetable", "suitable area", "moisture", "SSA" (Sub-Saharan Africa), and "validation", among others. This links to the food and nutrition security issues in SSA, which are highly impacted by climate variability, such that NUS crops ("African leafy vegetable") are the only suitable crops because of their drought tolerance. The fifth cluster, which is light blue, features co-occurring terms, such as "species", "reflectance", "density", "waveband", "nir", "classification", and "palmer amaranth". This cluster relates to the impact of species variability and plant or foliage density, which causes the spectral signatures of NUS crops to vary at different wavebands of the EM spectrum, facilitating optimal classification accuracies. The sixth cluster is characterised by the co-occurrence of terms such as "plant", "amaranth", "phenological", and "growth stage", which relates to the assessment of NUS crops' phenological characteristics.

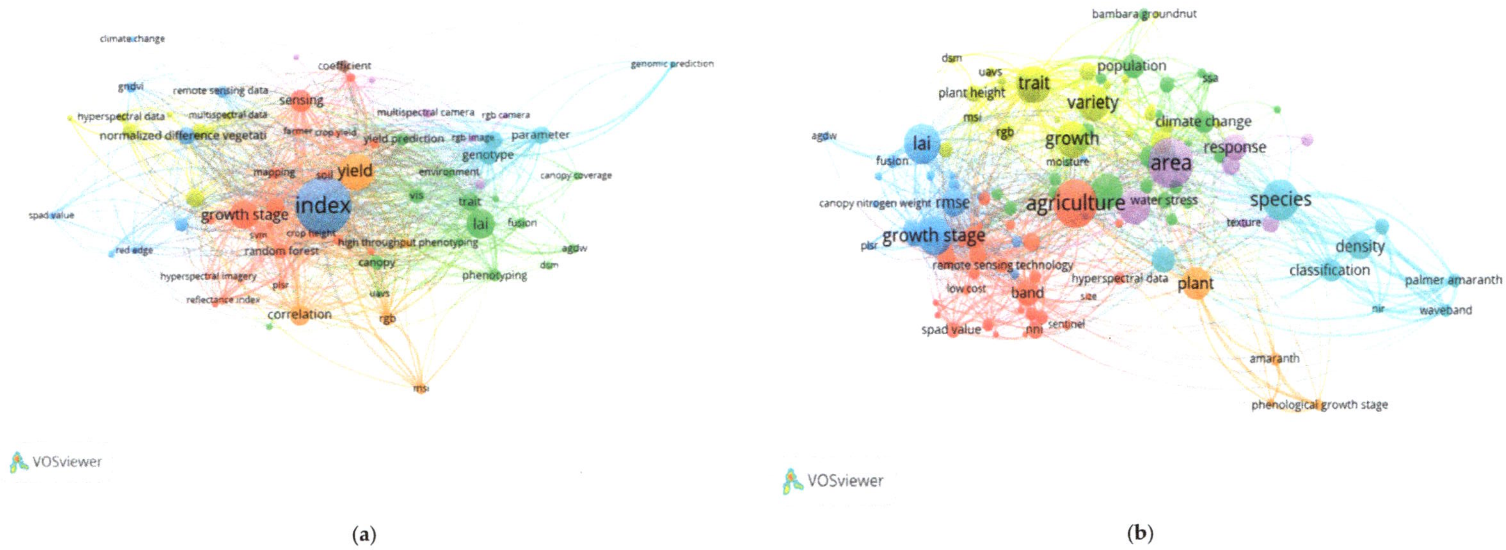

Figure 2. Topical concepts identified using a bibliometric analysis of titles and abstracts of articles that utilised (**a**) all remote sensing sensors and (**b**) only drone-acquired data in mapping the NUS. Various coloured lines establish connections between keywords that co-occurred within the same documents.

Figure 2b illustrates seven topical clusters in dark blue, light blue, red, green, purple, yellow, and orange, which were derived using abstracts and titles that utilised drone remotely sensed data in mapping NUS crop spatial extent and healthiness. The key terms from the red clusters were "mapping", "crop height", "growth stage", "yield" and, "crop yield", "reflectance index", "hyperspectral imagery", "random forest", and "svm". This cluster relates to using high-resolution remotely sensed data in mapping and monitoring NUS productivity elements, such as growth stage, using machine learning. The key terms in the green cluster are "phenotyping", "LAI", "canopy coverage", "trait", "agdw", "prediction", "uavs", "dsm", "fusion", "VIS", and "environment". This cluster relates to using UAV remotely sensed data that could be fused with other data for assessing the structural attributes of NUS crops (phenotyping). The third cluster in dark blue has the key terms of "remotely sensed data", "gndvi", "normalised difference vegetation", "red-edge", "spad value", and "climate change". This cluster relates to the utility of the widely used spectral variables ("gndvi", "normalised difference vegetation", and "red-edge") in monitoring the health of NUS crops. The fourth cluster is orange and has co-occurrence terms, such as "high throughput phenotyping", "rgb", "msi", "yield", and "correlation". This cluster can be attributed to applying the high-throughput phenotyping of NUS through RGB and multispectral spectrums in estimating crop yield. The fifth cluster in light blue had "genotype", "parameter", and "genomic parameters". This cluster suggests examining the genetic composition of NUS crops through the assessment of optimal parameters to assess and infer their crop health. The sixth cluster has "multispectral camera", "rgb camera", "rgb image". This cluster relates to the use of general colour imagery (in red, green, and blue spectrums) acquired using drones for mapping NUS. The last cluster has "hyperspectral data" and "multispectral data". This cluster relates to the utility of multispectral and hyperspectral data in mapping and monitoring NUS crop spatial distributions and health.

3.1. Progress in Mapping the Spatial Distribution and Health Status of Neglected and Underutilised Crop Species

Significant progress has been attained in detecting, mapping, and monitoring NUS crops' spatial distribution and health status using remotely sensed data (Figure 3). However, it should be noted that this progress relates to the collective of NUS, but not each individual species. The period between 2003 and 2013 is marked by a low frequency of published literature based on all Earth-observation sensors (Figure 3). Between 2014 and 2022, there was a rapid increase in articles that mapped the health and spatial distribution of NUS using all Earth-observation sensors. Again, from 2014 to 2022, there was a rapid growth in the literature that utilised UAV-acquired remotely sensed data to characterise NUS attributes. Despite substantial progress, only a few studies have demonstrated the effectiveness of remote sensing technologies in NUS crop classification, characterising their suitability ranges, or discriminating their varieties based on their phenotypic traits. Specifically, only 3% and 4% of the retrieved studies utilised drone and satellite-borne remotely sensed data in mapping the spatial distribution of NUS crops, respectively. However, most of the retrieved literature evaluated NUS phenological and phenotypic characteristics.

Regarding geographic distribution, the studies included in the meta-analysis were conducted in 47 countries. Most of the retrieved literature was conducted in Asia, America, and Africa. On a national scale, most of the studies were conducted in the United States of America (n = 18) [11,12], followed by South Africa (n = 18) [13,14] and China (n = 17) [15,16], (Figure 4). Interestingly, most of the African studies were conducted in southeast Africa.

The most prevalent NUS in the retrieved literature included sweet potato (Ipomoea *batatas*), sorghum (*Sorghum bicolor*), amaranth (*Amaranthus cruentus*), cassava (*Manihot esculenta*), cowpea (*Vigna unguiculata*), millet (i.e., pearl (*Pennisetum glaucum L.R. Br.*), finger (*Eleusine coracana*), and proso millet (*Panicum milliaceum*)) (Figure 5a). UAV-based remotely sensed data was utilised in mapping the attributes of amaranth (n= 6) [17], Legume (*Fabaceae*) (n = 2) [18], Sweet potato (n = 2) [20], sorghum (n = 2) [21], and bambara groundnut (*Vigna subterranea*) (n = 2) [22] (Figure 5b). Most of the retrieved literature remotely sensed various NUS using hyperspectral data, hence the decline in the frequency of NUS studies that utilised drone and satellite-acquired remotely sensed data (Figure 5).

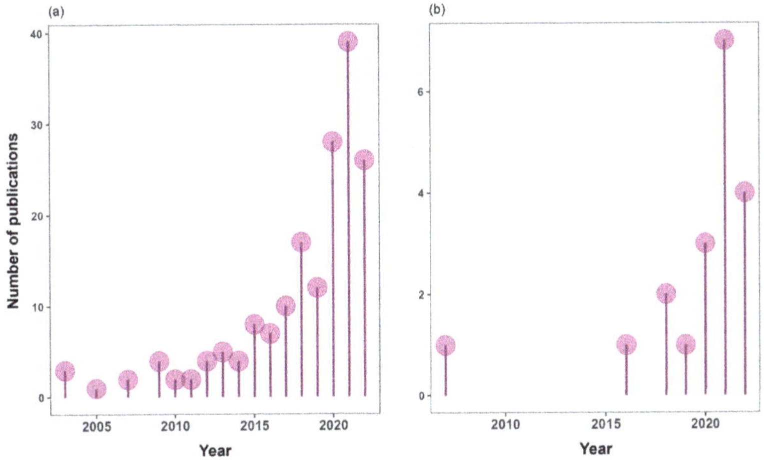

Figure 3. Frequency of published articles on remote sensing applications of NUS based on (**a**) all sensors and (**b**) drones.

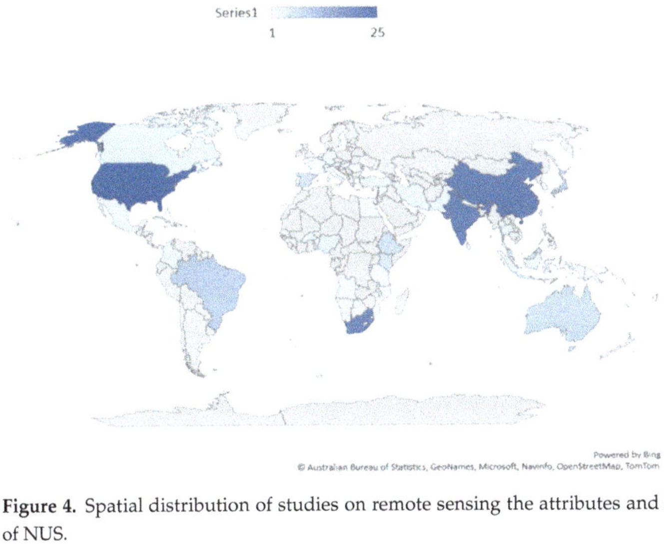

Figure 4. Spatial distribution of studies on remote sensing the attributes and spatial distribution of NUS.

Eight key broad research themes emerged from the reviewed literature on NUS. These include phenotyping, crop genetics, crop productivity, crop physiology, phenology, crop adaptation, classification, and land suitability (Figure 6). Most reviewed studies focused on quantifying NUS' physiological and phenological crop traits. A relatively small number of studies characterised crop classification and spatial extent. Five studies discriminated crops based on drone-acquired data, while six articles were based on satellite-borne remotely sensed data (Figure 6). The most extensively researched areas based on all satellite-borne sensors included phenology (n = 142), crop physiology (n = 110), and crop productivity (n = 104). When considering only the drone-borne sensors, 16 studies focused on NUS phenology. These studies mainly assessed external crop attributes, such as crop LAI, biomass, crop height, and crop yield periods across the growing season. For example,

Jewan, Pagay [23] monitored six distinct phenological stages of the Bambara groundnut, estimating its yield. Only nine studies assessed NUS crop productivity, while seven research studies examined NUS crop health and physiology. For instance, Lati, Avneri [24] assessed the utility of a UAV imaging platform coupled with an RGB sensor to monitor chickpea's physiological and morphological parameters, such as LAI, biomass, and yield, during irrigation periods.

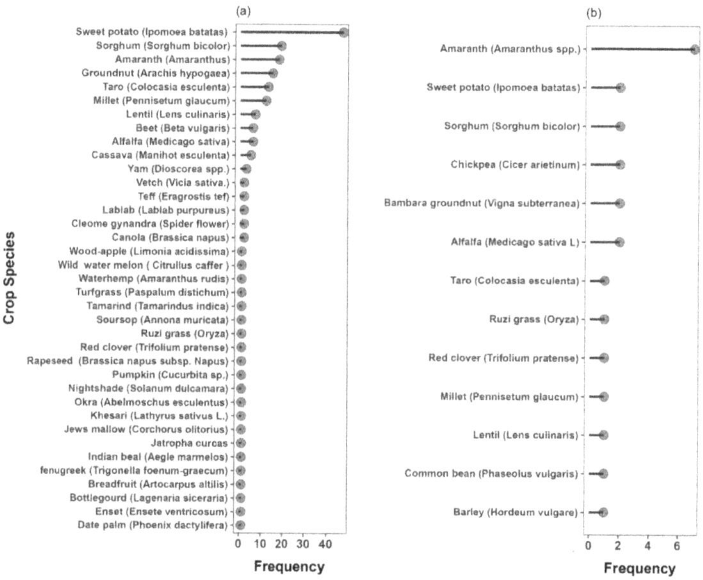

Figure 5. Frequency of NUS in the literature remotely sensed using (**a**) all various sensors and (**b**) exclusively drone-borne sensors.

Moreover, five studies focused on characterizing NUS crop spatial distribution using classification methods, and four concentrated on NUS's climatic suitability and adaptation. For example, Ramírez, Grüneberg [20] utilised vegetation and temperature indices to characterise various sweet potato genotypes based on productivity and resilience under drought treatments. Only five studies utilised crop phenotyping and breeding techniques. In many instances, crop phenotyping and genotyping included single-nucleotide polymorphism (SNP) markers. SNP markers provide a broad range of applications in various crops, including plant variety and cultivar characterization, quantitative trait loci (QTL) analysis, the production of a high-density genetic map, and genome-wide association analysis [25]. And lastly, four studies focused on land suitability.

Many of the retrieved studies mapped the productivity and water stress-related elements. Specifically, the most extensively researched NUS health attributes included crop yield, growth attributes, crop health, chlorophyll content, leaf water content, biomass, photosynthesis, LAI, stomatal conductance, canopy height, plant weight, leaf nitrogen, and canopy temperature (Figure 7a). The most researched NUS attributes in the context of UAV-based remote sensing were crop yield (n = 15), growth attributes (n = 13), biomass (n = 11), crop health (n = 10), chlorophyll content (n = 9), canopy cover (n = 7), plant/canopy height (n = 7), LAI (n = 6), leaf nitrogen (n = 5), leaf size attributes (n = 5), leaf water content (n = 3), and leaf temperature (n = 3). As aforementioned, there were very few studies that classified and characterised the spatial distribution of NUS (Figure 7b) [4].

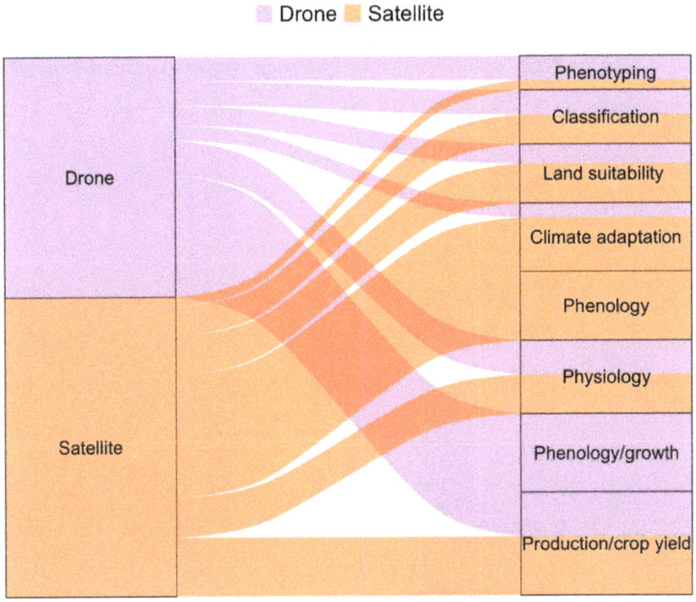

Figure 6. Frequency of articles that utilised remotely sensed data from drones and satellites to map NUS attributes. The thickness of the lines represents the frequency in the retrieved literature. (See Supplementary Materials spreadsheet 454 for frequency values).

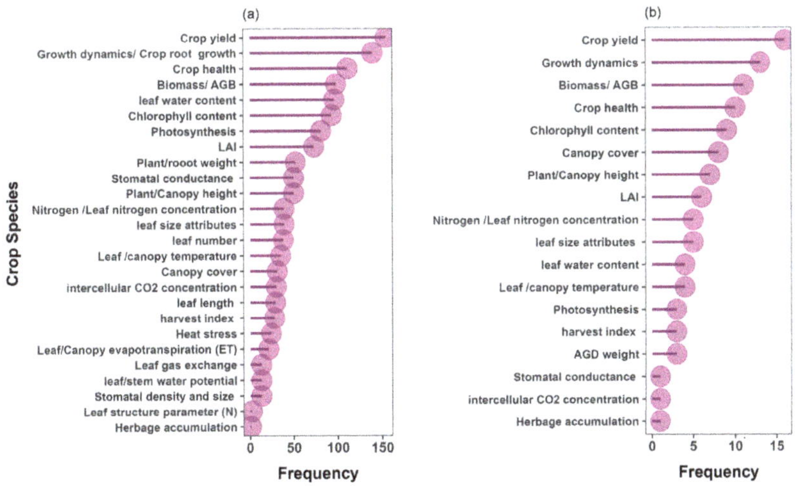

Figure 7. Frequency of studies that remotely sensed a specific crop attribute based on (**a**) all satellite and drone-borne sensors and (**b**) drone-borne sensors only.

3.2. Assessing Literature on Classification and Stomatal Conductance Estimation of Taro and Sweet Potato Crops

Based on the findings, 12 articles characterised the spatial distribution of various NUS (Figure 8). Sweet potato, lentil, and chickpea were the crops that received substantial attention in the literature (Figure 8). Overall, a limited number of studies have utilised UAV remotely sensed data to classify NUS. There is a gap in the research focusing on UAV classifications from 2008 to 2019. Furthermore, 2020 was the most predominant year for NUS classification studies. Nevertheless, research on the spatial distribution of NUS remains sparse and limited to developed regions. The increase in literature could point to an increase in the interest in NUS and the general application of UAV-acquired remotely sensed data.

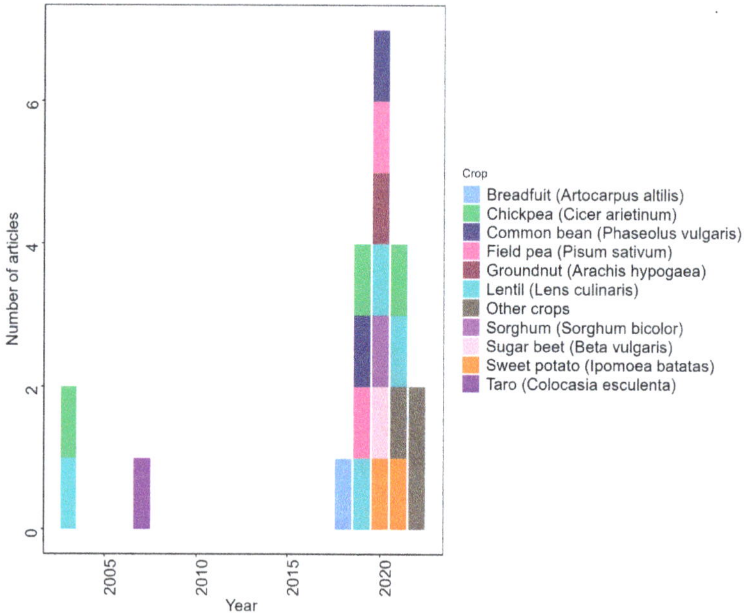

Figure 8. Frequency of published articles on NUS classifications.

Fifty articles estimated the stomatal conductance of NUS based on satellite-borne remotely sensed data. Specifically, studies in the USA and China mapped these NUS mostly using sensors, such as Planet scope, analytical spectral devices, UAVs, MODIS, and Sentinel 2 MSI (Supplementary Figures S1 and S2). There was a modest amount of research on stomatal conductance between 2003 and 2011. However, the literature related to farm-scale stomatal conductance increased from 2013 to 2022 (Figure 9). In remote sensing, the stomatal conductance of NUS, sweet potato received more research attention (17 studies), followed by taro (5 studies), cowpea (5 studies), sorghum (4 studies), and amaranth (4 studies) (Figure 9). Meanwhile, very few studies have been conducted concerning the stomatal conductance of crops such as cassava and millet. These crops were less frequent in the retrieved literature (Figure 9a,b) than previously stated. However, there were few studies on remote sensing applications for estimating the stomatal conductance of taro based on UAV-acquired remotely sensed data from the retrieved literature. The specific countries and sensors used in mapping these NUS are detailed in Supplementary Table S6.

3.3. Types of Sensors and Their Spectral Resolutions

The utilization of Earth-observation sensors in the remote sensing of NUS studies is significant. Thirteen different sensor types were noted in the reviewed literature (Figure 10). In terms of sensors, the findings of this review revealed that the spectrophotometer was the most widely used sensor for characterizing the health status of NUS, being used in 27 studies. Furthermore, research results indicate that various studies have used hand-held hyperspectral devices to acquire in situ remotely sensed data to detect and map NUS biophysical and phenological attributes. The most predominant sensors in the retrieved literature were spectrophotometers (27), UAV-borne sensors (n = 18), spectrometers (n = 18), radiometers (n = 11), Sentinel- 2 MSI (n = 9), MODIS (n = 3), and LiDAR (n = 3)

(Figure 10a). Meanwhile, the most frequently used drone-borne sensors were RGB cameras (in 11 studies) [26], RedEdge-MX (in three studies) [27], and Canon (in three studies) [23]. These were followed by thermal cameras [28], Parrot Sequoia, Micasense Altum [20], CMOS cameras [29], and MCA6 [17] in order of frequency in the retrieved literature (Figure 10b).

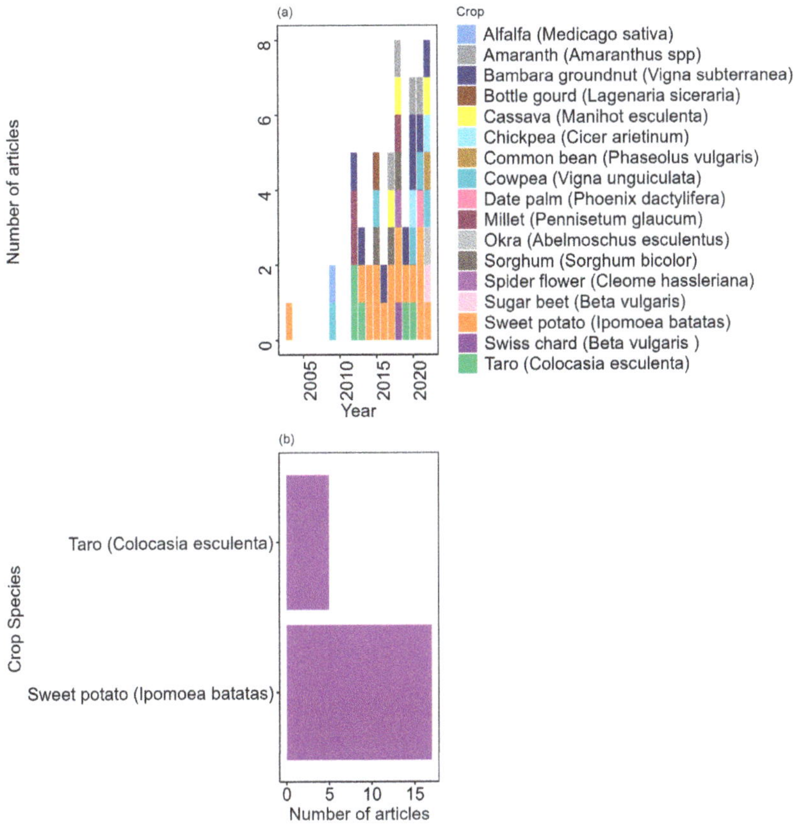

Figure 9. Frequency of published articles on mapping NUS stomatal conductance based on (**a**) all sensors for all NUS and (**b**) only on sweet potato and taro.

Across all platforms, the multispectral (broadbands) were highly utilised in the literature compared with hyperspectral (narrow) bands. The visible section of the electromagnetic spectrum, specifically the red, green, and blue (RGB) sections, are primarily the most utilised wavelengths in mapping the spatial distribution of NUS crops and their health attributes (Figure 11). Specifically, the RGB sections of the electromagnetic spectrum (EM) were utilised in 17, 18, and 49 studies based on drone, satellite-borne, and hyperspectral sensors, respectively (Figure 11). The second most widely used section of the electromagnetic spectrum in the literature was the NIR section, utilised in 12 studies with drone-acquired data, 18 studies with satellite-borne sensors, and 38 with spectroscopy. When considering only the drone-borne sensors, seven studies utilised the electromagnetic spectrum's red edge (RE) section. In comparison, 12 studies utilised the satellite remotely sensed RE section, while 44 studies utilised RE bands from spectroscopy. Few studies attempted to engage the thermal bands in characterising the spatial distribution of NUS and their health attributes. Four studies used drone-acquired thermal remotely sensed data and a similar number of studies used satellite-acquired thermal bands. Also, only five studies utilised the spectroscopy thermal section of the EM (Figure 11). When considering drone-borne sensors, only 2 studies utilised the ultra-violet (UV) and 38 studies used the spectroscopy UV section of the EM.

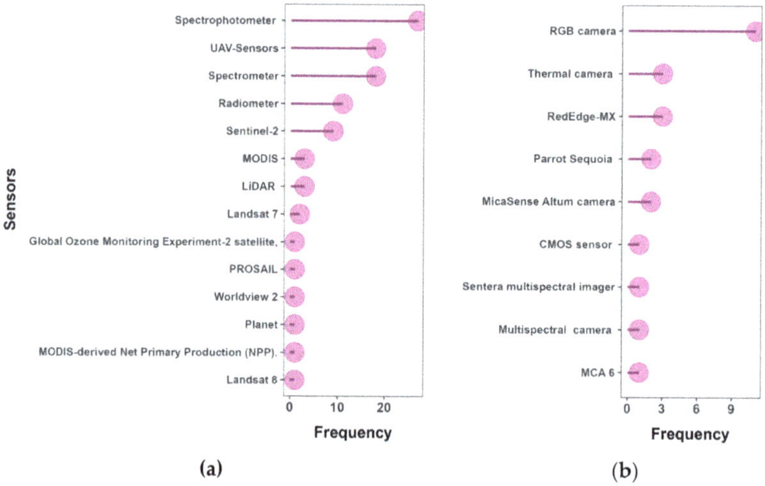

Figure 10. Frequency of (**a**) all sensor and (**b**) drone-borne sensors that have been used to map the spatial distribution of NUS and their attributes. (RGB represents red, green, and blue).

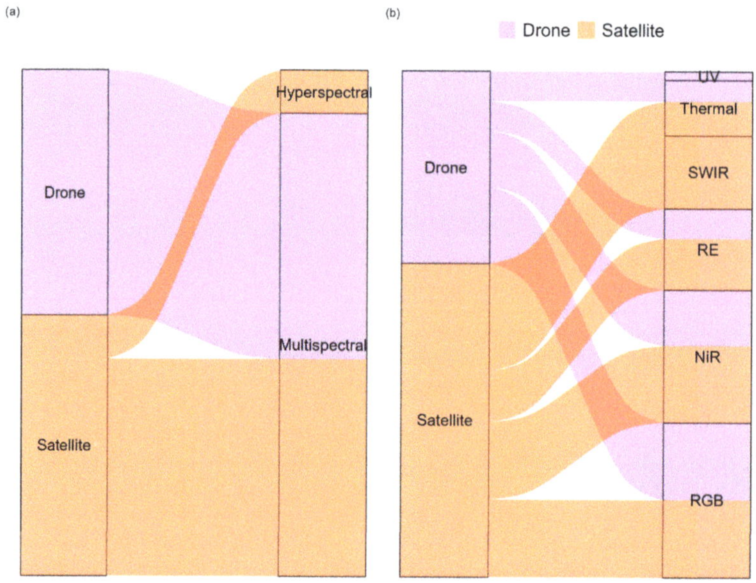

Figure 11. Visual distribution (frequency) of (**a**) Spectral characteristics of satellite and drone borne sensor, and (**b**) specific sections of the electromagnetic spectrum they covered in the literature (See Supplementary Table S2 for frequency values). UV is ultraviolet, SWIR is shortwave infrared, RE is red edge, NiR is near-infrared, and RGB is the red, green, and blue spectra. (See Supplementary Materials spreadsheet for frequency values.

3.4. UAV Platforms Utilised in the Literature

Regarding the drone platforms, the DJI fleet was utilised in a marginally higher number of studies in relation to all other platforms ($n = 12$). The Octocopter [30], mikrokopter [17], and Sensfly eBee [6] each featured in separate single studies (Figure 12b). When assessing the frequency of each specific sensor in the retrieved literature, it was observed that the DJI Phantom 4 Pro was the most frequently used platform across the board (appearing in six studies) [23], followed by the Octocopter UAV, which was utilised in one study [31] (Figure 12a). Furthermore, quadcopters were the most widely used drone platform type, followed by fixed-wing drones (Figure 12b). Quadcopter drones were utilised in 16 studies, and fixed-wing drones were used in only 2 studies. Furthermore, the DJI UAVs were popular in mapping a wider range of crops and research domains when compared with other platforms (Supplementary Tables S3 and S4). This could indicate that these platforms are more popular and versatile.

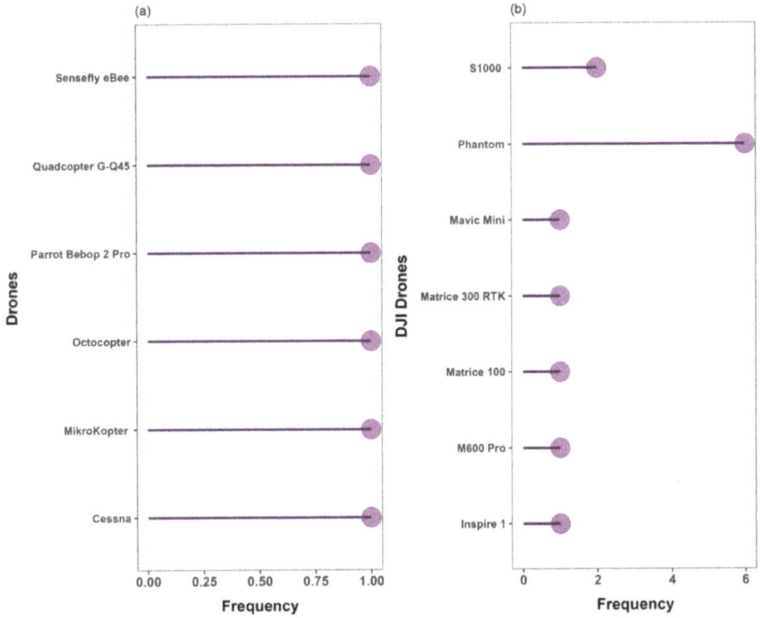

Figure 12. Frequency of (**a**) summarised drone platforms (**b**) and DJI drones utilised in the literature to map the spatial distribution of NUS and their health attributes.

3.5. Derived Vegetation Indices in Remote the Spatial Distribution and Health of NUS Crops

Algebraic combinations derived from multiple spectral bands, which are commonly known as vegetation indices (VIs), were used to estimate vegetation vigour and vegetative characteristics (canopy biomass, absorbed radiation, and chlorophyll content) in the retrieved literature [32]. Furthermore, the visible (green: 530–570 nm, red: 640–680 nm, and red edge: 730–740 nm), near-infrared (770–810 nm), and red edge (730–740 nm) sections of the electromagnetic spectrum were common in studies that assessed crop health. The reflectance values of these prominent wavelengths are generally used to calculate vegetation indices, such as the normalised difference vegetation index (NDVI), NDVI–red edge (NDRE) simple ratio (SR), green normalised difference vegetation index (GNDVI), green chlorophyll index (CIgreen), and soil-adjusted vegetation index (SAVI), which were frequently used in the retrieved literature (Figure 11). All VIs that were used in the literature are also listed in Supplementary Table S1. In this regard, there is still room to assess more image transformations in mapping the spatial distribution of NUS, such as sweet potato and taro, dominant in smallholder croplands.

3.6. Statistical and Machine Algorithms Were Utilised in Mapping the Spatial Distribution and Health of NUS Crops

This study's findings show that several basic statistical procedures, simple regression techniques, and machine learning techniques were used in mapping the spatial distribution and health of NUS crops. These algorithms can be further subdivided into three categories, which are (i) generic GIS classifications, (ii) machine learning and regression techniques, and (iii) multivariate techniques.

Based on Figure 13, the main machine learning algorithms utilised in conjunction with drone-acquired data were linear regression (39%), random forest (28%), support vector machine (SVM) (22%), and artificial neural network (ANN) (17%), in order of frequency in the literature. Linear regression techniques were the most frequent regression algorithms in assessing NUS crop health (Figure 13) [22,30]. The frequency of linear regression and machine learning-based algorithms was also detected during the bibliometric analysis, as illustrated in Figure 2b (red cluster). The RF was the second most widely used machine learning ensemble, followed by SVM, PLS, linear discriminant analysis, and artificial neural network (ANN), in order of frequency in the retrieved literature.

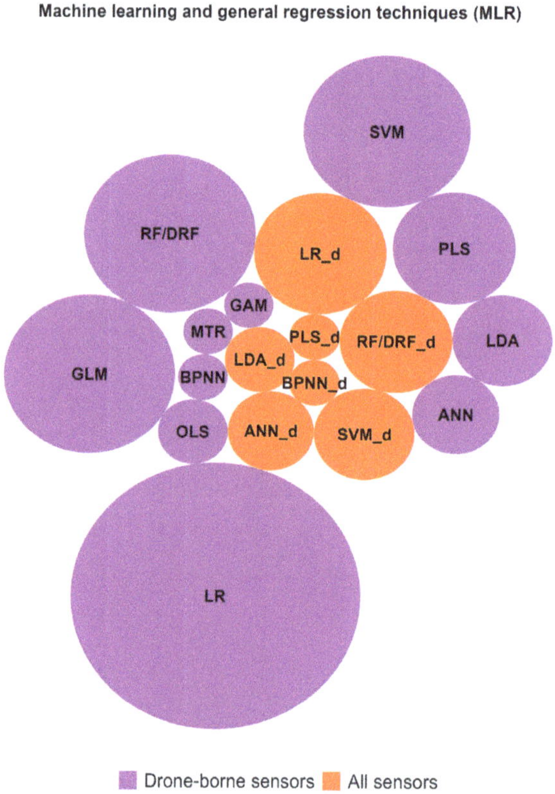

Figure 13. Frequency of machine learning and general regression techniques used in remote sensing NUS attributes based on all sensors and drone-borne sensors. (GLM is generalised linear model, RF/DRF is random forest, LR is linear regression, SVM is support vector machine, OLS is ordinary least squares regression, ANN is artificial neural network, BPNN is back propagation neural network, GAM is generative adversarial networks, LDA is linear discriminant analysis, and PLS is partial least squares regression. (See Supplementary Materials spreadsheet for frequency values).

In terms of generic statistics and classification techniques, Pearson correlation, ANOVA, maximum likelihood (ML), OBIA, and the empirical line method (ELM) were the most frequently utilised algorithms based on satellite and drone-borne remotely sensed data (Figure 14). The Mahalanobis distance, parallelepiped, k-means, and canny edge filtering were some of the popular generic classification algorithms used to map NUS based on satellite and drone-acquired remotely sensed data. Studies that were based on generic classification were relatively few for each algorithm (<5) when compared with all other algorithms utilised in the retrieved literature (Figure 14a). Generic classification procedures, such as the analytical hierarchical process (AHP), cluster analysis, fuzzy logic, and multi-layer perceptron, have not yet been utilised in conjunction with drone-acquired remotely sensed data for crop mapping (Figure 14b).

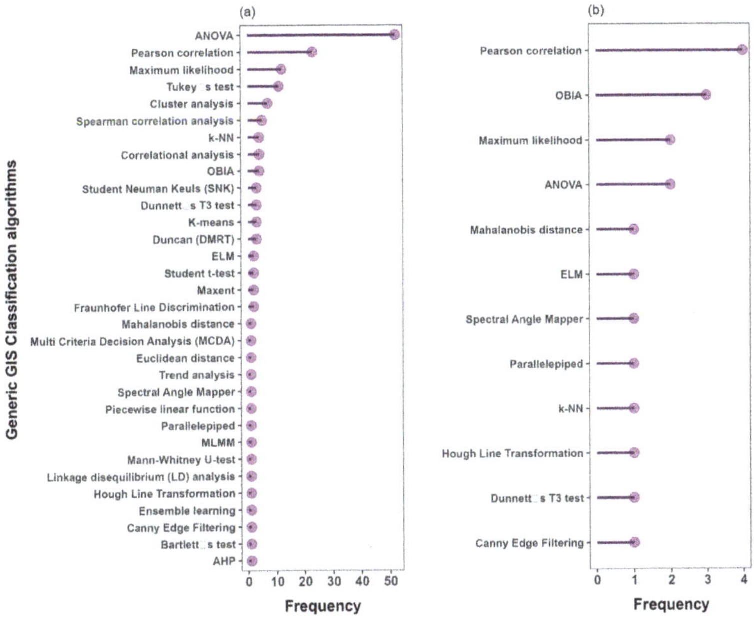

Figure 14. Frequency of generic GIS classification techniques used in remote sensing NUS attributes based on (**a**) all sensors and (**b**) drone-borne sensors.

Regarding multivariate techniques, PCA followed by cluster analysis and multiple regression were the most frequently used algorithms based on satellite and drone-acquired data combined (Figure 15). Studies based on linear mixed model and machine learning algorithm classifications were relatively few (<5) compared with PCA and cluster analysis. There seemed to be scanty literature (<5) that utilised multivariate techniques in conjunction with drone-acquired data for mapping the spatial distribution of NUS and their health attributes.

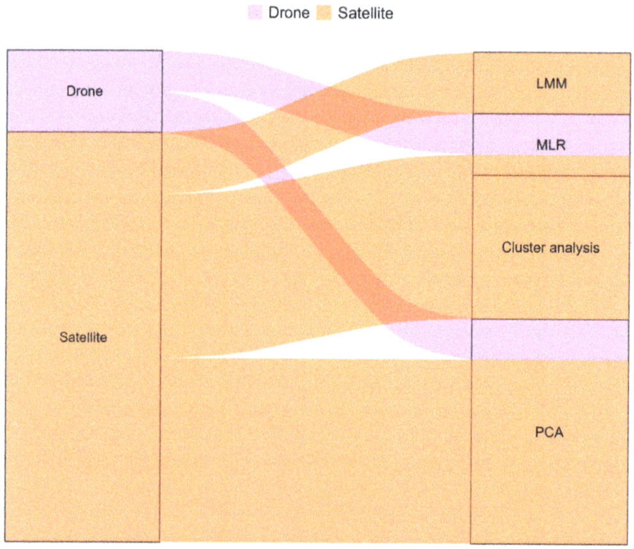

Figure 15. Frequency of multivariate techniques used in remote sensing NUS attributes based on all sensors and drone-borne sensors. (See Supplementary Materials spreadsheet for frequency values).

4. Discussion

4.1. Evolution of Drone Technology Applications in Remote Sensing

There have been numerous shifts in the key terms of the literature on the remote sensing of the distribution and health status of NUS crops. Specifically, there was a shift in the topical terms from mere correlations based on RGB remotely sensed data in mapping NUS attributes between 2018 and 2019 to using hyperspectral data in predicting and mapping NUS crop health attributes, such as yield and AGB, in 2020. Currently, research efforts are being exerted towards the fusion of drone-acquired data with satellite remotely sensed datasets in conjunction with robust machine learning algorithms, such as PLSR and SVM, in characterising yield genotype canopy coverages, amongst others. Progress is discussed in detail in the following sub-sections.

4.1.1. Frequency of Publication and Their Geographic Distribution

This study's findings revealed that the articles that utilised drone-based remotely sensed data in mapping the spatial distribution, health, and productivity elements of NUS crops increased gradually from 2009 to 2014 (Figure 3). This trend was similar to that of published studies that utilised satellite remotely sensed data (Figure 3b). The rapid changes, improvements, and increased accessibility of Earth-observation sensors and platforms could explain this. That period was characterised by limited access to high-spatial-resolution remotely sensed data for crop monitoring. These findings are echoed by numerous reviews, which include Mutanga, Dube [33] and Sibanda, Mutanga [34].

In the context of satellite-borne sensors, the research period from 2009 to 2014 was mainly dominated by the utility of Landsat and MODIS [35]. These sensors were incapable of capturing the land fragmentation and heterogeneity associated with NUS. Moreover, the only accessible fine-spatial-resolution images were those procured from commercial sensors, such as WorldView and QuickBird. These sensors are often associated with exorbitant costs that restrict research activities, especially in under-developed countries. The potential of NUS was still being researched during that period, and research investments were not channelled to the use of satellite-borne sensors. However, the period between 2015 and 2022 was marked by a rapid increase in drone platforms and associated sensors, hence the rapid increase. This would suggest an increase in the interest in NUS and the improved capabilities of drone and satellite sensors.

Interestingly, most of these drone-related studies were conducted by universities and agricultural institutions in China, America, South America, Europe, and Australia (Supplementary Table S6). This could be attributed to the fact that the earliest drone technologies emerged in these regions between 1849 and 1916 [34]. Also, some of the mentioned countries are pioneers in the research and preservation of NUS. Since then, technology has been spreading and advancing. As a result, there is an increasing need to improve the application of UAV technologies for precision agriculture in accordance with NUS spatial extent and health assessments. Furthermore, it was noted that most of the studies were conducted on experimental plots, in some instances with irrigation facilities at university experimental plots. Generally, commercial farmers who are endowed with resources are the most dormant users of these technologies in under-developed regions, such as southern Africa. No studies were conducted in rainfed smallholder croplands in the retrieved literature on remote sensing NUS. The findings of this study imply that these technologies are slowly being embraced and incorporated from developed countries to the Global South.

4.1.2. NUS Crop Attributes That Have Been Remotely Sensed Using Drone-Acquired Data

The limited number of studies focusing on NUS suggests that the utility of UAVs is still in its infancy in practice. In the retrieved literature, there were no studies that indicated whether there are any spectral libraries that have been generated for NUS. This could imply that, despite the significant efforts in advancing the remote sensing of NUS, more efforts are still required to match with other crops, such as maize. Although the results revealed that, from the year 2020 going onwards, there has been an increase in the number of studies, Merkert and Bushell [36] noted that there are still few research efforts directed towards NUS (Figure 8). In developing countries where NUS are beginning to gain recognition, the limited number of studies could be attributed to the expenses associated with drones and their accessories. Furthermore, the requirements to license and operate a drone in developing countries are still a challenge [37,38]. These findings underscore the importance of increasing knowledge and literature in the context of NUS to improve our insights into them as alternative crops.

The findings showed that 16 studies assessed crop phenology, 9 were on productivity, and 7 explored physiology using drone-acquired remotely sensed data. However, most of the retrieved studies focused on interrelated themes, which explains the high frequencies of studies on each crop attribute. Crop attributes, such as physiology, productivity, and phenology, are mutually dependent elements that determine crop development and health. According to Fageria, Baligar [39], crop physiology in particular is a useful crop attribute that could be used to quantify crop growth for optimising crop yields.

The findings of this study showed that 11 NUS crop productivity elements, namely, crop yield, crop growth, crop health, chlorophyll content, biomass, LAI, canopy height, cover, leaf nitrogen, leaf water content elements, and stomatal conductance, were the most widely researched [28,40]. These are the principal optical crop productivity and health elements; hence, they are anticipated to be covered extensively in the literature on agricultural remote sensing applications. The primary function of remote sensing applications is to optimise yield production and the supply of nutrient-rich foods [41]. Additionally, all research efforts have been exerted towards optimising agricultural productivity (increasing yields) and addressing sustainable goals 1 and 2 on hunger and poverty [23,33,42,43]. Subsequently, yield is expected to be the most intensely researched NUS crop attribute based on drone and satellite-borne remotely sensed data [23,42]. It must be noted that, in most instances, yield is synonymous with AGB. The findings of this study showed that 61% of drone-based remote sensing studies on NUS estimated crop AGB. AGB is an important parameter that can also be utilised to predict crop growth, yield. and productivity [44].

The findings of this study also showed that LAI is one of the most researched elements of NUS, featuring in 33% of the retrieved literature. This can be attributed to the fact that LAI is yet another accurate proxy of plant growth, as illustrated in Figure 7. It has been extensively proven to correlate with other plant growth indicators, including chlorophyll content, biomass, and, to some extent, yield [45,46]. LAI represents the structural attributes of the leaf components estimated by the area of the leaf per unit of ground surface area [47]. The area covered by leaves per unit of the ground surface changes with the changes in a crop's growth stage. This variable can accurately represent the space available for photon interception across a crop's phenological stages, which can affect yield [48]. Subsequently, LAI is a plausible indicator of canopy health and development that needs to be accurately mapped [49]. In addition, LAI can affect the surrounding canopy and the microclimate, e.g., radiation from the sun is intercepted by leaves, affecting transpiration and leaf surface temperature. This ultimately influences the photosynthetic nature of leaves and the stomatal conductance [50]. In this regard, LAI was one of the most important NUS attributes in the retrieved literature.

The chlorophyll content is another predominant NUS attribute covered by 50% of the retrieved literature in this study [51,52]. The chlorophyll content of leaf tissue indicates a plant's physiological structure, nutritional composition, and health [53,54]. The antenna pigments in chloroplasts absorb incoming solar radiation during photosynthetic activities [55]. The resulting radiation is then transferred to the reaction centre pigments, which discharge electrons to activate the photochemical process [55]. Specifically, chlorophyll is a very relevant and optical indicator of crop health, since chlorophyll highly absorbs energy in the red (650–700 nm) and blue (400–500 nm) regions of the electromagnetic spectrum to increase photosynthesis. The types of chlorophyll responsible for the higher absorption within the visible spectrum are chlorophylls a and b [55]. In this regard, there is a positive relationship between the leaf total chlorophyll content (chlorophyll a+b), the solar radiation absorbed by the leaf tissues, and the photosynthetic rate of the crop [55]. Subsequently, a leaf's biophysical pigments and biochemical photosynthetic processes are linked to a plant's health and productivity. Hence it is considered a suitable proxy for crop health in the light of agricultural remote sensing applications [54].

Meanwhile, the chlorophyll content has been widely proven to correlate positively with the nitrogen content in various crop species [56–59]. Indeed, approximately 75% of the total nitrogen is stored within the leaf chloroplasts [60]. In this regard, the findings of this study revealed that 28% of the retrieved studies estimated nitrogen concentration using drone remotely sensed data, thus rendering the nitrogen content another sought-after attribute of NUS. An elevated nitrogen concentration is associated with an increased CO2 assimilation rate and stomatal conductance in the crop, which aids in producing chlorophyll and green pigments. For example, Muhammad, Alam [61] conducted a related study to evaluate the impact of various nitrogen and phosphorous levels and beneficial microbes on enhancing canola productivity. The results revealed that nitrogen applied at a rate of 180 kg ha^{-1} increased plant pods, seed pods, the seed-filling duration, seed weight, biological yield, and seed yield.

The findings of this review also showed that crop structural parameters, such as crop height, crop growth, and canopy cover, were explored in 39%, 72%, and 39% of drone-based studies, respectively. These crop structural parameters are related to AGB, LAI, and the chlorophyll content, which are measured and used to predict the yield. Hence, these parameters are directly linked to food and nutrition security. The crop height and growth attributes directly interact with the incident electromagnetic energy that is typically measured and used to model productivity. Furthermore, these physiological crop variables are sensitive to variations in environmental conditions. These include environmental (precipitation, temperature, soil type, etc.) and biochemical conditions (fertility, weeds, pests, and diseases). In this regard, these structural attributes are instrumental in monitoring the health of crops to optimise crop production; hence, significant research efforts were devoted to them.

This study showed that stomatal conductance was another optimal NUS health parameter assessed by 11% of the retrieved studies that utilised drone data. The stomatal conductance measures the degree of stomatal opening and can indicate leaf gas exchange [62]. The stomatal conductance is strongly associated with leaf transpiration rates, photosynthetic efficiency, chlorophyll concentration, and nitrogen concentration [63]. Specifically, periods of crop water stress or drought stress will be limited by the CO_2 concentration at carboxylation sites (Cc) inside the chloroplast. This is determined by the CO_2 diffusion components, i.e., stomatal conductance (gs) and mesophyll conductance (gm) [64]. Particularly, the stomata control the CO_2 diffusion into the leaf tissue and water diffusion out of the plant. Therefore, it has been proven that, under water deficit conditions, plant stomata will close to prevent major water loss. This consequently decreases photosynthesis via the decreased influx of CO_2 [64].

In this regard, higher stomatal conductance and high photosynthetic efficiency are associated with limited moisture stress. Hence, these plants will be healthier, with a higher chlorophyll content and green pigment. In this regard, stomatal conductance was among the most researched crop attributes in the retrieved literature because it is an accurate proxy of moisture stress and the health status of crops [28,62]. In fact, the stomatal conductance aids in water use and irrigation scheduling as a pathway towards optimising food production. A study by Chai, Massawe [65] measured the morpho-physiological traits of Bambara groundnut exposed to progressive mild drought in a controlled environment. Drought stress reduced stomatal conductance significantly ($p < 0.01$). Furthermore, higher stomatal density and reduced leaf area were observed in drought-treated plants ($p < 0.01$). This suggests that NUS crops are more resistant to biotic and abiotic stresses, such as drought and water stress. Above all, this indicates the prospects of using stomatal conductance as a proxy for understanding crop water stress.

In most studies from the retrieved literature, there were strong correlations between the stomatal conductance and water relation status or leaf water content [28,66,67]. Specifically, 17% of studies based on drone remote sensing researched the NUS leaf water content. Water content is an important indicator of crop health [28,66,67]. A plant with higher water potential will produce greener pigmentation and have increased crop productivity. According to Ouyang, Struik [64], under water deficit or conditions of mild water stress, the stomatal conductance, internal CO_2 concentration, and net assimilation rate within a plant will decrease, and the A/G ratio will increase. Various studies assessed and reported that crop water stress allows for a decline in the total chlorophyll of various crops and a decline in plant productivity [68–70]. This includes non-stomatal regulation of photosynthesis, a decline in light and the CO_2 concentration, a reduction in photochemistry, declining activity of photosynthetic enzymes, and lowered mesophyll conductance. For instance, Chibarabada [71] assessed the stomatal conductance of three grain legume crops (groundnut, dry bean, and Bambara groundnut) grown under three water treatments. Their results indicated that, under varying water regimes, NUS crops adjusted to constrained soil water through stomatal regulation and reduced canopy size. Furthermore, Bambara groundnut showed a positive attribute under water-limited conditions. It had the lowest stomatal conductance under all watering regimes compared with the other crops. In the context of remote sensing, these plant conditions then impact the interaction between different sections of the EM spectrum and the various magnitudes of crop water stress [72]. This example illustrates why stomatal conductance and foliar temperature were significantly considered in NUS production.

4.1.3. Sensors and Platforms That Were Used in Remote Sensing NUS

The dominance of drone-based remote sensing studies on NUS could be explained by the fact that most of these crop species are orphaned, neglected, and underutilised. In this regard, they are generally planted in smaller areas than mainstream crops, such as maize. On the other hand, very few freely accessible satellite-borne sensors offer finer-spatial-resolution data suitable for capturing the variety of crops in fragmented smallholder croplands. Subsequently, the freely available datasets from moderate-spatial-resolution sensors, such as Landsat and Sentinel 2 MSI, cannot capture the dynamics of crops in smallholder farms when compared with UAVs. Drones offer rapid, ultra-fine-spatial-resolution data, often to a sub-metre resolution, in a cost-effective manner [73].

Furthermore, the user determines the spatial and spectral resolution of the drone remotely sensed data, offering endless opportunities in crop phenotyping from the stand level to field scale. Sentinel-2 MSI was the most widely used satellite-borne sensor in mapping the spatial distribution, health, and productivity parameters of NUS crops. Sentinel 2 MSI has a minimum spatial resolution of 10 m and a spectral resolution that covers the red edge section of the electromagnetic spectrum, which is sensitive to crop health [74]. These attributes make Sentinel 2 MSI the second most suitable sensor among UAV-borne sensors in mapping NUS, often grown in small, fragmented fields.

In terms of the drone sensors, our findings showed that the RGB and multispectral cameras (>three bands) were the most frequently used in the retrieved literature when compared with hyperspectral sensors (Figure 11).This could be explained by the exorbitant expenses associated with hyperspectral sensors to multispectral cameras. RGB and multispectral cameras generally cover the visible, including the red edge and the NIR, sections of the EM spectrum. This is usually in not more than six broader spectral band resolutions (i.e., MicaSense Altum) [75]. Meanwhile, most hyperspectral sensors, such as the Cubert S185 hyperspectral sensors, cover a wavelength range between 450 and 950 nm. These are composed of 125 channels at a resolution of 8 nm at 532 nm, ranging from visible to near-infrared (450–950 nm) with a sampling interval of 4 nm [76]. Furthermore, the RGB type of sensors offers models with less accuracy when compared with multispectral or hyperspectral images covering a wider spectral range [77]. Hyperspectral sensors have been widely proven to offer robust narrow, relatively contiguous bands, although multicollinearity issues often impact them. Despite their exorbitant prices, these bands are more sensitive to subtle crop variations compared with broadband sensors [78–80].

The findings of this study also showed that very limited studies utilised cameras that captured data in the red edge section of the EM spectrum. The multispectral sensors that captured the red edge section in the retrieved studies were the MicaSense series, and Red-edge MX utilised in two and three studies, respectively. The red edge section has been vastly proven to be sensitive to minute variations in plant attributes associated with health and productivity, such as the LAI, chlorophyll content, stomatal conductance, AGB, and nitrogen content [49]. For instance, the increased chlorophyll content, LAI, and biomass generally result in increased absorption in the red region, pushing the red edge to longer wavelengths [81]. Considering this phenomenon, the red edge has become one of the most sought-after sections of the EM spectrum in crop monitoring [82]. In this regard, there is a need for more studies that assess the robustness of hyperspectral and red-edge sensors in mapping NUS attributes. This will improve the assessment and monitoring of the health elements of NUS if their productivity is optimised based on the information derived from remotely sensed data.

The findings of this study also showed that the multirotor drones were the most widely used platforms in the retrieved literature compared with fixed-wing drones (Figure 12). This could be explained by the fact that, despite the endurance associated with fixed-wing drones, they often require take-off and landing space. However, these requirements are not always available in experimental sites, where most retrieved studies have been conducted [83]. Multirotor drones are characterised by vertical take-off and landing (VTOL), which makes them more suitable for utilisation in research areas with limited take-off space [83]. These drones are relatively cheaper than fixed-wing drones [34]. This could also explain the high utilisation frequency of DJI drones in the retrieved literature (Figure 12). The advantage of DJI drones is that some are associated with an automated image acquisition procedure, making it easy to fly them, as they require less expertise in drone piloting. According to Sibanda, Mutanga [34], the DJI platforms are generally compatible with many types of sensors from other platforms, which could also explain their wide utilisation in studies on the remote sensing of crops.

4.1.4. Performance of Vegetation Indices, Classification, and Estimation Algorithms

This review's findings also showed that vegetation indices (VIs), such as NDVI, NDRE, SR, GNDVI, CIgreen, and SAVI frequently mapped the spatial distribution of NUS and their health parameters. For instance, Li, Zheng [84] deployed a quadcopter UAV to investigate its utility in crop identification from different land cover types based on VIs. The maximum likelihood classifier, in combination with optimal VIs, exhibited high classification accuracies in that study. This approach was accurate because VIs could partially overcome the influence of shadows and other noise in the background [85]. A growing body of literature demonstrates the optimal performance of VIs [22,24,29]. Specifically, VIs are robust because they can suppress background effects compared with bands [79]. Furthermore, VIs derive their strength from two or more sections of the EM spectrum. For instance, Vis, such as NDVI, NDRE, SR, GNDVI, CIgreen, and CIred-edge, derive their strength from sections of the EMS (i.e., red edge, NIR, and SWIR) that are more sensitive to crop elements, such as the chlorophyll content, LAI, ABG, nitrogen concentration, and stomatal conductance [53]. Moreover, a large body of literature has proven that red-edge-based VIs are more robust than conventional vegetation indices, such as NDVI and SR [20,86]. NDVI tends to be insensitive to increases in chlorophyll, LAI, and biomass [85,87]. As aforementioned, the red edge is highly associated with plant physiological traits, such as the leaf angle distribution, chlorophyll concentration, and LAI, which directly influence vegetation spectral reflectance. Subsequently, red edge-derived vegetation harnesses this robust sensitivity to variations in LAI biomass and chlorophyll content, among others [85,88–91].

As was suggested in many studies, the combinations of sensitive spectral variables with robust algorithms produce models with relatively optimal accuracies [92]. The findings of this study showed that the machine learning algorithms that were utilised in conjunction with drone-acquired remotely sensed data were primarily linear regressions (LR) (39%), followed by RF (28%), SVM (22%), and ANN (17%). Machine learning regression algorithms present a potential approach for generating adaptive, robust, and fast crop estimates. A growing body of literature demonstrates machine learning algorithms' efficiency and optimal performance in estimating crop biophysical parameters [47,93,94]. The high frequency of using regression models, such as RF, LR, and SVM, can be credited to the fact that they are simple to implement using various platforms. This ranges from Microsoft Excel to complex programming platforms, such as R Statistical Software and the Google Earth Engine platform. Despite their high frequency in the retrieved literature of this current study, LR models are parametric algorithms associated with data normality assumptions. However, these assumptions are often challenging to attain due to high spatial varieties in crop fields. Furthermore, these models are susceptible to overfitting and outliers [95]. Subsequently, more efforts have been channelled towards machine learning algorithms, such as SVM and RF. These algorithms are non-parametric, robust, and less prone to outliers, as they are not extremely affected by the data frequency distributions [95]. Specifically, RF has (i) hyperparameters that are easy to adjust, (ii) is robust to outliers, overfitting, and data dimensionality, (iii) low bias and moderate to minimal variance due to the averaged trees, (vi) works well both continuous and categorical variables, (v) has a capability of discerning the importance of predictor variables, and (iv) it is relatively resistant to multicollinearity when tree depths are greater [91,96,97]. This makes this algorithm suitable for mapping cropping attributes.

Other than machine learning algorithms, numerous generic classification algorithms were utilised in the retrieved literature, such as maximum likelihood and minimum distance to mean classification algorithms [98–100]. Despite their optimal performance in the literature, these algorithms are less robust when compared with machine learning algorithms. For instance, maximum likelihood is also a traditional parametric algorithm that is often difficult to parametrise, despite the fact it is generally accessible through freely available GIS platforms.

4.2. Challenges in Mapping the Spatial Distribution and Health of NUS Using UAVs

Even though there is progress regarding the utility of drones in crop mapping and health quantification of NUS, several challenges impede their propagation, especially in Africa. Many African countries are battling to address various drone-related issues, such as privacy, public safety, and preventing the possession of malicious drones [101]. Moreover, restrictive UAV or drone regulations across many developing regions, including Africa, hinder their utilisation. To utilise UAVs, users must seek permission from landowners and municipalities, and only operate in certain areas, among several other issues [102].

For example, the civil aviation authorities (CAAs) in many countries aim to prevent UAVs from entering the flight paths of manned aircraft. Moreover, CAAs are attempting to construct an inclusive system that accommodates UAVs into their respective air navigation and surveillance systems [103]. To the best of our knowledge, only a few African countries, including Ghana, Kenya, South Africa, Rwanda, Zimbabwe, and Tanzania, have established complex legal requirements governing the use of UAVs in varied airspace practices. Some drone restrictions in South Africa (SA), for example, state that UAVs weighing more than 7 kg (15.4 pounds) are not permitted to operate [102]. In this regard, the regulation on the mass of UAVs at taking off tends to indirectly restrict the areal extent and the size of the camera to be mounted for research purposes [34]. Specifically, due to the weight restrictions, many of the sensor types that are frequently used tend to be lightweight, small, and general consumer grades with limited spectral resolution [34]. Also. this could explain why more studies are based on the VIS (RGB) spectra than other sections of the electromagnetic spectrum.

Furthermore, SACAA asserts that drone operators should maintain a continuous visual line of sight (VLOS) with their drones during flights. Moreover, remote-piloted aircraft (RPAs) are not permitted to fly beyond visual-line-of-sight in designated places (BVLOS) [102]. Sibanda, Mutanga [34] stated that supporting regulations and the operationalisation of BVLOS drone technology applications will facilitate coverage of greater areas on a single mission. Covering a greater area in a single mission improves the cost-effectiveness of acquiring VHR imagery. Other restrictions include requiring drone operators to obtain an RPAS operator certificate (ROC) from the CAA before flying [104]. UAV operators should obtain insurance to cover their liability in the case of committing physical or bodily harm to another individual while operating their drone [104].

Aside from these regions, numerous African countries are still attempting to establish the necessary regulations that endorse UAV operations. As a result, these regulations are becoming increasingly complicated, while attaining RPAS is associated with high expenses. More notably, the cost of obtaining a drone pilot license is excessive, with estimates ranging from USD 1500 to USD 2000 in 2021 [34]. Furthermore, the value of drone platforms, including sensors, is prohibitively expensive for several minority groups or scholars, making these technologies unattainable for research purposes in the vast bulk of Sub-Saharan African countries [34].

4.3. Research Gaps and Opportunities

The following gaps were noted in assessing the utility of drone-based remotely sensed data in mapping the spatial distribution and health of NUS crops:

- The observation of NUS crop health has garnered minimal research attention and interest from the scientific community. Further, few studies have sought to evaluate the utility of drone technology for characterizing crop dynamics, especially in the Global South. The limited research within this region means there are opportunities to innovate;
- Although NUS crops reportedly resist abiotic stresses, such as drought and heat stress, most of this information is anecdotal and inconsistent [1]. This incomplete body

of knowledge around drought and heat stress makes applying and validating RS techniques challenging. Hence, there is a requirement to generate more empirical information on the ecophysiology and morphology of NUS;

- Only a few research studies have sought to evaluate the effectiveness of robust ML algorithms in conjunction with VIs in predicting the spatial distribution and health of NUS crops. Further to this, few studies have attempted to assess and leverage the potential synergies between drone and satellite-borne datasets, especially considering the release of the freely accessible Planet Scope Sentinel 2 MSI and Landsat series;
- The application of UAV-based technology for estimating NUS' spatial extent and health has not attracted significant attention from the geospatial research community in practice. The spatial extent of NUS crops can be predicted at a granular scale using UAV-based modelling and classification. Such models will be useful for predicting crop yield, crop monitoring, predicting soil quality, and modelling evapotranspiration, precipitation, drought, and pest outbreaks;
- Modelling and predicting vegetation key variables, such as LAI, stomatal conductance, and AGB, are critical to understanding and quantifying NUS' morphological and phenological processes in the face of climate change;
- Optimal VIs, such as NDVI NDRE, and VARI, can aid smallholder farmers in analysing trends in plant health. Moreover, NDRE is useful in determining vegetative vigour late in the growing season.

It is, therefore, essential to research the prospects of these remote sensing technologies in monitoring these crops to promote their inclusion as mainstream crops within the agriculture sector.

4.4. Way Forward: Closing the Gaps in the Utilisation of Drone Technology in Mapping Spatial Distribution and Health Status of NUS Crops

Scholarly attention needs to be paid to promoting the application of UAVs for assessing NUS crops' spatial extent and health at the field scale. There is a need to increase and extend research efforts towards rainfed smallholder croplands in remotely sensing NUS. Hence, this will optimise their productivity while strengthening the livelihood and food security of marginalised subsistence farmers.

UAVs are becoming increasingly common in agricultural research as relevant sources of high-temporal-resolution data for agricultural health monitoring, amongst other applications. This has been previously limited to a farm scale. UAVs provide immense opportunities by incorporating climate-smart and precision agriculture into smallholder farming. They deliver high-resolution imagery at user-defined temporal resolutions, which benefit crop health monitoring and independent decision-making. For example, the integration of multi-rotor drones, such as the DJI series, in conjunction with multispectral sensors, such as the Mica Sense RedEdgeMX Parrot Sequoia and MicaSense Altum cameras, has been demonstrated to deliver accurate models of crop health characteristics at the field scale.

Additionally, the increasing adoption of cutting-edge narrow-band hyperspectral and LiDAR remote sensing technologies could provide the opportunity to assess diverse crop health parameters at high optical resolutions. This assessment would also allow for creating various optimal VIs. Hence, these are directly associated with NUS crop morphological and physiological characteristics, allowing for the quantification of their health status.

Moreover, to improve research based on the application of UAVs to estimate NUS crop health, it is essential to identify and implement robust and reliable non-parametric ML algorithms. Algorithms that enhance prediction accuracies with fewer assumptions, such as the RF, seem to hold endless prospects. Using multi-fusion techniques based on the combination of ideal VIs and ML algorithms will achieve the best optimal accuracies. As a result, additional research is required to evaluate the utility of UAVs with distinct spectral characteristics for NUS. It would be valuable to research whether UAV sensors that measure spectral reflectance along the electromagnetic spectrum's thermal, SWIR, and NIR sections enhance the prediction of NUS water stress at the farm scale. Therefore, near-real-time-temporal-resolution spatially explicit information on NUS spatial extent and health can certainly assist smallholder farmers. This will assist in detecting variations in crop morphology to optimise yields. Near-real-time fine-resolution NUS information could be used to draw up early warning systems for smallholder farmers to prevent any damage or reduction to crop yields.

5. Limitations of This Study

Many possible sources of information on the spatial extent and health of NUS were disregarded throughout our search approach. We primarily searched for English-language sources and peer-reviewed articles included in Scopus, Web of Science, and Google Scholar. In conducting the literature review, some studies were inaccessible in full length and others were not written in English. This could have had a potentially negative effect on quantifying all the studies that focus on remote sensing the spatial distribution and health status of NUS crops. More so, the exclusion of these studies has an impact on the spatial distribution of NUS studies. Since NUS crops research is highly under-represented and anecdotal, it was challenging to obtain research that focused on the health and spatial extent of NUS crops. Despite missed contributions, we believe that our findings are based on a sample that is adequate and diverse enough to offer meaningful insight and implications in the application of UAVs in the remote sensing of NUS crops in smallholder croplands. Another, limitation of the study was the fact that we limited our scope of research materials for the systematic review to only the Scopus, Web of Science, and Google Scholar databases. As such, we recommend that future work extend the sources to a larger number of databases.

6. Conclusions

The objective of this study was to systematically assess the literature on the progress, challenges, gaps, and opportunities in using drone-derived remotely sensed data to map the spatial distribution and health status of NUS. This study placed a special focus on classifying and measuring stomatal conductance in smallholder croplands in the Global South. From the reviewed literature, UAVs emerged as cutting-edge technology with a high potential to provide spatially explicit, timely, and reliable data for assessing the health of NUS. However, there is a noticeable scarcity of studies that have attempted to map the spatial distribution and stomatal conductance, along with other health attributes, of NUS in comparison with mainstream crops in smallholder croplands in the Global South. Several factors contribute to this scarcity, including the limited popularity of NUS, the prohibitive cost of drones and pilot licenses, a shortage of personnel with the necessary skills, and stringent regulations governing drone procurement and operation, among others. While acknowledging the substantial ownership costs associated with drones, our study advocates for the communal ownership of UAVs among smallholder farmers, which can significantly reduce operational expenses. This approach offers promising opportunities to integrate climate-smart agriculture into local farming systems through enhanced monitoring and mapping of crop health. It will equip farmers and, through extension, workers with valuable information for making informed decisions in the field. Additionally, it will enhance stress detection, improve irrigation scheduling, and boost crop productivity on a farm scale. Furthermore, improving the productivity of NUS and other crops presents an opportunity to revive local food culture and food systems, while simultaneously empowering smallholder farmers, who are often the stewards of this agrobiodiverse resource system. We also advocate for the use of UAVs to engage younger generations in agriculture, a critical priority in Sub-Saharan Africa. By delivering high-quality imagery and automating data collection, processing, and analysis at a low cost, the application of UAVs in local food systems can serve as a pathway to optimise food production among disadvantaged smallholder farmers.

Supplementary Materials: The following supporting information can be downloaded at: https://www.mdpi.com/article/10.3390/rs15194672/s1. A list of the articles used in this study is provided as supplementary data, Frequences. References [6,12,14,15,17,18,20–24,26–28,30,31,52,66,67, 84,86,94,99,105–123] are cited in the supplementary materials

Author Contributions: Conceptualization, M.A. and M.S.; methodology, M.A.; software, M.A. and M.S.; validation, M.A., M.S., T.D. and T.M.; formal analysis, M.A.; investigation, M.A. and M.S.; resources, M.S.; data curation, M.A.; writing—original draft preparation, M.A.; writing—review and editing, T.D., M.S., V.G.P.C. and T.M.; visualization, M.A. and M.S.; supervision, M.S., T.D. and T.M.; project administration, M.S.; funding acquisition, M.S. All authors have read and agreed to the published version of the manuscript.

Funding: This work was funded by the Water Research Commission of South Africa (WRC) through Project C2022/2023-00930 titled "Unmanned aerial vehicle high-throughput phenotyping of neglected and underutilized crops species (NUS) for improved water use and productivity in smallholder farms".

Data Availability Statement: The data will be availed upon request from the first author. The first author's contact details are detailed in the names section of tis manuscript.

Conflicts of Interest: The authors declare no conflict of interest.

References

1. Food and Agriculture Organisation of the United Nations. *Small Family Farmers Produce a Third of the World's Food*; Food and Agriculture Organisation of the United Nations: Rome, Italy, 2021.
2. Fan, S.; Rue, C. The role of smallholder farms in a changing world. In *The Role of Smallholder Farms in Food and Nutrition Security*; Springer Nature: Cham, Switzerland, 2020; pp. 13–28.
3. Mabhaudhi, T.; Chimonyo, V.G.; Modi, A.T. Status of underutilised crops in South Africa: Opportunities for developing research capacity. *Sustainability* **2017**, *9*, 1569. [CrossRef]
4. Mugiyo, H.; Chimonyo, V.G.; Sibanda, M.; Kunz, R.; Nhamo, L.; Masemola, C.R.; Dalin, C.; Modi, A.T.; Mabhaudhi, T. Multi-criteria suitability analysis for neglected and underutilised crop species in South Africa. *PLoS ONE* **2021**, *16*, e0244734. [CrossRef] [PubMed]
5. Joshi, B.K.; Shrestha, R.; Gauchan, D.; Shrestha, A. Neglected, underutilized, and future smart crop species in Nepal. *J. Crop Improv.* **2020**, *34*, 291–313. [CrossRef]
6. Li, K.-Y.; Burnside, N.G.; Sampaio de Lima, R.; Villoslada Peciña, M.; Sepp, K.; Yang, M.-D.; Raet, J.; Vain, A.; Selge, A.; Sepp, K. The application of an unmanned aerial system and machine learning techniques for red clover-grass mixture yield estimation under variety performance trials. *Remote Sens.* **2021**, *13*, 1994. [CrossRef]
7. Muruganantham, P.; Wibowo, S.; Grandhi, S.; Samrat, N.H.; Islam, N. A systematic literature review on crop yield prediction with deep learning and remote sensing. *Remote Sens.* **2022**, *14*, 1990. [CrossRef]
8. Butilă, E.V.; Boboc, R.G. Urban traffic monitoring and analysis using unmanned aerial vehicles (UAVs): A systematic literature review. *Remote Sens.* **2022**, *14*, 620. [CrossRef]
9. Duarte, A.; Borralho, N.; Cabral, P.; Caetano, M. Recent advances in forest insect pests and diseases monitoring using UAV-based data: A systematic review. *Forests* **2022**, *13*, 911. [CrossRef]
10. Ankrah, J.; Monteiro, A.; Madureira, H. Shoreline Change and Coastal Erosion in West Africa: A Systematic Review of Research Progress and Policy Recommendation. *Geosciences* **2023**, *13*, 59. [CrossRef]
11. Opole, R.A. *Effect of Environmental Stress and Management on Grain and Biomass Yield of Finger Millet [Eleusine coracana (L.) Gaertn.]*; Kansas State University: Manhattan, KS, USA, 2012.
12. Everitt, J.; Yang, C.; Davis, M.; Everitt, J.; Davis, M. Mapping wild taro with color-infrared aerial photography and image processing. *J. Aquat. Plant Manag.* **2007**, *45*, 106–110.
13. Mabhaudhi, T.; Modi, A.; Beletse, Y. Growth, phenological and yield responses of a bambara groundnut (Vigna subterranea L. Verdc) landrace to imposed water stress: II. Rain shelter conditions. *Water SA* **2013**, *39*, 191–198. [CrossRef]
14. Mazarire, T.; Ratshiedana, P.; Nyamugama, A.; Adam, E.; Chirima, G. Exploring machine learning algorithms for mapping crop types in a heterogeneous agriculture landscape using Sentinel-2 data. A case study of Free State Province, South Africa. *S. Afr. J. Geomat.* **2020**, *9*, 333–347. [CrossRef]
15. Malinao, R.M.L.; Hernandez, A.A. Classifying Breadfruit Tree using Artificial Neural Networks. In Proceedings of the 6th ACM/ACIS International Conference on Applied Computing and Information Technology, Kunming, China, 13–15 June 2018; pp. 27–31.
16. Shao, Y.; Liu, Y.; Xuan, G.; Wang, Y.; Gao, Z.; Hu, Z.; Han, X.; Gao, C.; Wang, K. Application of hyperspectral imaging for spatial prediction of soluble solid content in sweet potato. *RSC Adv.* **2020**, *10*, 33148–33154. [CrossRef] [PubMed]
17. Che'Ya, N.N.; Dunwoody, E.; Gupta, M. Assessment of Weed Classification Using Hyperspectral Reflectance and Optimal Multispectral UAV Imagery. *Agronomy* **2021**, *11*, 1435. [CrossRef]

18. Grüner, E.; Astor, T.; Wachendorf, M. Prediction of biomass and N fixation of legume–grass mixtures using sensor fusion. *Front. Plant Sci.* **2021**, *11*, 603921. [CrossRef]
19. Azizan, F.A.; Kiloes, A.M.; Astuti, I.S.; Abdul Aziz, A. Application of optical remote sensing in rubber plantations: A systematic review. *Remote Sens.* **2021**, *13*, 429. [CrossRef]
20. Ramírez, D.A.; Grüneberg, W.; Andrade, I.M.; De Boeck, B.; Loayza, H.; Makunde, S.G.; Ninanya, J.; Rinza, J.; Heck, S.; Campos, H. Phenotyping of productivity and resilience in sweetpotato under water stress through UAV-based multispectral and thermal imagery in Mozambique. *J. Agron. Crop Sci.* **2021**, *209*, 41–55. [CrossRef]
21. Shi, Y.; Murray, S.C.; Rooney, W.L.; Valasek, J.; Olsenholler, J.; Pugh, N.A.; Henrickson, J.; Bowden, E.; Zhang, D.; Thomasson, J.A. Corn and sorghum phenotyping using a fixed-wing UAV-based remote sensing system. In Proceedings of the Autonomous air and ground sensing systems for agricultural optimization and phenotyping, Baltimore, MD, USA, 17 May 2016; pp. 46–53.
22. Suhairi, T.A.S.T.M.; Sinin, S.S.M.; Wimalasiri, E.M.; Nizar, N.M.M.; Tharmandran, A.S.; Jahanshiri, E.; Gregory, P.J.; Azam-Ali, S.N. Use of Unmanned Aerial Vehicles (UAVs) Imagery in Phenotyping of Bambara Groundnut. *J. Agric. Sci.* **2020**, *12*, 12–22. [CrossRef]
23. Jewan, S.Y.Y.; Pagay, V.; Billa, L.; Tyerman, S.D.; Gautam, D.; Sparkes, D.; Chai, H.H.; Singh, A. The feasibility of using a low-cost near-infrared, sensitive, consumer-grade digital camera mounted on a commercial UAV to assess Bambara groundnut yield. *Int. J. Remote Sens.* **2022**, *43*, 393–423. [CrossRef]
24. Lati, R.; Avneri, A.; Aharon, S.; Atsmon, G.; Smirnov, E.; Sadeh, R.; Abbo, S.; Peleg, Z.; Herrmann, I.; Bonfil, D.J. Uav-Based Imaging for Prediction of Chickpea Crop Biophysical Parameters and Yield. *Comput. Electron. Agric.* **2023**, *205*, 107581. [CrossRef]
25. Xia, W.; Luo, T.; Zhang, W.; Mason, A.S.; Huang, D.; Huang, X.; Tang, W.; Dou, Y.; Zhang, C.; Xiao, Y. Development of high-density SNP markers and their application in evaluating genetic diversity and population structure in Elaeis guineensis. *Front. Plant Sci.* **2019**, *10*, 130. [CrossRef] [PubMed]
26. Parra, L.; Mostaza-Colado, D.; Yousfi, S.; Marin, J.F.; Mauri, P.V.; Lloret, J. Drone RGB images as a reliable information source to determine legumes establishment success. *Drones* **2021**, *5*, 79. [CrossRef]
27. Pereira, F.d.S.; de Lima, J.; Freitas, R.; Dos Reis, A.A.; do Amaral, L.R.; Figueiredo, G.K.D.A.; Lamparelli, R.A.; Magalhães, P.S.G. Nitrogen variability assessment of pasture fields under an integrated crop-livestock system using UAV, PlanetScope, and Sentinel-2 data. *Comput. Electron. Agric.* **2022**, *193*, 106645. [CrossRef]
28. Sobejano-Paz, V.; Mikkelsen, T.N.; Baum, A.; Mo, X.; Liu, S.; Köppl, C.J.; Johnson, M.S.; Gulyas, L.; García, M. Hyperspectral and thermal sensing of stomatal conductance, transpiration, and photosynthesis for soybean and maize under drought. *Remote Sens.* **2020**, *12*, 3182. [CrossRef]
29. Liu, Y.; Hatou, K.; Aihara, T.; Kurose, S.; Akiyama, T.; Kohno, Y.; Lu, S.; Omasa, K. A robust vegetation index based on different UAV RGB images to estimate SPAD values of naked barley leaves. *Remote Sens.* **2021**, *13*, 686. [CrossRef]
30. Huang, Y.; Reddy, K.N.; Fletcher, R.S.; Pennington, D. UAV low-altitude remote sensing for precision weed management. *Weed Technol.* **2018**, *32*, 2–6. [CrossRef]
31. Sankaran, S.; Zhou, J.; Khot, L.R.; Trapp, J.J.; Mndolwa, E.; Miklas, P.N. High-throughput field phenotyping in dry bean using small unmanned aerial vehicle based multispectral imagery. *Comput. Electron. Agric.* **2018**, *151*, 84–92. [CrossRef]
32. Candiago, S.; Remondino, F.; De Giglio, M.; Dubbini, M.; Gattelli, M. Evaluating multispectral images and vegetation indices for precision farming applications from UAV images. *Remote Sens.* **2015**, *7*, 4026–4047. [CrossRef]
33. Mutanga, O.; Dube, T.; Galal, O. Remote sensing of crop health for food security in Africa: Potentials and constraints. *Remote Sens. Appl. Soc. Environ.* **2017**, *8*, 231–239. [CrossRef]
34. Sibanda, M.; Mutanga, O.; Chimonyo, V.G.P.; Clulow, A.D.; Shoko, C.; Mazvimavi, D.; Dube, T.; Mabhaudhi, T. Application of Drone Technologies in Surface Water Resources Monitoring and Assessment: A Systematic Review of Progress, Challenges, and Opportunities in the Global South. *Drones* **2021**, *5*, 84. [CrossRef]
35. Funk, C.; Budde, M.E. Phenologically-tuned MODIS NDVI-based production anomaly estimates for Zimbabwe. *Remote Sens. Environ.* **2009**, *113*, 115–125. [CrossRef]
36. Merkert, R.; Bushell, J. Managing the drone revolution: A systematic literature review into the current use of airborne drones and future strategic directions for their effective control. *J. Air Transp. Manag.* **2020**, *89*, 101929. [CrossRef]
37. Goel, R.K.; Yadav, C.S.; Vishnoi, S.; Rastogi, R. Smart agriculture–Urgent need of the day in developing countries. *Sustain. Comput. Inform. Syst.* **2021**, *30*, 100512. [CrossRef]
38. Velusamy, P.; Rajendran, S.; Mahendran, R.K.; Naseer, S.; Shafiq, M.; Choi, J.-G. Unmanned Aerial Vehicles (UAV) in Precision Agriculture: Applications and Challenges. *Energies* **2021**, *15*, 217. [CrossRef]
39. Fageria, N.K.; Baligar, V.C.; Clark, R. *Physiology of Crop Production*; CRC Press: Boca Raton, FL, USA, 2006.
40. Aboutalebi, M.; Torres-Rua, A.F.; Allen, N. Multispectral Remote Sensing for Yield Estimation Using High-Resolution Imagery from an Unmanned Aerial Vehicle. In Proceedings of the Autonomous Air and Ground Sensing Systems for Agricultural Optimization and Phenotyping III, Orlando, FL, USA, 15–19 April 2018; SPIE: Bellingham, WA, USA, 2018; Volume 10664, pp. 140–149.
41. Sishodia, R.P.; Ray, R.L.; Singh, S.K. Applications of remote sensing in precision agriculture: A review. *Remote Sens.* **2020**, *12*, 3136. [CrossRef]
42. Van Wart, J.; Kersebaum, K.C.; Peng, S.; Milner, M.; Cassman, K.G. Estimating crop yield potential at regional to national scales. *Field Crops Res.* **2013**, *143*, 34–43. [CrossRef]

43. Duku, C.; Zwart, S.J.; van Bussel, L.G.; Hein, L. Quantifying trade-offs between future yield levels, food availability and forest and woodland conservation in Benin. *Sci. Total Environ.* **2018**, *610*, 1581–1589. [CrossRef] [PubMed]
44. Liu, Y.; Feng, H.; Yue, J.; Fan, Y.; Jin, X.; Song, X.; Yang, H.; Yang, G. Estimation of Potato Above-Ground Biomass Based on Vegetation Indices and Green-Edge Parameters Obtained from UAVs. *Remote Sens.* **2022**, *14*, 5323. [CrossRef]
45. Ali, A.; Imran, M. Evaluating the potential of red edge position (REP) of hyperspectral remote sensing data for real time estimation of LAI & chlorophyll content of kinnow mandarin (Citrus reticulata) fruit orchards. *Sci. Hortic.* **2020**, *267*, 109326.
46. Buthelezi, S.; Mutanga, O.; Sibanda, M.; Odindi, J.; Clulow, A.D.; Chimonyo, V.G.P.; Mabhaudhi, T. Assessing the Prospects of Remote Sensing Maize Leaf Area Index Using UAV-Derived Multi-Spectral Data in Smallholder Farms across the Growing Season. *Remote Sens.* **2023**, *15*, 1597. [CrossRef]
47. Wu, S.; Deng, L.; Guo, L.; Wu, Y. Wheat leaf area index prediction using data fusion based on high-resolution unmanned aerial vehicle imagery. *Plant Methods* **2022**, *18*, 1–16. [CrossRef]
48. Blessing, C.; Nhamo, M.; Rangarirai, M. The impact of plant density and spatial arrangement on light interception on cotton crop and seed cotton yield: An overview. *J. Cotton Res.* **2020**, *3*, 18.
49. Xue, J.; Su, B. Significant remote sensing vegetation indices: A review of developments and applications. *J. Sens.* **2017**, *2017*, 1353691. [CrossRef]
50. Williams, T.B.; Dodd, I.C.; Sobeih, W.Y.; Paul, N.D. Ultraviolet radiation causes leaf warming due to partial stomatal closure. *Hortic. Res.* **2022**, *9*, uhab066. [CrossRef] [PubMed]
51. Raji, S.N.; Aparna, G.N.; Mohanan, C.N.; Subhash, N. Proximal remote sensing of herbicide and drought stress in field grown colocasia and sweet potato plants by sunlight-induced chlorophyll fluorescence Imaging. *J. Indian Soc. Remote Sens.* **2017**, *45*, 463–475. [CrossRef]
52. Singhal, G.; Bansod, B.; Mathew, L.; Goswami, J.; Choudhury, B.; Raju, P. Estimation of leaf chlorophyll concentration in turmeric (Curcuma longa) using high-resolution unmanned aerial vehicle imagery based on kernel ridge regression. *J. Indian Soc. Remote Sens.* **2019**, *47*, 1111–1122. [CrossRef]
53. Brewer, K.; Clulow, A.; Sibanda, M.; Gokool, S.; Naiken, V.; Mabhaudhi, T. Predicting the Chlorophyll Content of Maize over Phenotyping as a Proxy for Crop Health in Smallholder Farming Systems. *Remote Sens.* **2022**, *14*, 518. [CrossRef]
54. Tahir, M.N.; Naqvi, S.Z.A.; Lan, Y.; Zhang, Y.; Wang, Y.; Afzal, M.; Cheema, M.J.M.; Amir, S. Real time estimation of chlorophyll content based on vegetation indices derived from multispectral UAV in the kinnow orchard. *Int. J. Precis. Agric. Aviat.* **2018**, *1*, 24–31.
55. Monteoliva, M.I.; Guzzo, M.C.; Posada, G.A. Breeding for Drought Tolerance by Monitoring Chlorophyll Content. *Gene Technol.* **2021**, *10*, 165.
56. Bojović, B.; Marković, A. Correlation between nitrogen and chlorophyll content in wheat (*Triticum aestivum* L.). *Kragujev. J. Sci.* **2009**, *31*, 69–74.
57. Musa, M.; Massawe, F.; Mayes, S.; Alshareef, I.; Singh, A. Nitrogen fixation and N-balance studies on Bambara groundnut (*Vigna subterranea* L. Verdc) landraces grown on tropical acidic soils of Malaysia. *Commun. Soil Sci. Plant Anal.* **2016**, *47*, 533–542.
58. Blumenthal, J.; Megherbi, D.B.; Lussier, R. Unsupervised machine learning via Hidden Markov Models for accurate clustering of plant stress levels based on imaged chlorophyll fluorescence profiles & their rate of change in time. *Comput. Electron. Agric.* **2020**, *174*, 105064.
59. Liu, C.; Liu, Y.; Lu, Y.; Liao, Y.; Nie, J.; Yuan, X.; Chen, F. Use of a leaf chlorophyll content index to improve the prediction of above-ground biomass and productivity. *PeerJ* **2019**, *6*, e6240. [CrossRef] [PubMed]
60. Li, Y.; Ren, B.; Ding, L.; Shen, Q.; Peng, S.; Guo, S. Does chloroplast size influence photosynthetic nitrogen use efficiency? *PLoS ONE* **2013**, *8*, e62036. [CrossRef] [PubMed]
61. Muhammad, A.; Alam, M.; Ahmad, I.; Jalal, A. Role of beneficial microbes with nitrogen and phosphorous levels on canola productivity. *Braz. J. Biol.* **2021**, *82*, e227703.
62. Iseki, K.; Olaleye, O. A new indicator of leaf stomatal conductance based on thermal imaging for field grown cowpea. *Plant Prod. Sci.* **2020**, *23*, 136–147. [CrossRef]
63. Wijewardana, C.; Alsajri, F.A.; Irby, J.T.; Krutz, L.J.; Golden, B.; Henry, W.B.; Gao, W.; Reddy, K.R. Physiological assessment of water deficit in soybean using midday leaf water potential and spectral features. *J. Plant Interact.* **2019**, *14*, 533–543. [CrossRef]
64. Ouyang, W.; Struik, P.C.; Yin, X.; Yang, J. Stomatal conductance, mesophyll conductance, and transpiration efficiency in relation to leaf anatomy in rice and wheat genotypes under drought. *J. Exp. Bot.* **2017**, *68*, 5191–5205. [CrossRef]
65. Chai, H.H.; Massawe, F.; Mayes, S. Effects of mild drought stress on the morpho-physiological characteristics of a bambara groundnut segregating population. *Euphytica* **2016**, *208*, 225–236. [CrossRef]
66. Awais, M.; Li, W.; Cheema, M.; Zaman, Q.; Shaheen, A.; Aslam, B.; Zhu, W.; Ajmal, M.; Faheem, M.; Hussain, S. UAV-based remote sensing in plant stress imagine using high-resolution thermal sensor for digital agriculture practices: A meta-review. *Int. J. Environ. Sci. Technol.* **2022**, *20*, 1135–1152. [CrossRef]
67. Awais, M.; Li, W.; Cheema, M.J.; Hussain, S.; Shaheen, A.; Aslam, B.; Liu, C.; Ali, A. Assessment of optimal flying height and timing using high-resolution unmanned aerial vehicle images in precision agriculture. *Int. J. Environ. Sci. Technol.* **2022**, *19*, 2703–2720. [CrossRef]
68. Majid, K.; Roza, G. The effect of drought stress on leaf chlorophyll content and stress resistance in maize cultivars (*Zea mays*). *Afr. J. Microbiol. Res.* **2012**, *6*, 2844–2848.

69. Maes, W.H.; Steppe, K. Estimating evapotranspiration and drought stress with ground-based thermal remote sensing in agriculture: A review. *J. Exp. Bot.* **2012**, *63*, 4671–4712. [CrossRef]
70. Pineda, M.; Baron, M.; Perez-Bueno, M.L. Thermal Imaging for Plant Stress Detection and Phenotyping. *Remote Sens.* **2021**, *13*, 68. [CrossRef]
71. Chibarabada, T.P. Water Use and Nutritional Water Productivity of Selected Major and Underutilised Grain Legumes. Ph.D. Dissertation, University of KwaZulu-Natal, Pietermartzburg, South Africa, 2018.
72. Bellvert, J.; Zarco-Tejada, P.J.; Girona, J.; Fereres, E. Mapping crop water stress index in a 'Pinot-noir'vineyard: Comparing ground measurements with thermal remote sensing imagery from an unmanned aerial vehicle. *Precis. Agric.* **2014**, *15*, 361–376. [CrossRef]
73. Gray, P.C.; Ridge, J.T.; Poulin, S.K.; Seymour, A.C.; Schwantes, A.M.; Swenson, J.J.; Johnston, D.W. Integrating drone imagery into high resolution satellite remote sensing assessments of estuarine environments. *Remote Sens.* **2018**, *10*, 1257. [CrossRef]
74. Clevers, J.G.; Gitelson, A.A. Remote estimation of crop and grass chlorophyll and nitrogen content using red-edge bands on Sentinel-2 and-3. *Int. J. Appl. Earth Obs. Geoinf.* **2013**, *23*, 344–351. [CrossRef]
75. Ndlovu, H.S.; Odindi, J.; Sibanda, M.; Mutanga, O.; Clulow, A.; Chimonyo, V.G.P.; Mabhaudhi, T. A Comparative Estimation of Maize Leaf Water Content Using Machine Learning Techniques and Unmanned Aerial Vehicle (UAV)-Based Proximal and Remotely Sensed Data. *Remote Sens.* **2021**, *13*, 4091. [CrossRef]
76. Zheng, Q.; Huang, W.; Ye, H.; Dong, Y.; Shi, Y.; Chen, S. Using continous wavelet analysis for monitoring wheat yellow rust in different infestation stages based on unmanned aerial vehicle hyperspectral images. *Appl. Opt.* **2020**, *59*, 8003–8013. [CrossRef]
77. Nyman, J. Pixel classification of hyperspectral images. In Proceedings of the 2022 5th International Conference on Machine Learning and Natural Language, Sanya, China, 23–25 December 2018.
78. Marshall, M.; Thenkabail, P. Advantage of hyperspectral EO-1 Hyperion over multispectral IKONOS, GeoEye-1, WorldView-2, Landsat ETM+, and MODIS vegetation indices in crop biomass estimation. *ISPRS J. Photogramm. Remote Sens.* **2015**, *108*, 205–218. [CrossRef]
79. Thenkabail, P.S.; Mariotto, I.; Gumma, M.K.; Middleton, E.M.; Landis, D.R.; Huemmrich, K.F. Selection of hyperspectral narrowbands (HNBs) and composition of hyperspectral twoband vegetation indices (HVIs) for biophysical characterization and discrimination of crop types using field reflectance and Hyperion/EO-1 data. *Sel. Top. Appl. Earth Obs. Remote Sens. IEEE J.* **2013**, *6*, 427–439. [CrossRef]
80. Thenkabail, P.S.; Smith, R.B.; De Pauw, E. Evaluation of narrowband and broadband vegetation indices for determining optimal hyperspectral wavebands for agricultural crop characterization. *Photogramm. Eng. Remote Sens.* **2002**, *68*, 607–622.
81. Mafuratidze, P. Discriminating Wetland Vegetation Species in an African Savanna Using Hyperspectral Data. Master's Dissertation, University of KwaZulu-Natal, Pietermartzburg, South Africa, 2010.
82. Zhang, L.; Zhang, H.; Niu, Y.; Han, W. Mapping Maize Water Stress Based on UAV Multispectral Remote Sensing. *Remote Sens.* **2019**, *11*, 605. [CrossRef]
83. Zaludin, Z.; Harituddin, A.S.M. Challenges and Trends of Changing from Hover to Forward Flight for a Converted Hybrid Fixed Wing VTOL UAS from Automatic Flight Control System Perspective. In Proceedings of the 2019 IEEE 9th International Conference on System Engineering and Technology (ICSET), Shah Alam, Malaysia, 7 October 2019; pp. 247–252.
84. Li, L.; Zheng, X.M.; Zhao, K.; Li, X.F.; Meng, Z.G.; Su, C.H. Potential Evaluation of High Spatial Resolution Multi-Spectral Images Based on Unmanned Aerial Vehicle in Accurate Recognition of Crop Types. *J. Indian Soc. Remote Sens.* **2020**, *48*, 1471–1478. [CrossRef]
85. Mutanga, O.; Masenyama, A.; Sibanda, M. Spectral saturation in the remote sensing of high-density vegetation traits: A systematic review of progress, challenges, and prospects. *ISPRS J. Photogramm. Remote Sens.* **2023**, *198*, 297–309. [CrossRef]
86. Sharifi, A. Remotely sensed vegetation indices for crop nutrition mapping. *J. Sci. Food Agric.* **2020**, *100*, 5191–5196. [CrossRef]
87. Cabrera-Bosquet, L.; Molero, G.; Stellacci, A.; Bort, J.; Nogués, S.; Araus, J. NDVI as a potential tool for predicting biomass, plant nitrogen content and growth in wheat genotypes subjected to different water and nitrogen conditions. *Cereal Res. Commun.* **2011**, *39*, 147–159. [CrossRef]
88. Filella, I.; Penuelas, J. The red edge position and shape as indicators of plant chlorophyll content, biomass and hydric status. *Int. J. Remote Sens.* **1994**, *15*, 1459–1470. [CrossRef]
89. Guyot, G.; Baret, F.; Jacquemoud, S. *Imaging Spectroscopy for Vegetation Studies*; Kluwer Academic Publishers: Norwell, MA, USA, 1992; Volume 2.
90. Mutanga, O.; Skidmore, A.K. Narrow band vegetation indices overcome the saturation problem in biomass estimation. *Int. J. Remote Sens.* **2004**, *25*, 3999–4014. [CrossRef]
91. Mutanga, O.; Adam, E.; Cho, M.A. High density biomass estimation for wetland vegetation using WorldView-2 imagery and random forest regression algorithm. *Int. J. Appl. Earth Obs. Geoinf.* **2012**, *18*, 399–406. [CrossRef]
92. Brewer, K.; Clulow, A.; Sibanda, M.; Gokool, S.; Odindi, J.; Mutanga, O.; Naiken, V.; Chimonyo, V.G.P.; Mabhaudhi, T. Estimation of Maize Foliar Temperature and Stomatal Conductance as Indicators of Water Stress Based on Optical and Thermal Imagery Acquired Using an Unmanned Aerial Vehicle (UAV) Platform. *Drones* **2022**, *6*, 169. [CrossRef]
93. Brewer, K.R. Assessment of Maize Crop Health and Water Stress Based on Multispectral and Thermal Infrared Unmanned Aerial Vehicle Phenotyping in Smallholder Farms. Master's Dissertation, University of KwaZulu-Natal, Pietermartzburg, South Africa, 2021.

94. Sapkota, B.; Singh, V.; Cope, D.; Valasek, J.; Bagavathiannan, M. Mapping and Estimating Weeds in Cotton Using Unmanned Aerial Systems-Borne Imagery. *Agriengineering* **2020**, *2*, 350–366. [CrossRef]
95. Gu, Y.; Wylie, B.K.; Boyte, S.P.; Picotte, J.; Howard, D.M.; Smith, K.; Nelson, K.J. An optimal sample data usage strategy to minimize overfitting and underfitting effects in regression tree models based on remotely-sensed data. *Remote Sens.* **2016**, *8*, 943. [CrossRef]
96. Singh, C.; Karan, S.K.; Sardar, P.; Samadder, S.R. Remote sensing-based biomass estimation of dry deciduous tropical forest using machine learning and ensemble analysis. *J. Environ. Manag.* **2022**, *308*, 114639. [CrossRef]
97. Ehlers, D.; Wang, C.; Coulston, J.; Zhang, Y.; Pavelsky, T.; Frankenberg, E.; Woodcock, C.; Song, C. Mapping Forest Aboveground Biomass Using Multisource Remotely Sensed Data. *Remote Sens.* **2022**, *14*, 1115. [CrossRef]
98. Zhang, Y.; Teng, P.; Aono, M.; Shimizu, Y.; Hosoi, F.; Omasa, K. 3D monitoring for plant growth parameters in field with a single camera by multi-view approach. *J. Agric. Meteorol.* **2018**, *74*, 129–139. [CrossRef]
99. Shirzadifar, A.; Bajwa, S.; Nowatzki, J.; Bazrafkan, A. Field identification of weed species and glyphosate-resistant weeds using high resolution imagery in early growing season. *Biosyst. Eng.* **2020**, *200*, 200–214. [CrossRef]
100. Sengupta, S.; Mohinuddin, S.; Arif, M.; Sengupta, B.; Zhang, W. Assessment of agricultural land suitability using GIS and Fuzzy Analytical Hierarchy Process approach in Ranchi District, India. *Geocarto Int.* **2022**, *37*, 13337–13368. [CrossRef]
101. Nguyen, H.P.D.; Nguyen, D.D. Drone application in smart cities: The general overview of security vulnerabilities and counter-measures for data communication. In *Development and Future of Internet of Drones (IoD): Insights, Trends and Road Ahead*; Springer: Cham, Switzerland, 2021; Volume 332, pp. 185–210.
102. Kemp, L.; Roux, M.; Kemp, M.; Kock, R. Application of drones and image processing for bridge inspections in South Africa. *Civ. Eng. Siviele Ingenieurswese* **2021**, *29*, 25–30.
103. Grote, M.; Pilko, A.; Scanlan, J.; Cherrett, T.; Dickinson, J.; Smith, A.; Oakey, A.; Marsden, G. Sharing airspace with Uncrewed Aerial Vehicles (UAVs): Views of the General Aviation (GA) community. *J. Air Transp. Manag.* **2022**, *102*, 102218. [CrossRef]
104. Stopforth, R. Drone licenses-neccesities and requirements. *Il Ponte* **2017**, *73*, 149–156.
105. Kutugata, M.; Hu, C.; Sapkota, B.; Bagavathiannan, M. Seed rain potential in late-season weed escapes can be estimated using remote sensing. *Weed Sci.* **2021**, *69*, 653–659. [CrossRef]
106. Biswas, A.; Andrade, M.H.M.L.; Acharya, J.P.; de Souza, C.L.; Lopez, Y.; De Assis, G.; Shirbhate, S.; Singh, A.; Munoz, P.; Rios, E.F. Phenomics-Assisted Selection for Herbage Accumulation in Alfalfa (*Medicago sativa* L.). *Front. Plant Sci.* **2021**, *12*, 756768. [CrossRef] [PubMed]
107. Kourouma, J.M.; Eze, E.; Negash, E.; Phiri, D.; Vinya, R.; Girma, A.; Zenebe, A. Assessing the spatio-temporal variability of NDVI and VCI as indices of crops productivity in Ethiopia: A remote sensing approach. *Geomat. Nat. Hazards Risk* **2021**, *12*, 2880–2903. [CrossRef]
108. Lima, J.G.; Espínola, J.; Medeiros, J.F.d.; Viana, P.C.; Maniçoba, R.M. Water requirement and crop coefficients of sorghum in Apodi Plateau. *Rev. Bras. Eng. Agrícola Ambient.* **2021**, *25*, 684–688. [CrossRef]
109. Zakarya, Y.M.; Metwaly, M.M.; AbdelRahman, M.A.; Metwalli, M.R.; Koubouris, G. Optimized land use through integrated land suitability and GIS approach in West El-Minia Governorate, Upper Egypt. *Sustainability* **2021**, *13*, 12236. [CrossRef]
110. Guo, Q.; Wu, F.; Pang, S.; Zhao, X.; Chen, L.; Liu, J.; Xue, B.; Xu, G.; Li, L.; Jing, H. Crop 3D—A LiDAR based platform for 3D high-throughput crop phenotyping. *Sci. China Life Sci.* **2018**, *61*, 328–339. [CrossRef]
111. Tedesco, D.; de Oliveira, M.F.; dos Santos, A.F.; Silva, E.H.C.; de Souza Rolim, G.; da Silva, R.P. Use of remote sensing to characterize the phenological development and to predict sweet potato yield in two growing seasons. *Eur. J. Agron.* **2021**, *129*, 126337. [CrossRef]
112. Tedesco, D.; de Almeida Moreira, B.R.; Júnior, M.R.B.; Papa, J.P.; da Silva, R.P. Predicting on multi-target regression for the yield of sweet potato by the market class of its roots upon vegetation indices. *Comput. Electron. Agric.* **2021**, *191*, 106544. [CrossRef]
113. Punalekar, S.M.; Verhoef, A.; Quaife, T.; Humphries, D.; Bermingham, L.; Reynolds, C. Application of Sentinel-2A data for pasture biomass monitoring using a physically based radiative transfer model. *Remote Sens. Environ.* **2018**, *218*, 207–220. [CrossRef]
114. Thenkabail, P. Biophysical and yield information for precision farming from near-real-time and historical Landsat TM images. *Int. J. Remote Sens.* **2003**, *24*, 2879–2904. [CrossRef]
115. Zhang, P.; Hu, S.; Li, W.; Zhang, C. Parcel-level mapping of crops in a smallholder agricultural area: A case of central China using single-temporal VHSR imagery. *Comput. Electron. Agric.* **2020**, *175*, 105581. [CrossRef]
116. Karimi, N.; Sheshangosht, S.; Eftekhari, M. Crop type detection using an object-based classification method and multi-temporal Landsat satellite images. *Paddy Water Environ.* **2022**, *20*, 395–412. [CrossRef]
117. Karlson, M.; Ostwald, M.; Bayala, J.; Bazié, H.R.; Ouedraogo, A.S.; Soro, B.; Sanou, J.; Reese, H. The potential of Sentinel-2 for crop production estimation in a smallholder agroforestry landscape, Burkina Faso. *Front. Environ. Sci.* **2020**, *8*, 85. [CrossRef]
118. Guan, K.; Berry, J.A.; Zhang, Y.; Joiner, J.; Guanter, L.; Badgley, G.; Lobell, D.B. Improving the monitoring of crop productivity using spaceborne solar-induced fluorescence. *Glob. Chang. Biol.* **2016**, *22*, 716–726. [CrossRef]
119. Wimalasiri, E.M.; Jahanshiri, E.; Suhairi, T.; Udayangani, H.; Mapa, R.B.; Karunaratne, A.S.; Vidhanarachchi, L.P.; Azam-Ali, S.N. Basic Soil Data Requirements for Process-Based Crop Models as a Basis for Crop Diversification. *Sustainability* **2020**, *12*, 7781. [CrossRef]
120. Teng, P.; Ono, E.; Zhang, Y.; Aono, M.; Shimizu, Y.; Hosoi, F.; Omasa, K. Estimation of ground surface and accuracy assessments of growth parameters for a sweet potato community in ridge cultivation. *Remote Sens.* **2019**, *11*, 1487. [CrossRef]

121. Waldner, F.; Chen, Y.; Lawes, R.; Hochman, Z. Needle in a haystack: Mapping rare and infrequent crops using satellite imagery and data balancing methods. *Remote Sens. Environ.* **2019**, *233*, 111375. [CrossRef]
122. Adams, T.; Bruton, R.; Ruiz, H.; Barrios-Perez, I.; Selvaraj, M.G.; Hays, D.B. Prediction of Aboveground Biomass of Three Cassava (Manihot esculenta) Genotypes Using a Terrestrial Laser Scanner. *Remote Sens.* **2021**, *13*, 1272. [CrossRef]
123. Hama, A.; Matsumoto, Y.; Matsuoka, N. Estimating Leaf Water Content through Low-Cost LiDAR. *Agronomy* **2022**, *12*, 1183. [CrossRef]

remote sensing

MDPI

Technical Note

Multispectral, Thermal, and Hyperspectral Sensing Data Depict Stomatal Conductance in Grapevine

Kesevan Veloo [1,2,†], Carlos Zúñiga Espinoza [1,3,†], Alberto Espinoza Salgado [3], Pete W. Jacoby [4,5] and Sindhuja Sankaran [1,5,*]

1 Department of Biological System Engineering, Washington State University, Pullman, WA 99164, USA; k.veloo@msstate.edu (K.V.); czuniga@inia.cl (C.Z.E.)
2 Department of Agricultural and Biological Engineering, Mississippi State University, Starkville, MS 39762, USA
3 Instituto de Investigaciones Agropecuarias (INIA), 2280454 La Cruz, Chile; alberto.espinoza@inia.cl
4 Department of Crop and Soil Science, Washington State University, Pullman, WA 99164, USA; jacoby@wsu.edu
5 Center for Precision and Automated Agricultural Systems, Washington State University, Prosser, WA 99350, USA
* Correspondence: s.sankaran@wsu.edu; Tel.: +1-(509)-335-8828
† These authors contributed equally to this work.

Abstract: Climate-driven water challenges in the Pacific Northwest necessitate precise irrigation for sustainable vineyard management. In such scenarios, conservation of water using different approaches, including subsurface irrigation, becomes critical. Detecting crop water status becomes key to evaluating and managing such approaches. This study examines how multispectral, thermal, and hyperspectral proximal sensing data depict irrigation-induced variations in stomatal conductance in Cabernet Sauvignon vineyards during 2016 and 2017. The roles of individual and combined sensing modalities were analyzed, with key contributions including the identification of indices that characterize stomatal conductance. Data were collected at the following growth stages: 80 and 44 days before harvest (DBH) in 2016; and 64, 44, and 8 DBH in 2017. The vegetation indices analyzed included the green normalized difference vegetation index (GNDVI) and leaf area index (LAI) from multispectral data, crop water stress index (CWSI) from thermal data, and normalized difference spectral indices (NDSI) from hyperspectral data. Pearson's correlations at 80 and 44 DBH (2016) showed significant relationships between normalized stomatal conductance and multispectral indices (LAI: $r = 0.59$ to 0.66, GNDVI: $r = 0.41$ to 0.50, both $p < 0.01$). NDSI pairs (1380 nm with 1570 nm, 1570 nm with 1810 nm) at 80 DBH showed significant correlations ($r = -0.27$, 0.31, both $p < 0.05$). In 2017, the thermal data showed the strongest correlation with normalized stomatal conductance ($r = -0.83$) at 44 DBH. In the same year, NDSI pairs exhibited stronger correlations than multispectral indices as the DBH decreased (1380 nm with 1570 nm: $r = -0.58$ to -0.69, 1570 nm with 1810 nm: $r = 0.64$ to 0.48, both $p < 0.05$). Combining LAI with these NDSI pairs improved stomatal conductance predictions (2016: $R^2 = 0.37$–0.50; 2017: $R^2 = 0.51$–0.63, both $p < 0.01$). These results demonstrate the precision of a multimodal sensing approach, particularly integrating multispectral and hyperspectral data, to improve irrigation strategies and promote sustainable viticulture.

Keywords: proximal sensing; NDSI; water stress; vineyard monitoring

1. Introduction

Viticulture faces mounting challenges from climate change, particularly in the western United States, where a shift towards warmer, more arid conditions is becoming increasingly prevalent [1,2]. Traditionally, grape-growing climates were classified as warm or hot if the average temperature during the seven-month growing season (April to October) ranged between 17 °C and 21 °C, while temperatures above 24 °C were considered too extreme for optimal grape cultivation [3]. However, recent observations from 2021 and 2022 align with the temperature extremes once projected for the late 21st century, with days surpassing 35 °C as no longer an anomaly [4,5]. This trend towards more frequent exceedingly hot days during the growing season is severely impacting the viability of wine grape cultivation in numerous United States regions [6].

Compounding these temperature challenges is the viticulture sector's reliance on winter snow accumulation for summer irrigation, a system now jeopardized by climate-induced changes [1]. Diminished snowpack levels and premature melt-offs are altering the timing and availability of irrigation water, exacerbating drought vulnerability [7]. By addressing the twin challenges of climate change and water scarcity, the sector is actively seeking strategies to sustain wine grape cultivation's viability, productivity, and financial sustainability [8,9].

Research in viticulture has prioritized optimizing water use to address these challenges. Efforts include developing drought-tolerant varieties (both scions and rootstocks) [10–12], enhancing soil moisture through mulching and cover crops [13,14], and refining canopy management [15,16]. Advances in irrigation techniques, such as regulated deficit irrigation and partial root-zone drying, aim to minimize water use while maintaining vine vitality and grape quality [17]. Complementing these agronomic strategies, technological advancements have introduced decision-support systems that aim to optimize irrigation scheduling by integrating environmental data and soil moisture measurements, often obtained from automated weather stations and soil moisture sensors [18–20].

While these strategies provide valuable tools for managing water resources, they often overlook critical physiological indicators of stress, particularly those occurring at the leaf level. Soil moisture sensors focus on root water uptake but do not capture the spatial variability of plant physiological states, leading to gaps in understanding how grapevines respond to water scarcity. Factors such as soil type, sensor placement, and management practices further limit the accuracy and applicability of these tools [21,22]. To bridge these gaps, focusing on physiological indicators such as stomatal conductance is essential.

Stomatal conductance, a fundamental physiological parameter, provides precise insights into plant water stress responses. It quantifies the exchange of carbon dioxide and water vapor through the stomata [23]. During a water deficit, abscisic acid (ABA) production induces stomatal closure, reducing stomatal conductance to conserve water. This reduction directly affects photosynthesis, transpiration, and leaf temperature. As a dynamic indicator of water stress, stomatal conductance effectively links physiological responses to environmental conditions, offering scalable and accurate measure of grapevine water status.

Recent advancements in remote sensing have introduced effective tools for assessing vine water stress by capturing spatial variability in stomatal conductance. Thermal imaging, in particular, has demonstrated its value by leveraging the inverse relationship between leaf temperature and stomatal conductance across numerous crops, including strawberry [24], olive [25,26], almond [27], and peach and nectarine [28]. The crop water stress index (CWSI), derived from unmanned aerial vehicle (UAV) thermal imagery, has been shown to correlate strongly with photosynthetic rate and stomatal conductance in grapevines, emphasizing its applicability across varieties [29]. Expanding on the use of thermal indices, Pagay and Kidman [30] evaluated the CWSI, the stomatal conductance index, and the canopy-to-ambient temperature difference in Cabernet Sauvignon and Shiraz. Their findings indicated that the CWSI was more strongly correlated with stem water potential and stomatal conductance during warmer, drier seasons, while the stomatal conductance index and canopy-to-ambient temperature difference were more effective under cooler, wetter conditions. The relevance of thermal imaging for evaluating vine water stress has been further substantiated by other studies exploring similar indices [31,32].

Building on the insights provided by thermal imaging, multispectral imaging has emerged as another valuable tool for assessing vine water stress. Espinoza et al. [33] employed UAVs equipped with multispectral cameras and found significant correlations of normalized stomatal conductance with the normalized difference vegetation index (NDVI, $r = 0.56$) and the green normalized difference vegetation index (GNDVI, $r = 0.65$). These findings affirm the pivotal roles of the NDVI and GNDVI in monitoring water stress. López–García et al. [34] expanded upon this by correlating visible and multispectral vegetation indices, including the NDVI, GNDVI, green cover canopy, green leaf index, renormalized difference vegetation index, and visible atmospherically resistant index, with the stem water potential. Their results highlighted strong correlations for the green leaf index ($R^2 = 0.80$), green cover canopy ($R^2 = 0.73$), and visible atmospherically resistant index ($R^2 = 0.72$), emphasizing these indices' effectiveness in assessing water stress.

While multispectral imaging focuses on canopy reflectance, hyperspectral imaging provides granular spectral data across broader wavelengths, detecting subtle physiological changes linked to water stress. Loggenberg et al. [35] reported 83% accuracy with the random forest algorithm for water stress detection. In a complementary study, Zovko et al. [36] harnessed hyperspectral imaging alongside partial least squares-support vector machines to assess drought stress in grapevines. Conducted over various irrigation levels, this research achieved remarkable accuracy, with a drought stress classification accuracy of over 97%.

Despite the extensive use of multispectral, thermal, and hyperspectral imaging in monitoring water stress in vineyards, these methods are often applied individually, providing only fragmented insights into vine physiological responses. Furthermore, most studies rely on remote sensing that, while effective for large-scale monitoring, lacks the spatial and temporal resolution necessary to capture fine-scale plant–water interactions. Proximal sensing addresses these limitations by enabling detailed, close-range monitoring of stress indicators under variable conditions. This study introduces a novel framework that integrates multispectral, thermal, and hyperspectral data collected over two years under conditions with and without predominant water stress. Using an agricultural utility vehicle (AUV) for multispectral and thermal imaging and a spectroradiometer for hyperspectral measurements, this approach evaluates the individual and combined contributions of these methods to characterize water stress and its relationship to stomatal conductance. By bridging the gaps between single-method approaches and large-scale remote sensing, this framework offers a comprehensive understanding of water-stress dynamics, advancing the precision and effectiveness of vineyard water management strategies.

2. Materials and Methods

2.1. Experimental Design

The experiment was conducted over the 2016 and 2017 growing seasons (April to September) in a commercial block of Cabernet Sauvignon (*Vitis vinifera* L.) at Kiona Vineyards, located on Red Mountain near Benton City, WA, USA (46°16′59.1″N, 119°26′31.9″W). Spanning 0.44 ha, the experimental site featured Hezel loamy fine sandy soil. A split-plot design (Supplementary Materials, Figure S1) was implemented in 2016 (and revised in 2017), as reported by Espinoza et al. [33], to assess the effects of irrigation method (subsurface irrigation, continuous/pulse irrigation), irrigation rate (15% to 100% of evapotranspiration, ET), and depth (surface to 90 cm) on vine water stress. The two irrigation patterns evaluated were continuous (constant flow) or pulsed (delivered every two hours) irrigation. Subplots were subdivided into four irrigation levels, relative to a grower-defined baseline rate: 100% (Control), 60% (High), 30% (Medium), and 15% (Low).

Sub-subplots were defined by irrigation depth, with subsurface drip lines installed at depths of 30 cm, 60 cm, and 90 cm. These depths capture variability in water availability, with 30 cm reflecting shallow moisture (greater variability, higher evaporation potential), and 60 cm and 90 cm representing more stable moisture availability at depth, requiring deeper root access. The irrigation rates were to introduce stress gradually, while the maximum depth was selected in accordance with studies indicating that most grapevine roots are concentrated within 80 cm [37–39]. The experimental layout consisted of 60 plots arranged in a 6 × 10 grid, each plot containing three rows of five vines. Treatment-specific data were collected from the central three vines in the middle row, which served as representative vines for analysis.

The experimental framework in 2017 followed a similar split-plot design with two modifications (Supplementary Materials, Figure S2). Irrigation levels were set to 100% (Control), 80% (High), 60% (Medium), and 40% (Low) to simulate a broader range of water stress. An additional irrigation depth of 0 cm (surface application) was included alongside the previous depths of 30 cm, 60 cm, and 90 cm. This addition was made to assess the influence of surface-applied water, particularly in scenarios where evaporation is more pronounced. All plots received continuous irrigation throughout the growing season. Further details are provided in Ma et al. [40,41].

2.2. Data Collection

Table 1 describes the dates of data acquisition and corresponding days before harvest (DBH) alongside the growth stages.

Table 1. Data acquisition dates and growth stages during data collection.

Year	Date	DBH	Growth Stages
2016	6 July	80	Bunch closure
	16 August	44	Veraison (onset of ripening)
2017	2 August	64	Berry development (10 mm diameter)
	22 August	44	Veraison (onset of ripening)
	27 September	8	Ripening

High-resolution multispectral and thermal images were captured using an AUV, a modified John Deere Gator™ XUV590i (John Deere, Moline, IL, USA) equipped with an imaging sensor setup as described in Figure 1. The AUV was outfitted with a retractable mast (FM50-25, Floatagraph Technologies, Santa Barbara, CA, USA), elevating the imaging sensors 7 m above the ground. The imaging suite comprised a customized multispectral digital camera (NiteCanon ELPH110, LDP LLC, Carlstadt, NJ, USA) and a thermal infrared camera, Tau 2 640 (Teledyne FLIR LLC, Wilsonville, OR, USA). The multispectral camera was configured to capture images in the near infrared (NIR), green (G), and blue (B) bands. This setup facilitated geo-referenced interval shooting, manually triggered to optimize image acquisition timing. The multispectral camera produced 8-bit JPG images at a resolution of 16.1 megapixels (4608 × 3456), while the thermal camera generated 8-bit JPG images at a resolution of 32,000 pixels (640 × 512). To ensure radiometric consistency across data collection periods, multispectral reflectance was corrected using a white reference panel (Spectralon Reflectance Standard, Labsphere, North Sutton, NH, USA) that was present within the field of view of each image. Moreover, in 2017, dry cloth and wet cloth (cotton—black for maximum heat absorption and wet cloth was made damp with water) were used within the field of view of each thermal image for estimating the CWSI. All data were stored on an onboard SD card.

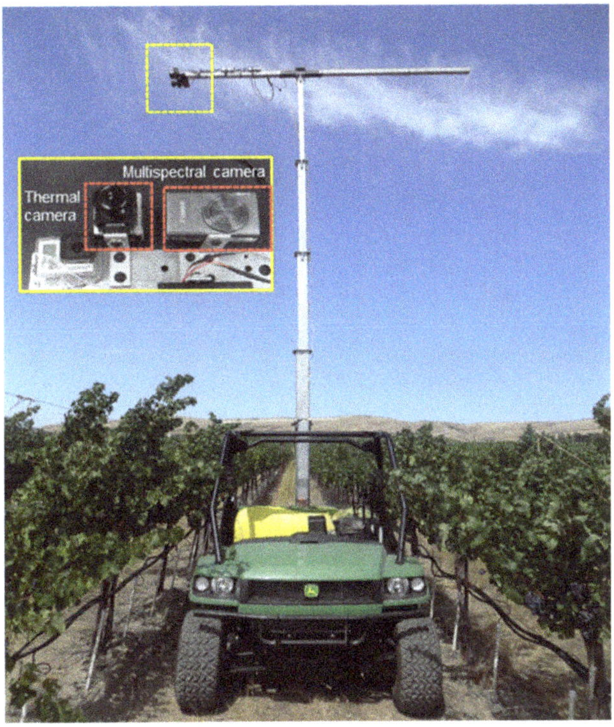

Figure 1. Imaging sensor setup and field data collection using an AUV deployed in the vineyard for image acquisition. The inset picture comprised of sensor configuration on the retractable mast, featuring a thermal infrared camera and a modified multispectral digital camera.

Concurrent with image capture, stomatal conductance and leaf hyperspectral data were collected between 11:00 AM and 2:00 PM on the same days. Stomatal conductance was measured using a porometer (Porometer SC1, Decagon Devices, Pullman, WA, USA) on two leaves per plot. Hyperspectral reflectance data were collected from three leaves per plot using a spectroradiometer (SVC HR-1024i, Spectra Vista Corporation, Poughkeepsie, NY, USA) to maintain consistency across years. Data collection used a leaf clip (LC-RP PRO, Spectra Vista Corporation) and a fiber optic cable to capture reflectance from 350 to 2500 nm. The system measured reflectance data across the 350 to 2500 nm range, with spectral resolutions of $\leq$3.5 nm at 700 nm, $\leq$9.5 nm at 1500 nm, and $\leq$6.5 nm at 2100 nm. The system has an in-built white reference panel for automated white reference correction prior to the hyperspectral reflectance measurement. Calibration was performed often to address sensor drift, ensuring sensor accuracy across data collection.

3. Data Extraction and Analysis

3.1. Multispectral and Thermal Data Processing

Figure 2 illustrates the data processing workflow for multispectral and thermal imaging, encompassing steps from image acquisition with an AUV to the correlation of computed indices with normalized stomatal conductance. Stomatal conductance was normalized by calculating the ratio of each stomatal conductance reading to the maximum stomatal conductance recorded on the respective data collection date (Equation (1)).

$$Norm\_g_s = \frac{g_s}{max\_g_s} \quad (1)$$

where

$Norm\_g_s$ = Normalized stomatal conductance;
g_s = Measured stomatal conductance (mmol. $m^{-2}s^{-1}$);
max_g_s = Maximum stomatal conductance on data acquisition date (mmol. $m^{-2}s^{-1}$).

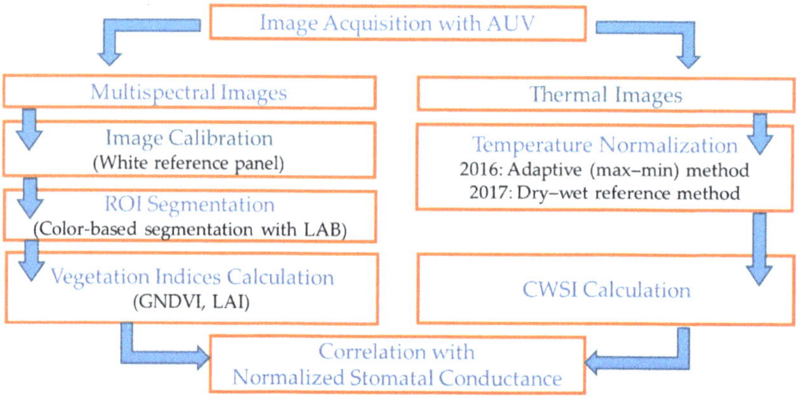

Figure 2. Workflow for multispectral and thermal data processing in vineyard monitoring. The process begins with image acquisition using an AUV, followed by image calibration, segmentation, and index calculation for multispectral data, temperature normalization, and CWSI calculation for thermal data. Both workflows converge to correlate computed indices with normalized stomatal conductance.

Multispectral imagery from each plot was processed using the MATLAB Image Processing Toolbox (MathWorks, Natick, MA, USA). Each image was calibrated against a reference panel to ensure consistency. Following calibration, the region of interest (ROI) in each image, primarily representing vine canopies, was segmented utilizing color-based thresholds in the CIE LAB color space. From these ROIs, two vegetation indices were derived: the green normalized difference vegetation index (GNDVI) (Equation (2)) [42,43] and the leaf area index (LAI) (Equation (3)) [44].

$$GNDVI = \frac{(NIR - Green)}{(NIR + Green)} \quad (2)$$

where

NIR = Reflectance in the near-infrared region;
$Green$ = Reflectance in the green region.

$$LAI = \frac{N_{canopy}}{N_{total}} \quad (3)$$

where

N_{canopy} = Number of pixels with GNDVI > 0.1 (canopy pixels);
N_{total} = Total number of pixels in the image region of interest (ROI).

Thermal imagery was utilized to compute the crop water stress index (CWSI) using two distinct normalization methods (Equation (4)). In 2016, the normalization used the observed maximum (T_{M1}) and minimum (T_{M2}) canopy temperatures within each thermal image. In 2017, a more refined method was applied, where the canopy temperature was normalized using fixed dry (T_{M1}) and wet (T_{M2}) reference surfaces, providing consistent temperature bounds. Canopy-averaged CWSI values were subsequently correlated with normalized stomatal conductance to evaluate the thermal response to irrigation treatments [33].

$$CWSI = \left(\frac{T_{canopy} - T_{M2}}{T_{M1} - T_{M2}} \right) \quad (4)$$

where

T_{canopy} = Canopy temperature;

T_{M1} = Maximum image canopy temperature (2016) or the temperature of the dry reference (2017);

T_{M2} = Minimum image canopy temperature (2016) or the temperature of the wet reference (2017).

3.2. Hyperspectral Data Processing

Hyperspectral data were processed using Matlab (Mathworks, Natick, MA, USA). The data were binned every 10 nm such that the spectral reflectance from 350 to 2500 nm was recorded with a 10 ($\pm$5) nm interval (216 total spectral features). Following this, the spectral data were normalized using Euclidean length or vector magnitude, where each vector refers to the hyperspectral data of an individual sample. Finally, three replicate sample spectral data from each canopy (treatment block) were averaged and further analyzed utilizing the corresponding stomatal conductance data. The spectral reflectance data were utilized to estimate the normalized difference spectral indices (NDSI) [45] using Equation (5), computed from every combination of two spectral bands:

$$NDSI = \frac{(R_k - R_n)}{(R_k + R_n)} \quad (5)$$

where

R_k and R_n represent the reflectance at the k-th and n-th spectral bands, respectively, with $k < n$.

This NDSI dataset computed from 2016 and 2017 datasets at different time points was correlated with normalized stomatal conductance data, and NDSI showing significant correlation ($p < 0.001$) with normalized stomatal conductance data for each time point were identified. The ANOVA test was performed to confirm that these NDSI showed significant differences in treatment (irrigation rate). All results were inferred at a 5% significance level.

3.3. Statistical Analysis

Comprehensive statistical analyses, including two–way ANOVA, Pearson correlation, and regression analysis, were conducted using R (4.3.3, R Foundation for Statistical Computing, Vienna, AT) and RStudio (2024.04.3, Posit Software PBC, MA, USA). Prior to analysis, data pre-processing was conducted to ensure accuracy and inter-annual consistency. Outliers were evaluated using scatterplot and interquartile range methods for each variable. Missing data points were assessed by reviewing replicate measurements within treatment plots. If partial replicates were missing, the remaining replicates were averaged; if all replicates were missing, the data point was excluded.

The main effects and interactions of irrigation depth and rate on stomatal conductance were assessed, and the relationships between vegetation indices, thermal metrics, NDSI and stomatal conductance were explored. Additionally, bidirectional stepwise regression was used to identify the best predictors of stomatal conductance within and across the hyperspectral, multispectral, and thermal datasets.

4. Result and Discussion

4.1. Ground Reference Data

Analysis using two-way ANOVA indicated that irrigation depth (comparing the control to depths of 30, 60, and 90 cm, with no significant differences detected among the latter depths) significantly influenced stomatal conductance in 2016 (80, 44 DBH: all $p < 0.05$) but not in 2017 (64, 44, 8 DBH: all $p > 0.10$). In contrast, irrigation rate consistently affected stomatal conductance in 2016 (80, 44 DBH: both $p < 0.01$) and 2017 (64, 8 DBH: both $p < 0.05$, 44 DBH: $p = 0.052$). The lack of significant interaction between depth and rate ($p > 0.36$ for all growth stages) underscores the independent and stable influence of irrigation rate, justifying its focus as the primary variable of interest in this study.

It was observed that stomatal conductance variability in terms of the coefficient of variation (CV) at 80 DBH (2016) increased significantly under reduced irrigation regimes (High irrigation rate CV: 17%, Medium irrigation rate CV: 26%, Low irrigation rate CV: 25%) relative to optimal irrigation conditions (Control CV: 12%). At 44 DBH (2016), the variability within the control treatment was 39%, while at other irrigation rates (low to high irrigation rates), the CV ranged from 23 to 30%. Despite the mitigating effects of natural rainfall, some variability in stomatal conductance under stress persisted. At 64 DBH (2017), the CV in stomatal conductance values in control and high irrigation levels was around 13%, while at medium and low irrigation rate treatments, it was about 26–34%. This trend continued into 44 DBH (2017) and 8 DBH (2017), where the control treatments showed the least variability in stomatal conductance values (7–12%), while the corresponding variability in the other irrigation rates was higher (High irrigation rate CV: 16–33%, Medium irrigation rate CV: 22–35%, Low irrigation rate CV: 20–26%). The higher variability in crop performance associated with drought stress has also been observed in other crops, such as wheat, where Veloo et al. [46] attributed this to differences in plant responses to drought effects. In grapevines, stomatal conductance exhibits a threshold-based response to water deficits, closing sharply within a narrow range of water potentials across genotypes [47]. Despite genotypic differences in stomatal sensitivity, this universal threshold behavior amplifies physiological variability under differing irrigation rates, as minor variations in water availability disproportionately affect stomatal regulation. Nevertheless, given the variability in stomatal conductance in the data, the datasets were suitable to evaluate the relationship between the remote sensing data with stomatal conductance values.

The effect of irrigation rates was further analyzed with box plots illustrating stomatal conductance distributions across various irrigation rates and time points as presented in Figure 3. It can be observed that the control and high irrigation rate had similar stomatal conductance values, while the medium and low irrigation rates had similar stomatal conductance values in 2016. In 2017, the median stomatal conductance values were similar across the control, high, and medium irrigation rates, except for the low irrigation rate, which had lower values, especially during 8 DBH (2017). The pronounced impact of irrigation rate in year 2016 is consistent with the findings from Espinoza et al. [33], while the absence of such effects in year 2017 is as reported by Chakraborty et al. [48]. The lower water stress in 2017 could be attributed to higher total precipitation during the growing season (approximately 70 mm in 2017 compared to 39 mm in 2016) (Supplementary Materials, Figure S3). This increased precipitation may have reduced the effectiveness of various irrigation strategies by maintaining elevated soil moisture levels. These results are consistent with those reported by Ma et al. [40,41].

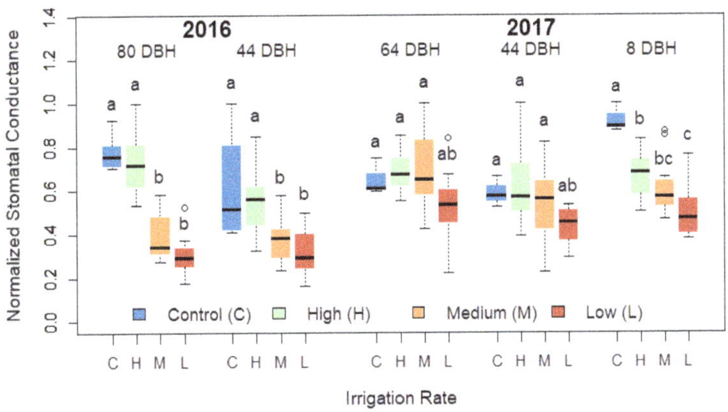

Figure 3. Boxplot of normalized stomatal conductance data with respect to different irrigation rates at different growth stages across seasons 2016 and 2017. The letters a, b, and c are from ANOVA posthoc test, where same letters indicate similar mean normalized stomatal conductance. Note: *DBH* = days before harvest.

4.2. Multispectral, Thermal, and Hyperspectral Data Analysis

4.2.1. Multispectral and Thermal Data Analysis

The relationships between the LAI, GNDVI, and CWSI with normalized stomatal conductance were assessed. Table 2 summarizes the year-to-year differences in correlation coefficients between these traits.

Table 2. Correlation coefficients (r) of the LAI, GNDVI, and CWSI with normalized stomatal conductance across 2016 and 2017.

Year	DBH	LAI (r)	GNDVI (r)	CWSI † (r)
2016	80	0.66 *	0.50 *	−0.70 *
	44	0.59 *	0.41 *	-
2017	64	0.07	0.13	-
	44	0.00	0.24	−0.83 *
	8	−0.07	0.27	-

Note: Significance levels are indicated as follows: * $0.01 \leq p < 0.05$. No asterisk is used where $p \geq 0.05$, indicating non-significant correlations; *DBH* = days before harvest. † CWSI normalization using max–min temperature in 2016 and dry–wet reference temperature in 2017.

Multispectral data analysis reveals distinct trends in the relationship between the LAI, GNDVI, and stomatal conductance. In 2016, the LAI exhibited strong positive correlations with normalized stomatal conductance at 80 and 44 DBH (both $p < 0.01$), while the GNDVI showed moderate correlations at 80 DBH ($r = 0.50$, $p < 0.01$) that weakened at 44 DBH ($r = 0.41$, $p < 0.01$) (Table 2, Figure 4). In 2017, the correlations for both the LAI and GNDVI were weak and statistically non-significant at 64, 44, and 8 DBH. This decline in correlation strength aligns with other observations, where increased rainfall likely masked treatment-induced differences in vine water stress.

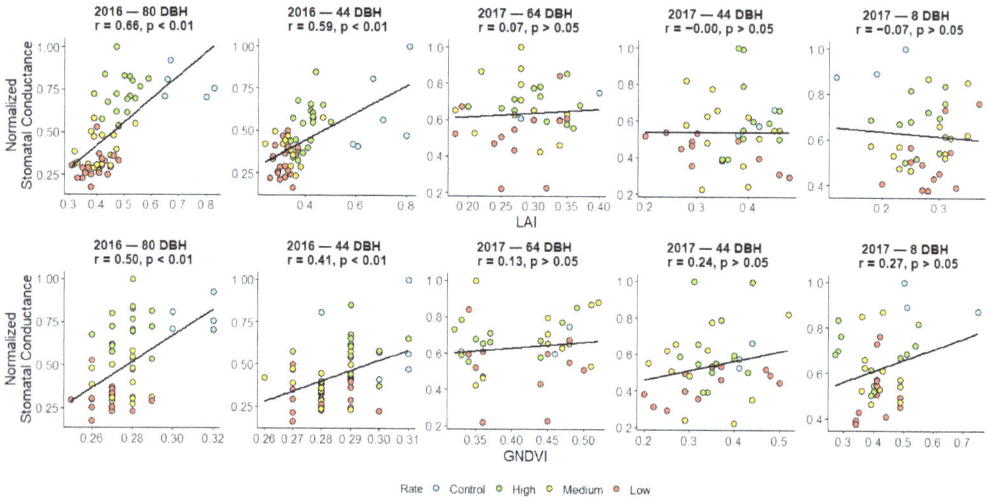

Figure 4. Scatterplots showcasing Pearson's correlation coefficients of features extracted from the AUV multispectral data with normalized stomatal conductance. Note: *DBH* = days before harvest.

Unlike prior studies that emphasized the GNDVI as a key index for water stress detection [34,49], this study identifies the LAI as a more robust indicator of stomatal conductance, particularly under controlled treatment conditions. The LAI reflects the leaf surface area, a direct determinant of transpiration and photosynthesis, which may explain its stronger relationship with stomatal regulation. By contrast, the GNDVI, which assesses chlorophyll content, captures overall vine health but is more sensitive to environmental factors such as rainfall and canopy shading. These findings highlight the distinct functional roles of the LAI and GNDVI, with the LAI providing a more direct physiological link to water stress, especially under conditions with clear treatment differentiation [50,51].

Thermal data analysis revealed differences in the CWSI calculation methods: adaptive method (max–min pixel normalization within images) (2016) and dry–wet reference normalization (2017). The shift to the dry–wet method was informed by prior studies, including Gonzalez-Dugo and Zarco-Tejada [52], who highlighted the influence of relative humidity and sensor precision on CWSI variability, and Katimbo et al. [53], who demonstrated that CWSI sensitivity increases as soil water depletion surpasses 80%, emphasizing the need for stable reference points. This shift yielded a stronger negative correlation with stomatal conductance in 2017 ($r = -0.83$) compared to 2016 ($r = -0.70$) (Table 2, Figure 5), demonstrating the improved sensitivity and reliability of the dry–wet reference method. In general, it was observed that lower irrigation rates resulted in higher canopy temperature (thus higher CWSI data) and lower normalized stomatal conductance.

Our comparative analysis highlights the limitations of adaptive normalization methods. Zhou et al. [54] observed no significant correlation between thermal data and stem water potential when using an adaptive approach, underscoring its instability. In contrast, the dry–wet reference method employed in our study yielded stronger associations with stomatal conductance, indicating that reference-based normalization offers greater consistency in assessing vine water status. The inclusion of fixed dry and wet references provided more stable CWSI estimates, particularly during the wetter 2017 growing season, where the maximum and minimum canopy temperatures may not have been fully captured within individual images.

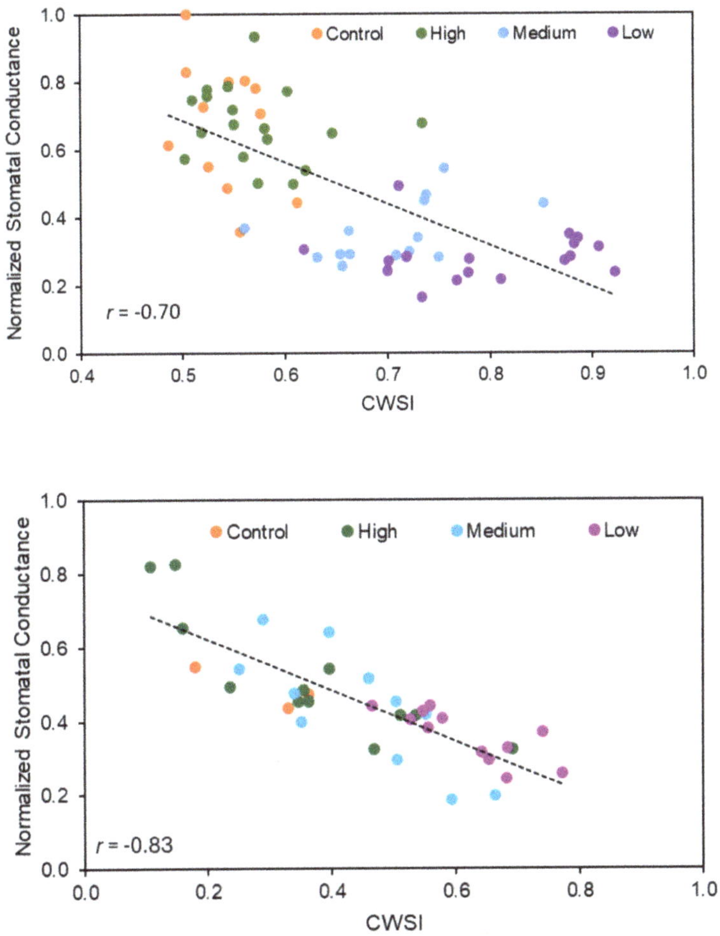

Figure 5. Correlation of the crop water stress index (CWSI) extracted from the AUV thermal data with normalized stomatal conductance. Note: Data are from 80 days before harvest (DBH) in 2016 (**top**) and 44 days before harvest in 2017 (**bottom**).

Canopy temperature, a widely recognized indicator of water stress, was negatively correlated with stomatal conductance in this study, consistent with the findings from Park et al. [28], who reported a determination coefficient ($R^2 = 0.82$) for this relationship in nectarine and peach orchards. Similar trends have been observed in viticulture, as Araújo-Paredes et al. [55] found moderate correlations ($R^2 = 0.58$) between canopy temperature and stem water potential using UAV-mounted thermal cameras. Unlike UAV-based methods, this study employed an AUV-mounted thermal camera, achieving higher spatial resolution (0.56 cm/pixel) and consistent observation angles. This proximal sensing approach reduces the wind-induced image instability observed in UAV systems and supports high-frequency monitoring, improving the precision of canopy temperature measurements for stomatal conductance assessment.

4.2.2. Hyperspectral Data Analysis

The analysis of various NDSI computed from diverse datasets demonstrated a robust correlation with normalized stomatal conductance data. There was a high variability in the number of statistically significant NDSI correlated with stomatal conductance across datasets (Supplementary Materials, Figures S4 and S5). At 80 and 44 DBH (2016), 194 and 6 NDSI exhibited significance at $p < 0.05$, respectively. A marked increase was observed in 2017, with 493, 794, and 1036 NDSI at 64, 44, and 8 DBH (2017), respectively. While the number of NDSI fluctuated between years, the number of NDSI features evidencing significant treatment differences ($p < 0.001$) remained relatively consistent in 2016, with only five NDSI significant at 44 DBH (2017). This trend continued in 2017, where approximately 270, 735, and 1025 NDSI at 64, 44, and 8 DBH (2017) exhibited significant treatment effects at $p < 0.001$.

From each dataset (five in total), the top 20 NDSI with the highest correlations were selected for further analysis. The spectral wavelengths occurring at high frequencies (>5) in different regions of the spectra (to avoid collinearity) were selected from the wavelength combinations. The five prominent spectral bands were 1380 nm, 1420 nm, 1570 nm, 1810 nm, and 1890 nm, situated within the short-wave infrared (SWIR) spectral region. These bands coincide with regions of significant water absorption between 1300 and 2500 nm, which exhibit a pronounced sensitivity to variations in leaf water content [36,56]. The relationship between these spectral bands and stomatal conductance underscores the influence of leaf water status on vine physiological responses.

This refined selection facilitated a more targeted quantitative analysis. By computing NDSI for each possible pair using these selected spectral bands, ten NDSI were analyzed: NDSI1 [1380 nm, 1420 nm], NDSI2 [1380 nm, 1570 nm], NDSI3 [1380 nm, 1810 nm], NDSI4 [1380 nm, 1890 nm], NDSI5 [1420 nm, 1570 nm], NDSI6 [1420 nm, 1810 nm], NDSI7 Our comparative analysis highlights the limitations of adaptive normalization methods. Zhou et al. [54] observed no significant correlation between thermal data and stem water potential when using an adaptive approach, underscoring its instability. In contrast, the dry–wet reference method employed in our study yielded stronger associations with stomatal conductance, indicating that reference-based normalization offers greater consistency in assessing vine water status. The inclusion of fixed dry and wet references provided more stable CWSI estimates, particularly during the wetter 2017 growing season, where the maximum and minimum canopy temperatures may not have been fully captured within individual images.[1420 nm, 1890 nm], NDSI8 [1570 nm, 1810 nm], NDSI9 [1570 nm, 1890 nm], and NDSI10 [1810 nm, 1890 nm]. Pearson's correlation analysis revealed dynamic relationships between various NDSI and stomatal conductance at different time points (Figure 6).

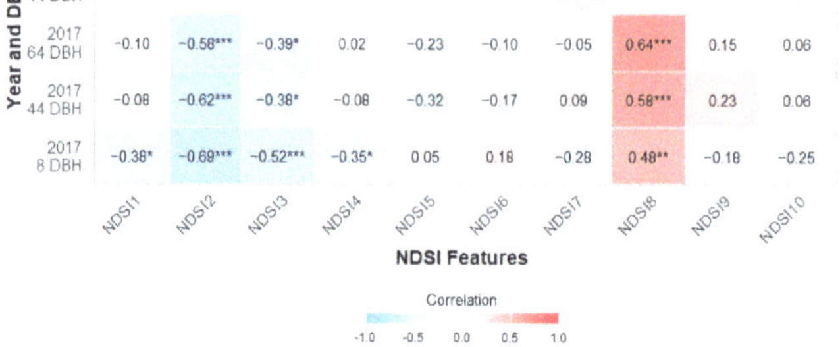

Figure 6. Heatmap showcasing Pearson's correlation coefficients of NDSI features with normalized stomatal conductance. Note: Significance levels are indicated as follows: *** $p < 0.001$, ** $0.001 \leq p < 0.01$, * $0.01 \leq p < 0.05$. No asterisk is used where $p \geq 0.05$, indicating non-significant correlations; *DBH* = days before harvest.

At 80 DBH (2016), significant correlations were observed for most NDSI, suggesting initial vine adaptation to reduced irrigation through decreased stomatal conductance [57]. This phase indicates proactive water loss management strategies by the vines. By 44 DBH (2016), the absence of significant correlations suggests a stabilization of physiological responses, supported by a reduced sensitivity of both stomatal conductance and chlorophyll fluorescence to prolonged water stress [58].

During 64, 44, and 8 DBH (2017), increased rainfall significantly impacted the correlations. Notably, NDSI8 showed a strong positive correlation at 64 DBH (2017) ($r = 0.64$, Our comparative analysis highlights the limitations of adaptive normalization methods. Zhou et al. [54] observed no significant correlation between thermal data and stem water potential when using an adaptive approach, underscoring its instability. In contrast, the dry–wet reference method employed in our study yielded stronger associations with stomatal conductance, indicating that reference-based normalization offers greater consistency in assessing vine water status. The inclusion of fixed dry and wet references provided more stable CWSI estimates, particularly during the wetter 2017 growing season, where the maximum and minimum canopy temperatures may not have been fully captured within individual images.$p < 0.001$), indicative of an immediate adaptive response to environmental changes, whereas NDSI2 demonstrated a pronounced negative correlation by 8 DBH (2017) ($r = -0.69$, $p < 0.001$), reflecting a delayed physiological adjustment over the season. Additionally, the significant negative correlations for NDSI3 ($r = -0.38$ to -0.52, $p < 0.05$) during these periods suggest that increased rainfall compensated for reduced irrigation, maintaining vine hydration. The consistent physiological responses across various irrigation treatments, particularly for NDSI2 and NDSI8, underscore the uniform adaptability of vines to both natural rainfall and managed irrigation practices. This suggests that vines modulate their physiological responses dynamically, integrating the combined effects of rainfall and irrigation [58].

The 1570 nm spectral band, central to both NDSI2 and NDSI8, demonstrates a strong correlation with stem water potential within the pivotal 1500–1700 nm range, highlighting its relevance in evaluating vine hydration [59]. This wavelength, alongside the 1380 nm and 1890 nm bands in NDSI2 and NDSI8, respectively, enables detailed monitoring of vine physiological adaptations, particularly in stomatal conductance. Additionally, Rapaport et al. [60] emphasize the critical nature of the 1500 nm region for reflecting water availability, a key factor in regulating stomatal activity and assessing grapevine water status.

4.3. Comparative Analysis of Multispectral, Hyperspectral, and Combined Data

Bidirectional stepwise regression, optimized with the Akaike Information Criterion (AIC), was used to evaluate the predictive capacity of hyperspectral and multispectral indices for modeling stomatal conductance across the 2016 and 2017 growing seasons. This approach combines the canopy level measures with leaf level measures. Predictors included NDSI1–NDSI10, LAI, and GNDVI, both individually and in combination. Table 3 summarizes the predictive performance of key indices (LAI, GNDVI, NDSI2, NDSI3 and NDSI8) and the best predictor set (LAI + NDSI2 + NDSI8) across both years.

In 2016, the LAI was the strongest predictor from the multispectral indices at both 80 DBH ($R^2 = 0.44$, adjusted $R^2 = 0.43$; $p < 0.001$) and 44 DBH ($R^2 = 0.35$, adjusted $R^2 = 0.34$; In 2016, the LAI was the strongest predictor from the multispectral indices at both 80 DBH ($R^2 = 0.44$, adjusted $R^2 = 0.43$; $p < 0.001$) and 44 DBH ($R^2 = 0.35$, adjusted $R^2 = 0.34$; $p < 0.001$). The GNDVI, though weaker, remained statistically significant at both 80 DBH ($R^2 = 0.25$, adjusted $R^2 = 0.23$; $p < 0.01$) and 44 DBH ($R^2 = 0.17$, adjusted $R^2 = 0.15$; $p < 0.01$). When the LAI and GNDVI were combined, the model showed moderate predictive strength at both time points (80 DBH: $R^2 = 0.44$, adjusted $R^2 = 0.42$; 44 DBH: $R^2 = 0.36$, adjusted $R^2 = 0.33$; both $p < 0.001$). The hyperspectral indices showed mixed results: at 80 DBH, NDSI8 ($R^2 = 0.09$, adjusted $R^2 = 0.08$; $p = 0.02$) and NDSI2 ($R^2 = 0.08$, adjusted $R^2 = 0.06$; $p = 0.04$) were borderline significant, but neither were significant at 44 DBH ($p > 0.05$). NDSI3 and all NDSI combinations were not significant at either time point (all $p > 0.05$).$p < 0.001$). The GNDVI, though weaker, remained statistically significant at both 80 DBH ($R^2 = 0.25$, adjusted $R^2 = 0.23$; $p < 0.01$) and 44 DBH ($R^2 = 0.17$, adjusted $R^2 = 0.15$; $p < 0.01$). When the LAI and GNDVI were combined, the model showed moderate predictive strength at both time points (80 DBH: $R^2 = 0.44$, adjusted $R^2 = 0.42$; 44 DBH: $R^2 = 0.36$, adjusted $R^2 = 0.33$; both $p < 0.001$). The hyperspectral indices showed mixed results: at 80 DBH, NDSI8 ($R^2 = 0.09$, adjusted $R^2 = 0.08$; $p = 0.02$) and NDSI2 ($R^2 = 0.08$, adjusted $R^2 = 0.06$; $p = 0.04$) were borderline significant, but neither were significant at 44 DBH ($p > 0.05$). NDSI3 and all NDSI combinations were not significant at either time point (all $p > 0.05$).

Table 3. Predictive power (R^2) of key indices for modeling stomatal conductance across 2016 and 2017.

Year	DBH	LAI (R^2)	GNDVI (R^2)	NDSI2 (R^2)	NDSI3 (R^2)	NDSI8 (R^2)	LAI + NDSI2 + NDSI8 (R^2)
2016	80	0.44 **	0.25 *	0.08 *	0.03	0.09 *	0.50 **
	44	0.35 **	0.17 *	0.03	0.04	0.00	0.37 **
2017	64	0.01	0.02	0.34 **	0.15 *	0.42 **	0.63 **
	44	0.00	0.06	0.39 **	0.15 *	0.34 **	0.51 **
	8	0.00	0.07	0.47 **	0.27 *	0.23 *	0.52 **

Note: Significance levels are indicated as follows: ** $0.001 \leq p < 0.01$, * $0.01 \leq p < 0.05$. No asterisk is used where $p \geq 0.05$, indicating non-significant; *DBH* = days before harvest.

For the 2017 growing season, hyperspectral indices revealed distinct trends in the predictive power of NDSI. NDSI2 increased as DBH decreased (64 DBH: $R^2 = 0.34$, adjusted $R^2 = 0.32$; 44 DBH: $R^2 = 0.39$, adjusted $R^2 = 0.37$; 8 DBH: $R^2 = 0.47$, adjusted $R^2 = 0.46$; all $p < 0.001$), while NDSI8 showed the opposite trend (64 DBH: $R^2 = 0.42$, adjusted $R^2 = 0.40$; 44 DBH: $R^2 = 0.34$, adjusted $R^2 = 0.32$; 8 DBH: $R^2 = 0.23$, adjusted $R^2 = 0.21$; all $p < 0.01$). NDSI3 maintained moderate predictive power across all time points (64 DBH: $R^2 = 0.15$, adjusted $R^2 = 0.13$; 44 DBH: $R^2 = 0.15$, adjusted $R^2 = 0.12$; 8 DBH: $R^2 = 0.27$, adjusted $R^2 = 0.25$, all $p < 0.05$). Among the combinations, NDSI2 + NDSI8 consistently showed the strongest predictive power at all stages (64 DBH: $R^2 = 0.57$, adjusted $R^2 = 0.55$; 44 DBH: $R^2 = 0.44$, adjusted $R^2 = 0.41$; 8 DBH: $R^2 = 0.51$, adjusted $R^2 = 0.47$; all $p < 0.01$). Multispectral indices (both LAI and GNDVI, and their combinations) were not significant at 64, 44, or 8 DBH (all $p > 0.10$).

The integration of hyperspectral and multispectral data enhanced model accuracy across growth stages. At 80 DBH (2016), the LAI + NDSI2 + NDSI8 combination demonstrated strong predictive power ($R^2 = 0.50$, adjusted $R^2 = 0.47$; $p < 0.001$), remaining the best predictor at 44 DBH (2016) ($R^2 = 0.37$, adjusted $R^2 = 0.34$; $p < 0.001$). In 2017, the same model showed high performance at 64 DBH ($R^2 = 0.63$, adjusted $R^2 = 0.60$; $p < 0.01$), with strong results continuing at 44 DBH ($R^2 = 0.51$, adjusted $R^2 = 0.47$; $p < 0.01$) and 8 DBH ($R^2 = 0.52$, adjusted $R^2 = 0.48$; $p < 0.001$). Furthermore, reinforcing our findings, Maimaitiyiming et al. [61] highlighted the effective use of NDSI for estimating stomatal conductance, particularly noting that NDSI calculated with bands at 603 nm and 558 nm exhibited a high correlation ($R^2 = 0.72$) with stomatal conductance, surpassing traditional indices. This aligns with our results, suggesting that the strategic integration of specific NDSI can significantly enhance predictive capabilities.

Our findings, supported by previous studies, show that, while individual hyperspectral or multispectral datasets provide valuable insights under specific conditions, combining these data sources consistently improves model stability and predictive accuracy. This improvement is particularly notable during the earlier growth stages in 2017, despite the lack of treatment effects.

5. Summary and Discussion

This study demonstrates that integrating multispectral, thermal, and hyperspectral data enabled stomatal conductance assessment in Cabernet Sauvignon vineyards over two growing seasons in the Pacific Northwest, USA. Proximal sensing facilitated spatial and temporal monitoring of water stress, with the LAI, NDSI2 [1380 nm, 1570 nm], and NDSI8 [1570 nm, 1810 nm] identified as key predictive indices. In 2016, the LAI was the strongest predictor, while in 2017, NDSI2 and NDSI8 outperformed other indices under minimal treatment effects. Combining these indices (LAI + NDSI2 + NDSI8) enhanced early-season water stress detection, while thermal data (CWSI) captured stress under severe drought, supporting precision irrigation strategies. These indices enable viticulturists to adjust irrigation schedules based on early stress detection, supporting water conservation and yield stability.

The AUV-based platform offers a practical alternative to UAV-based vineyard monitoring, enabling continuous, weather-independent data collection with consistent observation angles. The system integrates multispectral (LAI, GNDVI) and thermal (CWSI) sensors for real-time stress detection throughout the growing season, while hyperspectral data (NDSI2, NDSI8) are collected via a portable spectroradiometer for leaf-level insights. Implementation requires a standard utility vehicle with mounting brackets for sensor attachments and minimal hardware modifications. Cost components include sensor procurement, mounting brackets, and optional vehicle modifications (medium-to-high initial investment relative to UAVs with comparable sensor payloads). Unlike UAVs, AUVs avoid airspace restrictions and recurring operational fees, reducing regulatory burdens and long-term costs. Routine system checks and periodic calibration are recommended to maintain sensor accuracy, with ongoing operational demands limited to data collection and storage. By enabling in-field, real-time stress detection, this platform supports dynamic irrigation scheduling, minimizing water use and operational costs.

Despite promising results, certain methodological limitations must be addressed. The shift from max–min normalization (2016) to dry–wet reference normalization (2017) for thermal data improved sensitivity but introduced variability in baseline temperature references, affecting inter-annual comparability. Rainfall in 2017 likely masked irrigation effects, reducing the detectability of changes in stomatal conductance. Sub-canopy variability remains a challenge, as canopy architecture and leaf shading may bias sensor-derived indices. Addressing these issues may require segmentation methods to improve sub-canopy resolution and standardized normalization protocols to ensure consistency across seasons. Moreover, there is a potential for other stress conditions to mask the spectral responses. This may require more sophisticated machine learning techniques for image-based classification. Future work should prioritize validation of the framework across diverse climates (arid, semi-arid, Mediterranean) and grape cultivars with differing physiological traits (e.g., leaf morphology, transpiration dynamics) to assess generalizability. Broader testing is recommended to ensure robustness across these conditions. Research should further investigate the role of the LAI, NDSI2, NDSI8, and CWSI in defining dynamic irrigation thresholds and integrating these indices into decision-support systems for automated irrigation control. Embedding this approach into automated systems with real-time environmental data could further optimize irrigation scheduling, supporting sustainable viticulture under climate variability.

Supplementary Materials: The following supporting information can be downloaded at: https://www.mdpi.com/article/10.3390/rs17010137/s1, Figure S1. Plot design describing the irrigation treatments in the field during season 2016. Figure S2. Plot design describing the irrigation treatments in the field during season 2017. Figure S3. Total precipitation during the vineyard growing season (April to September) for 2016 and 2017. The higher total precipitation in 2017 (approximately 70 mm) compared to 2016 (39 mm) likely contributed to the observed reduction in water stress (wrcc.dri.edu). Figure S4. Correlation between the NDSIs computed from hyperspectral data with normalized stomatal conductance across time points in the year 2016. Note: DBH = days before harvest. Figure S5. Correlation between the NDSIs computed from hyperspectral data with normalized stomatal conductance across time points in the year 2017. Note: DBH = days before harvest.

Author Contributions: Conceptualization, S.S. and P.W.J.; methodology, C.Z.E. and S.S.; software, K.V., C.Z.E. and A.E.S.; validation, K.V., C.Z.E. and A.E.S.; formal analysis, K.V., C.Z.E., S.S. and A.E.S.; resources, S.S. and P.W.J.; data curation, K.V., C.Z.E., S.S. and A.E.S.; writing—original draft preparation, K.V. and C.Z.E.; writing—review and editing, all authors; visualization, K.V., C.Z.E. and S.S.; supervision, S.S.; funding acquisition, S.S. and P.W.J. All authors have read and agreed to the published version of the manuscript.

Funding: This activity was funded, in part, with an Emerging Research Issues Internal Competitive Grant from the Washington State University, College of Agricultural, Human, and Natural Resource Sciences, the US Department of Agriculture National Institute of Food and Agriculture (USDA-NIFA) hatch and multistate projects (Accession Number WNP00011, NC1212), and USDA Agricultural Research Services.

Data Availability Statement: Data will be available upon request.

Acknowledgments: We sincerely thank Lav R Khot for his guidance and support during the project. We are very grateful to Sanaz Jarolmasjed, Chongyuan Zhang, and Rajeev Sinha for their help in data collection. Additionally, we acknowledge Kiona Winery and Vineyards that provided the land, vines, irrigation, and overall management of the vineyard site and appreciate their collaboration.

Conflicts of Interest: The authors declare no conflicts of interest.

References

1. Keller, M.; Mills, L.J.; Kawakami, A.K. Optimizing irrigation for mechanized Concord juice grape production. *Am. J. Enol. Vitic.* **2023**, *74*, 0740008. [CrossRef]
2. van Leeuwen, C.; Sgubin, G.; Bois, B.; Ollat, N.; Swingedouw, D.; Zito, S.; Gambetta, G.A. Climate change impacts and adaptations of wine production. *Nat. Rev. Earth Environ.* **2024**, *5*, 258–275. [CrossRef]
3. Jones, G.V.; Duff, A.A.; Hall, A.; Myers, J.W. Spatial analysis of climate in winegrape growing regions in the Western United States. *Am. J. Enol. Vitic.* **2010**, *61*, 313–326. [CrossRef]
4. White, M.A.; Diffenbaugh, N.S.; Jones, G.V.; Pal, J.S.; Giorgi, F. Extreme heat reduces and shifts United States premium wine production in the 21st century. *Proc. Natl. Acad. Sci. USA* **2006**, *103*, 11217–11222. [CrossRef] [PubMed]
5. Keller, M. Climate change impacts on vineyards in warm and dry areas: Challenges and opportunities. *Am. J. Enol. Vitic.* **2023**, *74*, 0740033. [CrossRef]
6. Jones, G.V.; Edwards, E.J.; Bonada, M.; Sadras, V.O.; Krstic, M.P.; Herderich, M.J. Climate change and its consequences for viticulture. *Manag. Wine Qual. Vol. One Vitic. Wine Qual.* **2022**, *1*, 727–778. [CrossRef]
7. Wagner, A.M.; Bennett, K.E.; Liston, G.E.; Hiemstra, C.A.; Cooley, D. Multiple indicators of extreme changes in snow-dominated streamflow regimes, Yakima River Basin Region, USA. *Water* **2021**, *13*, 2608. [CrossRef]
8. Mirás-Avalos, J.M.; Araujo, E.S. Optimization of vineyard water management: Challenges, strategies, and perspectives. *Water* **2021**, *13*, 746. [CrossRef]
9. Finco, A.; Bentivoglio, D.; Chiaraluce, G.; Alberi, M.; Chiarelli, E.; Maino, A.; Mantovani, F.; Montuschi, M.; Raptis, K.G.C.; Semenza, F.; et al. Combining precision viticulture technologies and economic indices to sustainable water use management. *Water* **2022**, *14*, 1493. [CrossRef]

10. Carbonneau, A. The early selection of grapevine rootstocks for resistance to drought conditions. *Am. J. Enol. Vitic.* **1985**, *36*, 195–198. [CrossRef]
11. Iacono, F.; Peterlunger, E. Rootstock-scion interaction may affect drought tolerance in *vitis vinifera* cultivars. Implications in selection programs. *Acta Hortic.* **2000**, *528*, 543–549. [CrossRef]
12. Serra, I.; Strever, A.; Myburgh, P.A.; Deloire, A. Review: The interaction between rootstocks and cultivars (*Vitis vinifera* L.) to enhance drought tolerance in grapevine. *Aust. J. Grape Wine Res.* **2014**, *20*, 1–14. [CrossRef]
13. Gil, P.M.; Lobos, P.; Durán, K.; Olguín, J.; Cea, D.; Schaffer, B. Partial root-zone drying irrigation, shading, or mulching effects on water savings, productivity and quality of 'Syrah' grapevines. *Sci. Hortic.* **2018**, *240*, 478–483. [CrossRef]
14. Delpuech, X.; Metay, A. Adapting cover crop soil coverage to soil depth to limit competition for water in a Mediterranean vineyard. *Eur. J. Agron.* **2018**, *97*, 60–69. [CrossRef]
15. Palliotti, A.; Tombesi, S.; Silvestroni, O.; Lanari, V.; Gatti, M.; Poni, S. Changes in vineyard establishment and canopy management urged by earlier climate-related grape ripening: A review. *Sci. Hortic.* **2014**, *178*, 43–54. [CrossRef]
16. Silvestroni, O.; Lanari, V.; Lattanzi, T. Canopy management strategies to control yield and grape composition of Montepulciano grapevines. *Aust. J. Grape Wine Res.* **2018**, *25*, 30–42. [CrossRef]
17. Chaves, M.M.; Zarrouk, O.; Francisco, R.; Costa, J.M.; Santos, T.; Regalado, A.P.; Rodrigues, M.L.; Lopes, C.M. Grapevine under deficit irrigation: Hints from physiological and molecular data. *Ann. Bot.* **2010**, *105*, 661–676. [CrossRef]
18. Tsirogiannis, I.L.; Malamos, N.; Baltzoi, P. Application of a generic participatory decision support system for irrigation management for the case of a wine grapevine at Epirus, Northwest Greece. *Horticulturae* **2023**, *9*, 267. [CrossRef]
19. Kang, C.; Diverres, G.; Karkee, M.; Zhang, Q.; Keller, M. Decision-support system for precision regulated deficit irrigation management for wine grapes. *Comput. Electron. Agric.* **2023**, *208*, 107777. [CrossRef]
20. King, B.A.; Shellie, K.C. A crop water stress index based internet of things decision support system for precision irrigation of wine grape. *Smart Agric. Technol.* **2023**, *4*, 100202. [CrossRef]
21. RoTimi Ojo, E.; Bullock, P.R.; Fitzmaurice, J. Field performance of five soil moisture instruments in heavy clay soils. *Soil Sci. Soc. Am. J.* **2015**, *79*, 20–29. [CrossRef]
22. Huang, Y.; Chen, Z.; Yu, T.; Huang, X.; Gu, X. Agricultural remote sensing big data: Management and applications. *J. Integr. Agric.* **2018**, *7*, 1915–1931. [CrossRef]
23. Osakabe, Y.; Osakabe, K.; Shinozaki, K.; Tran, L.S.P. Response of plants to water stress. *Front. Plant Sci.* **2014**, *5*, 86. [CrossRef]
24. Grant, O.M.; Davies, M.J.; James, C.M.; Johnson, A.W.; Leinonen, I.; Simpson, D.W. Thermal imaging and carbon isotope composition indicate variation amongst strawberry (*Fragaria* × *ananassa*) cultivars in stomatal conductance and water use efficiency. *Environ. Exp. Bot.* **2012**, *76*, 7–15. [CrossRef]
25. Egea, G.; Padilla-Díaz, C.M.; Martinez-Guanter, J.; Fernández, J.E.; Pérez-Ruiz, M. Assessing a crop water stress index derived from aerial thermal imaging and infrared thermometry in super-high density olive orchards. *Agric. Water Manag.* **2017**, *187*, 210–221. [CrossRef]
26. Marques, P.; Pádua, L.; Sousa, J.J.; Fernandes-Silva, A. Assessing the water status and leaf pigment content of olive trees: Evaluating the potential and feasibility of unmanned aerial vehicle multispectral and thermal data for estimation purposes. *Remote Sens.* **2023**, *15*, 4777. [CrossRef]
27. Gutiérrez-Gordillo, S.; García-Tejero, I.F.; Durán Zuazo, V.H.; García Escalera, A.; Ferrera Gil, F.; Amores-Agüera, J.J.; Cárceles Rodríguez, B.; Hernández-Santana, V. Assessing the water-stress baselines by thermal imaging for irrigation management in almond plantations under water scarcity conditions. *Water* **2020**, *12*, 1298. [CrossRef]
28. Park, S.; Ryu, D.; Fuentes, S.; Chung, H.; Hernández-Montes, E.; O'Connell, M. Adaptive estimation of crop water stress in nectarine and peach orchards using high-resolution imagery from an unmanned aerial vehicle (UAV). *Remote Sens.* **2017**, *9*, 828. [CrossRef]
29. Matese, A.; Baraldi, R.; Berton, A.; Cesaraccio, C.; Di Gennaro, S.F.; Duce, P.; Facini, O.; Mameli, M.G.; Piga, A.; Zaldei, A. Estimation of water stress in grapevines using proximal and remote sensing methods. *Remote Sens.* **2018**, *10*, 114. [CrossRef]
30. Pagay, V.; Kidman, C.M. Evaluating remotely-sensed grapevine (*Vitis vinifera* L.) water stress responses across a viticultural region. *Agronomy* **2019**, *9*, 682. [CrossRef]
31. Baluja, J.; Diago, M.P.; Balda, P.; Zorer, R.; Meggio, M.; Morales, F.; Tardaguila, J. Assessment of vineyard water status variability by thermal and multispectral imagery using an unmanned aerial vehicle (UAV). *Irrig. Sci.* **2012**, *30*, 511–522. [CrossRef]
32. Buunk, T.; Vélez, S.; Ariza-Sentís, M.; Valente, J. Comparing nadir and oblique thermal imagery in UAV-Based 3D crop water stress index applications for precision viticulture with LiDAR validation. *Sensors* **2023**, *23*, 8625. [CrossRef] [PubMed]
33. Espinoza, C.Z.; Khot, L.R.; Sankaran, S.; Jacoby, P.W. High Resolution Multispectral and Thermal Remote Sensing-Based Water Stress Assessment in Subsurface Irrigated Grapevines. *Remote Sens.* **2017**, *9*, 961. [CrossRef]
34. López-García, P.; Intrigliolo, D.S.; Moreno, M.A.; Martínez-Moreno, A.; Ortega, J.F.; Pérez-Álvarez, E.P.; Ballesteros, R. Assessment of vineyard water status by multispectral and RGB imagery obtained from an unmanned aerial vehicle. *Am. J. Enol. Vitic.* **2021**, *72*, 285–297. [CrossRef]

35. Loggenberg, K.; Strever, A.; Greyling, B.; Poona, N. Modelling water stress in a Shiraz vineyard using hyperspectral imaging and machine learning. *Remote Sens.* **2018**, *10*, 202. [CrossRef]
36. Zovko, M.; Žibrat, U.; Knapič, M.; Kovačić, M.B.; Romić, D. Hyperspectral remote sensing of grapevine drought stress. *Precis. Agric.* **2019**, *20*, 335–347. [CrossRef]
37. Morano, L.; Kliewer, W.M. Root distribution of three grapevine rootstocks grafted to Cabernet Sauvignon Grown on a very gravelly clay loam soil in Oakville, California. *Am. J. Enol. Vitic.* **1994**, *45*, 345–348. [CrossRef]
38. Rijal, J.P.; Bergh, J.C. Food-Finding Capability of Grape Root Borer (Lepidoptera: Sesiidae) Neonates in Soil Column Bioassays. *J. Entomol. Sci.* **2016**, *51*, 54–68. [CrossRef]
39. Ferlito, F.; Distefano, G.; Gentile, A.; Allegra, M.; Lakso, A.N.; Nicolosi, E. Scion–rootstock interactions influence the growth and behaviour of the grapevine root system in a heavy clay soil. *Aust. J. Grape Wine Res.* **2020**, *26*, 68–78. [CrossRef]
40. Ma, X.; Sanguinet, K.A.; Jacoby, P.W. Performance of direct root-zone deficit irrigation on *Vitis vinifera* L. cv. Cabernet Sauvignon production and water use efficiency in semi-arid southcentral Washington. *Agric. Water Manag.* **2019**, *221*, 47–57. [CrossRef]
41. Ma, X.; Sanguinet, K.A.; Jacoby, P.W. Direct root-zone irrigation outperforms surface drip irrigation for grape yield and crop water use efficiency while restricting root growth. *Agric. Water Manag.* **2020**, *231*, 105993. [CrossRef]
42. Diago, M.P.; Tardaguila, J.; Barrio, I.; Fernández-Novales, J. Combination of multispectral imagery, environmental data and thermography for on-the-go monitoring of the grapevine water status in commercial vineyards. *Eur. J. Agron.* **2022**, *140*, 126586. [CrossRef]
43. Gitelson, A.A.; Merzlyak, M.N. Remote sensing of chlorophyll concentration in higher plant leaves. *Adv. Space Res.* **1998**, *22*, 689–692. [CrossRef]
44. Gowdy, M.; Pieri, P.; Suter, B.; Marguerit, E.; Destrac-Irvine, A.; Gambetta, G.; van Leeuwen, C. estimating bulk stomatal conductance in grapevine canopies. *Front. Plant Sci.* **2022**, *13*, 839378. [CrossRef]
45. Marzougui, A.; Ma, Y.; Zhang, C.; McGee, R.J.; Coyne, C.J.; Main, D.; Sankaran, S. Advanced Imaging for Quantitative Evaluation of Aphanomyces Root Rot Resistance in Lentil. *Front. Plant Sci.* **2019**, *10*, 383. [CrossRef]
46. Veloo, K.; Valencia-Ortiz, M.; Pumphrey, M.; Carter, A.; Garland-Campbell; Sankaran, S. AGIcam+: Performance comparison of an in-field multispectral and thermal sensor system with UAV imaging for wheat drought stress monitoring. *Comput. Electron. Agric.* **2024**. *submitted for publication*.
47. Gambetta, G.A.; Herrera, J.C.; Dayer, S.; Feng, Q.; Hochberg, U.; Castellarin, S.D. The physiology of drought stress in grapevine: Towards an integrative definition of drought tolerance. *J. Exp. Bot.* **2020**, *71*, 4658–4676. [CrossRef] [PubMed]
48. Chakraborty, M.; Khot, L.R.; Sankaran, S.; Jacoby, P.W. Evaluation of mobile 3D light detection and ranging based canopy mapping system for tree fruit crops. *Comput. Electron. Agric.* **2019**, *158*, 284–293. [CrossRef]
49. Brandmeier, M.; Heßdörfer, D.; Siebenlist, P.; Meyer-Spelbrink, A.; Kraus, A. Time Series Analysis of Multisensor Data for Precision Viticulture—Assessing Microscale Variations in Plant Development with Respect to Irrigation and Topography. *Remote Sens.* **2024**, *16*, 1419. [CrossRef]
50. Ben-Asher, J.; Tsuyuki, I.; Bravdo, B.-A.; Sagih, M. Irrigation of grapevines with saline water: I. leaf area index, stomatal conductance, transpiration and photosynthesis. *Agric. Water Manag.* **2006**, *83*, 13–21. [CrossRef]
51. Sperling, O.; Perry, A.; Ben-Gal, A.; Yermiyahu, U.; Hochberg, U. Potassium deficiency reduces grapevine transpiration through decreased leaf area and stomatal conductance. *Plant Physiol. Biochem.* **2024**, *208*, 108534. [CrossRef]
52. Gonzalez-Dugo, V.; Zarco-Tejada, P.J. Assessing the impact of measurement errors in the calculation of cwsi for characterizing the water status of several crop species. *Irrig. Sci.* **2022**, *42*, 431–443. [CrossRef]
53. Katimbo, A.; Rudnick, D.R.; DeJonge, K.C.; Lo, T.H.; Qiao, X.; Franz, T.E.; Nakabuye, H.N.; Duan, J. Crop water stress index computation approaches and their sensitivity to soil water dynamics. *Agric. Water Manag.* **2022**, *266*, 107575. [CrossRef]
54. Zhou, Z.; Diverres, G.; Kang, C.; Thapa, S.; Karkee, M.; Zhang, Q.; Keller, M. Ground-based thermal imaging for assessing crop water status in grapevines over a growing season. *Agronomy* **2022**, *12*, 322. [CrossRef]
55. Araújo-Paredes, C.; Portela, F.; Mendes, S.; Valín, M.I. Using Aerial Thermal Imagery to Evaluate Water Status in *Vitis vinifera cv.* Loureiro. *Sensors* **2022**, *22*, 8056. [CrossRef] [PubMed]
56. Zhang, Y.; Wu, J.; Wang, A. Comparison of various approaches for estimating leaf water content and stomatal conductance in different plant species using hyperspectral data. *Ecol. Indic.* **2022**, *142*, 109278. [CrossRef]
57. Benyahia, F.; Bastos Campos, F.; Ben Abdelkader, A.; Basile, B.; Tagliavini, M.; Andreotti, C.; Zanotelli, D. Assessing grapevine water status by integrating vine transpiration, leaf gas exchanges, chlorophyll fluorescence and sap flow measurements. *Agronomy* **2023**, *13*, 464. [CrossRef]
58. Trigo-Córdoba, E.; Bouzas-Cid, Y.; Orriols-Fernández, I.; Mirás-Avalos, J.M. Effects of deficit irrigation on the performance of grapevine (*Vitis vinifera* L.) cv. 'Godello' and 'Treixadura' in Ribeiro, NW Spain. *Agric. Water Manag.* **2015**, *161*, 20–30. [CrossRef]
59. Laroche-Pinel, E.; Albughdadi, M.; Duthoit, S.; Chéret, V.; Rousseau, J.; Clenet, H. understanding vine hyperspectral signature through different irrigation plans: A first step to monitor vineyard water status. *Remote Sens.* **2021**, *13*, 536. [CrossRef]

60. Rapaport, T.; Hochberg, U.; Shoshany, M.; Karnieli, A.; Rachmilevitch, S. Combining leaf physiology, hyperspectral imaging and partial least squares-regression (PLS-R) for grapevine water status assessment. *ISPRS J. Photogramm. Remote Sens.* **2015**, *109*, 88–97. [CrossRef]
61. Maimaitiyiming, M.; Ghulam, A.; Bozzolo, A.; Wilkins, J.L.; Kwasniewski, M.T. Early detection of plant physiological responses to different levels of water stress using reflectance spectroscopy. *Remote Sens.* **2017**, *9*, 745. [CrossRef]

MDPI AG
Grosspeteranlage 5
4052 Basel
Switzerland
Tel.: +41 61 683 77 34

Remote Sensing Editorial Office
E-mail: remotesensing@mdpi.com
www.mdpi.com/journal/remotesensing

www.ingramcontent.com/pod-product-compliance
Lightning Source LLC
Chambersburg PA
CBHW041017270926
51850CB00024B/1328
* 9 7 8 3 7 2 5 8 4 7 1 7 4 *